RESEARCHING OPEN INNOVATION IN SMEs

RESEARCHING OPEN INNOVATION IN SMEs

Editors

Wim Vanhaverbeke
Universiteit Hasselt, Belgium
ESADE Business School, Spain
National University of Singapore, Singapore

Federico Frattini
Politecnico di Milano, Italy

Nadine Roijakkers
Open Universiteit, the Netherlands

Muhammad Usman
Universiteit Hasselt, Belgium

World Scientific

NEW JERSEY • LONDON • SINGAPORE • BEIJING • SHANGHAI • HONG KONG • TAIPEI • CHENNAI • TOKYO

Published by

World Scientific Publishing Co. Pte. Ltd.
5 Toh Tuck Link, Singapore 596224
USA office: 27 Warren Street, Suite 401-402, Hackensack, NJ 07601
UK office: 57 Shelton Street, Covent Garden, London WC2H 9HE

Library of Congress Cataloging-in-Publication Data
Names: Vanhaverbeke, Wim, editor.
Title: Researching open innovation in SMEs / edited by Wim Vanhaverbeke
(Universiteit Hasselt, Belgium & ESADE Business School, Spain & NUS, Singapore)
[and three others].
Description: New Jersey : World Scientific, [2018]
Identifiers: LCCN 2017044676 | ISBN 9789813230965 (hc : alk. paper)
Subjects: LCSH: Small business--Management. | Technological innovations--Management. |
Diffusion of innovations--Management.
Classification: LCC HD62.7 .R467 2018 | DDC 658.4/063--dc23
LC record available at https://lccn.loc.gov/2017044676

British Library Cataloguing-in-Publication Data
A catalogue record for this book is available from the British Library.

For any available supplementary material, please visit
http://www.worldscientific.com/worldscibooks/10.1142/10733#t=suppl

Desk Editors: Herbert Moses/Alisha Nguyen

Typeset by Stallion Press
Email: enquiries@stallionpress.com

Printed in Singapore

About the Editors

Federico Frattini is a Full Professor of Strategic Management and Innovation at the School of Management of Politecnico di Milano (Italy) and Honorary Researcher at the Lancaster University Management School (UK). At the School of Management of Politecnico di Milano, he is also Director of the MBA and Executive MBA Division, Director of the ICT and Digital Learning Division, Coordinator of the Strategic Management Teaching Area, and Dean's Delegate for Rankings. His research area is innovation and technology management. On these topics, he has written more than 200 books and articles published in proceedings of international conferences and journals such as *Academy of Management Perspectives, Journal of Product Innovation Management, Family Business Review, Technovation, Technological Analysis & Strategic Management*, and many others. In 2013, he was nominated among the top 50 authors of Technology and Innovation Management worldwide by the International Association for Management of Technology (IAMOT).

Nadine Roijakkers is an Associate Professor of Open Innovation at the Open University in Heerlen, The Netherlands. She wrote her PhD thesis at the United Nations University/MERIT, Maastricht, the Netherlands, on interfirm collaborative innovation in the pharmaceutical biotechnology industry. For several years, she was a senior strategy consultant at KPMG Consulting in Utrecht (The Netherlands). Her articles have appeared in journals such as *Long Range Planning,*

Research Policy, Harvard Business History Review, British Journal of Management, European Management Journal, Technological Forecasting and Social Change, Small Business Economics, California Management Review, Journal of Product Innovation Management, and *Organizational Dynamics.*

Muhammad Usman is a Doctoral candidate with a research focus on strategy and open innovation. His academic research aims to explore open innovation phenomenon with a particular emphasis on small and medium enterprises. He has hands-on experience of project management prior to joining the doctoral program at the Hasselt University, Belgium. Drawing on this practical experience along with an MBA degree gives him an apt blend of understanding to focus on emerging innovation strategies and practices.

Wim Vanhaverbeke is a Professor at the University of Hasselt and a visiting professor at ESADE Business School and the National University of Singapore. He published in different international journals. He is co-editor with Henry Chesbrough and Joel West of two books *Open Innovation: Researching a New Paradigm* (OUP, 2006) and *New Frontiers in Open Innovation* (OUP, 2014), and he recently published *Open Innovation in SMEs* (CUP, 2017), an OI management guideline for small firms. His current research focuses on open innovation in SMEs, innovation ecosystems, and on the implementation of open innovation practices. He was recognized by the International Association of Management of Technology (IAMOT) as one of the top 50 authors of Technology and Innovation Management during the period 2008–2012.

About the Contributors

Nuran Acur is a Senior Lecturer in Innovation and Technology Management at the Adam Smith Business School, University of Glasgow, UK. Previously, Dr Acur worked at the University of Strathclyde (UK), Ozyegin University (Turkey), and Aalborg University (Denmark). She received her PhD in Strategic Management from the University of Strathclyde, UK. She has published award-winning, theory-driven, policy and practice-relevant articles across the fields of technology innovation, open innovation, social innovation, and operations strategy. Her work has appeared in the *Journal of Product Innovation Management, International Journal of Operations & Production Management, Creativity and Innovation Management, European Management Review, Supply Chain Management*, an international journal, and other scholarly journals. Currently, she is on the editorial board of *Journal of Product Innovation Management* and *Creativity and Innovation Journal.*

Joon Mo Ahn is an Assistant Professor at Graduate School of MOT (Management of Technology), Sogang University, Korea. Prior to joining Sogang University, he worked for Korean government agencies, such as the Small Medium Business Administration and the Ministry of Education, Science and Technology, and the Ministry of Science, ICT, and Future Planning where he engaged in the development of innovation policies. He has a BSc in Chemical Engineering from Seoul National University, Korea, and a PhD from the

University of Cambridge, UK. His current research is focused on open innovation in small firms, innovation management, innovation policy, and entrepreneurship.

Robert William Anderson has recently completed his PhD in Innovation Management at the University of Strathclyde. With an interest in open innovation, Robert researches crowdsourcing and business model in SMEs. Robert holds degrees in Product Design and Innovation and Supply Chain and Operations Management. Being a fully funded EPSRC scholar, he has presented his PhD work at many international conferences secured grants to help run open innovation projects for SMEs. His PhD work has given him experience in managing crowdsourcing for business model innovation projects for SMEs.

Anne Berthinier-Poncet is an Associate Professor at Cnam, Paris, where she teaches innovation management. She holds a PhD degree in Management Sciences obtained in 2012 at University Savoie Mont-Blanc. A member of the LIRSA Lab, her main research areas focus on collaborative innovation in the specific context of clusters (technopoles, competitiveness clusters) and makerspaces (fablabs). She is particularly interested in topics such as cluster governance, knowledge management at the collective level, and institutional work. Before her academic career, she worked for 15 years in the industrial sector as International Sales and Marketing Director.

Barbara Bigliardi is an Associate Professor at the Department of Engineering and Architecture of the University of Parma where she teaches Economics and Corporate Organization. Her expertise and key research interests concern the implementation of open innovation and technology transfer. On these topics, she has authored or co-authored more than 50 papers published in international journals, as well as in national and international conference proceedings. She acts as referee for more than 50 international scientific journals and conferences. She is also member of the editorial board for three international journals.

Marcel Bogers is a Professor of Innovation and Entrepreneurship at the Department of Food and Resource Economics (Unit for Innovation, Entrepreneurship and Management at the Section for Production, Markets and Policy), University of Copenhagen. He obtained a combined BSc and MSc in Technology and Society (Innovation Sciences) from Eindhoven University of Technology and a PhD in Management of Technology from Ecole Polytechnique Fédérale de Lausanne (Swiss Federal Institute of Technology). He previously held (visiting) positions at University of Southern Denmark (where this research was conducted), Chalmers University of Technology, and University of Trento. His main interests center on the design, organization and management of technology, innovation, and entrepreneurship in general, and on openness and participation in innovation and entrepreneurial processes in particular. More specifically, he has studied areas such as business models, open innovation, users as innovators, collaborative prototyping, family firms, improvisation, learning-by-doing, and university-industry relations.

Henry W. Chesbrough is best known as "the father of Open Innovation." He teaches at the Haas School of Business at the University of California-Berkeley, where he heads the Garwood Center for Open Innovation. His research focuses on managing technology and innovation. His first book was *Open Innovation* (Harvard Business School Press, 2003), where he discussed why companies must access external as well as internal technologies, and take them to market through internal and external paths. His next book, *Open Business Models* (Harvard Business School Press, 2006), extended his analysis of innovation to business model innovation, intellectual property management, and markets for innovation. *Open Services Innovation* (Jossey-Bass, 2011) studies open innovation in the services realm and examines the business model implications of shifting from products to services. He has been recognized as one of the leading business thinkers by Thinkers50. He received an Innovation Luminary award from the European Commission and Intel in 2014. He received the Industrial Research Institute Medal of Achievement in 2017, and has honorary doctorates from Hasselt University and the University of Vic.

Jonathan Corney is a Professor of Design and Manufacture at the University of Strathclyde. He graduated in Mechanical Engineering in 1983 and worked as a junior robot designer for the Westinghouse Electric Corp. He subsequently became a researcher at Edinburgh University's Department of Artificial Intelligence, before joining Heriot-Watt University as a lecturer, where he researched topics in mechanical CAD/CAM (e.g., feature recognition, 3D content-based retrieval). He has been principal investigator on over £1.1 million of EPSRC-funded research. He has published two books and over 70 papers on various aspects of CAD/CAM and advanced manufacturing. Since taking up the chair of "Design and Manufacture" at Strathclyde University in 2007, his research interests have ranged from remanufacturing and intelligent CAD/CAM to design innovation processes. His current research involves "Cloud Manufacturing Services," "Crowdsourcing for Industrial Applications," and "Intelligent CAD/CAM Interfaces."

Claudio Dell'Era is an Associate Professor in Design Strategy at the School of Management of Politecnico di Milano, where he also serves as Co-Founder of LEADIN'Lab, the Laboratory of LEAdership, Design and INnovation. His research activities are concentrated in the area of Design Strategy and Design Thinking, and he has published in key international journals including *Journal of Product Innovation Management, Long Range Planning, R&D Management, International Journal of Operations & Production Management, Industry & Innovation, International Journal of Innovation Management.*

Chiara Eleonora De Marco is a PhD candidate in Management of Innovation at Scuola Superiore Sant'Anna (Pisa, Italy). Her research interests focus on open innovation strategies and the innovation policy for the small business. Chiara investigates the downsides of implementing open innovation in public and private R&D to support innovation policy decision-makers. While visiting Haas

School of Business (UC Berkeley, CA), Chiara has conducted research on the knowledge transfer and the mobility of high-skilled workers, and the role of public policies in fostering private OI implementation. Before starting her PhD, Chiara was Junior Advisor to the Italian Minister of Education, University and Research on Innovation Policy and Technology. Chiara graduated in Law, gained two Master's degrees, respectively, in Diplomatic Studies and International Relations, and has a lot of experience in studying and working abroad.

Alberto Di Minin is an Associate Professor of Management at Scuola Superiore Sant'Anna (Pisa, Italy) and Research Fellow with the Berkeley Roundtable on the International Economy (BRIE), University of California, Berkeley, and Social Innovation Fellow with the Meridian International Center of Washington, DC. He is currently the Italian representative of the SMEs and Access to Finance Programme Committee, for Horizon 2020, with the European Commission; the Co-director of the Executive Doctorate in Business Administration Program at the Sant'Anna; the Director of the Confucius Institute of Pisa; and the Director of the Galilei Institute in Chongqing University. Alberto teaches Innovation Management and Innovation Policy, and his research deals with open innovation, appropriation of innovation, and science and technology policy. He also works on technology transfer, intellectual property, and R&D management. He is co-author of *Fiat: Open Innovation in a Downturn (1993–2003)* (Springer, 2010, with A. Di Minin and F. Frattini) and "Open Social Innovation" (*New Frontiers in Open Innovation*, Oxford, 2014, with H. Chesbrough).

Lawrence Dooley (MComm, PhD) is a College Lecturer in Enterprise and Innovation at University College Cork (UCC) since 2004. Prior to joining UCC, he was based at the Centre for Enterprise Management in the University of Dundee, Scotland. He undertook his doctoral thesis entitled "Systems Innovation Management" at the National University of Ireland, Galway. His core research interests focus on organizational innovation and issues related to interenterprise collaboration and value creation from the SME perspective.

Other related interests include knowledge exchange and discovery and the unique demands of lower-tech firms when engaging in innovation. He has published widely over recent years, actively liaises with industry both through applied research projects, consultancy and research seminars, and is also Associate Editor of the *R&D Management* journal.

Francesco Galati is an Assistant Professor at the Department of Engineering and Architecture of the University of Parma. He received his PhD in Industrial Engineering from the same university. His research activities mainly concern the management of technological innovation, and he also works on technology transfer, knowledge asset, and intellectual capital management and entrepreneurship. He has authored or co-authored more than 30 papers published in international journals, as well as in national and international conference proceedings. He acts as referee for more than 20 international scientific journals.

Giancarlo Giudici is an Associate Professor of Corporate Finance at Politecnico di Milano. He belongs to the faculty of MIP Graduate School of Business where he teaches Finance. He has written several publications in domestic and international journals on the topics of entrepreneurship, corporate financing, listings and IPOs, venture capital, and crowdfunding. He is the Director of the Italian Observatories on Mini-Bond and Crowdinvesting at Politecnico di Milano, School of Management. He led several projects' financed by public and private entities on the topics of competitiveness and firm financing. He is Adjunct Professor at the Ton Duc Thang University in Ho Chi Minh City (Vietnam).

Aineias Gkikas is a Research Fellow in SME Growth and Development, Birmingham City University Business School, and is currently completing his PhD in Economic Geography, Cardiff University. Aineias is interested in research concerning the growth and performance of small and medium-sized enterprises (SMEs) from an international perspective. His research concentration is growth and performance of SMEs, focusing on developing global insights into SME growth and performance helping to inform the way smaller businesses work, grow, and prosper in the economy. Aineias has taught at both Cardiff and University College London and presented research at national and international conferences. His role within the Centre for Enterprise, Innovation and Growth is progressing research on SME development in international contexts, collaborating with businesses and universities across the world including China, Greece, Slovakia, Slovenia, and Indonesia.

Simona Grama-Vigouroux is an Assistant Teacher at Business School of Troyes. She is teaching innovation and entrepreneurship lessons. She has a PhD degree from IAE de Lyon in Innovation Management. Her main research areas are focused on the human side of open innovation, the impact of support structures as technopoles and business incubators on entrepreneurs' performance, and the impact of entrepreneurial competences on the start-up success. Before integrating the educational field, she worked in the online recruitment industry in Romania.

Marjolein Hins is Founding Partner of Q-Search, a professional services network in the Netherlands for the development of people, organizations, and society. Marjolein works as a strategic connector, providing advice, inspiration, and collaboration to organizations large and small that seek sustainability and impact creation. Additionally, she specializes in developing and facilitating partnerships, networked companies, and/or open innovation ecosystems in various industries and domains. Throughout her career, Marjolein has initiated various social movements and has been a guest author for a.o. DuurzaamNieuws (sustainability) and Platform O. (government). Currently, she is part of networks of (practice) scientists in work and education, for example, CRN Career Research Network and MIG Social Intervision Group, and provides advice and guidance to policymakers in the Netherlands and elsewhere in Europe.

Mokter Hossain is an Assistant Professor at the Center for Industrial Production, Aalborg University, Denmark. He was a postdoctoral researcher at Imperial College London, after graduating with PhD in

Technology and Knowledge Management in 2016 from the Institute of Strategy and Venturing, Department of Industrial Engineering and Management, Aalto University. His research interests include innovation, strategy, and entrepreneurship. He has published articles on a range of research topics, such as Open Innovation, Crowdsourcing, Crowdfunding, Frugal Innovation, Reverse Innovation, Grassroots Innovation, and Business Model Innovation. He also has a good understanding of some emerging phenomena, such as Sharing Economy and Industry 4.0.

Dylan Jones-Evans O.B.E. is a Professor Entrepreneurship and Assistant Pro-Vice Chancellor for Enterprise at the University of South Wales, UK. He has published over 100 articles within edited books, academic conference proceedings, and refereed journals (including *R&D Management, Entrepreneurship Theory and Practice, International Small Business Journal, Technovation and Regional Studies*). Along with Professor Sara Carter, he is the author of the best-selling textbook *Enterprise and Small Business.*

Niall G. MacKenzie is a Senior Lecturer at the Hunter Centre for Entrepreneurship, University of Strathclyde, where he is also Director of Postgraduate Research and the principal investigator on a European Commission-funded technology commercialization project working in conjunction with Fraunhofer and the Swedish University of Agricultural Sciences among others. He is a research associate at the Centre for Business Research at the University of Cambridge and the Centre for Business History in Scotland at the University of Glasgow and served as treasurer of the Association of Business Historians in the United Kingdom for three years until July 2016. He previously worked at the universities of Glasgow, Cambridge, and Wales before joining Strathclyde. His work has been published in *Journal of Product Innovation Management, Small Business Economics, Asia Pacific Journal of Management, Business History*, and other journals.

Stefano Magistretti is a Research Fellow in Design and Innovation Management in the Department of Management, Economics and

Industrial Engineering of Politecnico di Milano, where he is also a PhD candidate. His research interests focus on technology innovation and innovation of meanings, in particular, how to foster radical innovation of meanings starting from the early stage of the development of a technology. His previous research focused on the relationship between innovation and corporate governance systems and particularly how family firms compared to nonfamily firms influence the innovation process.

Cristina Marullo is a Research Fellow at the Institute of Management, Scuola Superiore Sant'Anna (Pisa, Italy). She holds a Master's Degree in Economics from the University of Florence and received a PhD in Innovation Management from Scuola Superiore Sant'Anna upon completion of the dissertation: "The S&T antecedents of 'Small is Beautiful': testing R&D Management effectiveness in Small High-Technology Firms." Her research focuses on entrepreneurship and R&D and innovation management in SMEs. Actually she is interested in the impact of open innovation strategies and organizational learning processes on SMEs and start-up growth. Before the PhD, she had been working as a junior analyst in a public government research center supporting the creation of regional and national policy measures sustaining entrepreneurship and innovating SMEs.

Tim Minshall is the Head of the Institute for Manufacturing (IfM) as well as of the IfM's Centre for Technology Management. He is the inaugural Dr. John C. Taylor Professor of Innovation. He researches, teaches, writes, and consults on the topics of open innovation, technology enterprise, the financing of innovation, and university–industry knowledge exchange. He is a Nonexecutive Director of St John's Innovation Centre Ltd, Cambridge and a Visiting Professor at Doshisha University Institute for Technology, Enterprise and Competitiveness in Japan. He has a BEng from Aston University, and a PhD. from Cambridge University Engineering Department. His current research is focused on innovation collaborations, the emergence of additive manufacturing/3D printing, and the development of engineering skills.

Letizia Mortara is a Senior Research Associate at the University of Cambridge IfM's Centre for Technology Management. Her expertise and key research interests concern the understanding of how companies implement open innovation and technology intelligence systems. Letizia's research currently concentrates on understanding how Digital Fabrication technologies could have an impact on innovation and manufacturing. She has authored several papers and reports in these areas. She is also an Associate Editor for the *R&D Management* journal. Prior to joining the University of Cambridge in 2005, Letizia gained a first degree in Industrial Chemistry from the University of Bologna in Italy. After spending three years working as a process/product manager in the chemical industry, she moved to the UK where she obtained her PhD in processing and scale-up of advanced ceramic materials at Cranfield University.

David O'Sullivan is a Professor of Industrial Engineering at the School of Engineering and Informatics and also Director of Quality, both at the National University of Ireland Galway. His research interests are in the area of innovation management and quality in higher education. His most recent projects include innovation management within SMEs, distributed innovation management across extended enterprises, and the impact of quality assurance in higher education. David has over 150 publications including books — *Applying Innovation* (Sage); *Manufacturing Systems Redesign* (Prentice-Hall); *Reengineering the Enterprise* (Chapman & Hall), and *The Handbook of IS Management* (Auerbach). David works with leading organizations where innovation is a core value including IBM, Thermo King, Fujisawa, Hewlett-Packard, and Boston Scientific. David also plays a key role in strategic and operational planning, quality assurance and enhancement, and performance development at his university.

Andrea Piccaluga is a Full Professor of Innovation Management at the Institute of Management, Scuola Superiore Sant'Anna in Pisa, Italy where he is coordinator of the International PhD programme in Management (Innovation, Sustainability, and Healthcare). He is President of Netval (www.netval.it), the Italian network of University

Technology Transfer Offices. He holds a PhD from Scuola Superiore Sant'Anna and a Master's in Technology and Innovation Management from SPRU (University of Sussex, Brighton). He is Associate Editor of the *R&D Management Journal* and *Creativity and Innovation Management Journal*. He has published papers and books in the field of R&D management and technology transfer. He is co-author of *Fiat: Open Innovation in a Downturn (1993–2003)* (Springer, 2010) and *La gestione del trasferimento tecnologico* (Springer). He collaborates with large and medium-sized companies (Knauf, GE Oil&Gas, Loccioni, TT Venture) in the field of open innovation and with the regional governments of Puglia and Tuscany, Italy, in the field of policies for entrepreneurship and technology transfer. He is member of the board of SIAF (Scuola Internazionale di Alta Formazione) in Volterra, Italy.

Oana-Maria Pop is a PhD candidate at the Department of Marketing and Strategy at Hasselt University (BE). Her scholarly background is in innovation management, a field she became acquainted with during her Master's studies at Aarhus University in Denmark. Her research interests include innovation ecosystems, specifically: the emergence and evolution of ecosystems, ecosystems' potential for impact creation, and ecosystem actor capabilities. Prior to joining academia, Oana was an editor and marketing consultant in Denmark and Sweden, working closely with innovation management services firms, multinationals, public authorities, universities, as well individual innovators. Oana also writes regularly on the HYPE Innovation Blog and is involved in editing case studies and teaching materials.

Joaquin Proenca is a Business Management and Administration PhD student in Universitat Politècnica de València (UPV). His research interests are culture and innovation management where he examines open innovation strategy with a particular focus on the role played by users in new services development. He has a Master's degree in Scientific Culture and Innovation (UPV), Valencia, Spain, Tourism Activities Management (UNED), Madrid, Spain, and a

Marketing postgraduate in ISG Economics and Business School, Lisbon, Portugal. He has professional experience in marketing communications, branding assistance, and corporate and retail banking.

Agnieszka Radziwon is an Assistant Professor of Innovation at the Department of Business Development and Technology, Aarhus University. She obtained her PhD degree in Product Design and Innovation from University of Southern Denmark. She was a visiting researcher at Eindhoven University of Technology, Vienna University of Technology, Chalmers University of Technology, and University of Copenhagen. Her main interests center on the design, organization, and management of technology. In particular, her research focuses on open innovation, business models as well as regional ecosystems collaboration and alliances, which could help SMEs in getting more competitive. Agnieszka has published her work in journals like *Industry and Innovation, International Journal of Technology Management* and *Technological Forecasting and Social Change.*

Martin Rhisiart was a Professor of Strategy and Innovation at South Wales Business School where he was Director of the Centre for Research in Futures and Innovation. Martin designed and delivered a range of international research projects on innovation and strategic foresight. His work was funded by a range of national and international bodies, including the UK Commission for Employment and Skills, Arts and Humanities Research Council (UK), DG Research (European Union), Forfas (National Innovation Agency, Ireland), and the Welsh Government. Martin completed his undergraduate degree at the London School of Economics and his PhD at Cardiff University. Throughout his career, Martin worked on a range of international and national projects on innovation-related themes, including open innovation, new product development, and eco-innovation. Martin passed away during the production of this book.

Cristina Rossi-Lamastra is an Associate Professor at the Politecnico di Milano School of Management, where she serves as Director of the Executive MBA, part time. Her research interests are in the area of entrepreneurship, organizational economics, and user innovation. Cristina Rossi-Lamastra has published on these topics in *Science, Management Science, Entrepreneurship Theory & Practice, California Management Review, Research Policy, Long Range Planning, Small Business Economics*, and other scholarly journals.

Diana Rus is a Managing Partner at Creative Peas and a Senior Lecturer in Organizational Psychology at the University of Groningen (NL). She received her PhD from the Rotterdam School of Management at Erasmus University in the Netherlands. Her field of work is (open) innovation management and leadership development, and her current research interests center around leadership processes in open innovation management. Her work is published in various organizational behavior and psychology journals such as *Journal of Experimental Social Psychology, The Leadership Quarterly*, and *Social Psychological and Personality Science.*

Sana Saidi has a PhD in Management Sciences from the University of Toulouse 1 since 2011. She is actually an Assistant Professor in Finance at Business School of Troyes. Her main fields of research lie at the frontier between accounting and entrepreneurship. They focus on the sociocultural and politicoinstitutional practices of audit firms and accountants, the entrepreneurial motivation and the innovation performance of firms in the technopoles.

Marc Steen works as a Senior Research Scientist at TNO, an independent research and innovation organization in the Netherlands. He earned MSc, PDEng, and PhD degrees in Industrial Design Engineering at Delft University of Technology. He worked at Philips and KPN before joining TNO. His expertise is in human-centered design, open innovation, and applied ethics. He published articles on these topics, and on value-sensitive design, participatory design, the capability approach, and innovation management. He is interested in helping organizations to focus their innovation and design projects on promoting people's well-being and on creating sustainable impact (people, planet, profit). His current research is focused on ethical and societal issues that arise in (big) data applications and innovation (e.g., Machine Learning, Internet of Things), especially in ways to promote freedom, fairness, and accountability.

Roberto Verganti is a Full Professor of Leadership and Innovation at the Politecnico di Milano where he also serves as the Scientific Director of the MaDe In Lab, the laboratory for education in management of design and innovation. He has twice been a visiting scholar at Harvard Business School and a visiting professor of Design Management at the Copenhagen Business School. Roberto Verganti has published over 150 articles that lie at the intersection of strategy, design, and technology management in journals including *Management Science, Journal of Product Innovation Management, Harvard Business Review*. He has been featured in *The Wall Street Journal, The New York Times, Financial Times,* and *BusinessWeek*. He is also a regular contributor to the *Harvard Business Review* online magazine.

Contents

Introduction

Federico Frattini, Muhammad Usman,
Nadine Roijakkers and Wim Vanhaverbeke

Since the seminal book written by Henry Chesbrough in 2003 (Chesbrough, 2003), Open Innovation (OI) has become one of the most debated topics in management and innovation research (Durst and Ståhle, 2013; West and Bogers, 2014). Different conceptualizations of OI exist, and scholars may have a different understanding of what OI entails. A widely accepted definition of OI qualifies it as the "purposive inflows and outflows of knowledge to accelerate internal innovation, and expand the markets for external use of innovation, respectively" (Chesbrough 2006, p. 1).

OI is a strategic approach to industrial innovation management that has emerged as a response to a number of competitive challenges, such as cross-fertilization in knowledge generation, increasing costs for technology development, shortening of the life cycle of new technologies, shrinking returns of investments in R&D and technological innovation, and easier knowledge leakage and spillovers (Chesbrough, 2004; Chesbrough and Crowther, 2006; Feller *et al.*, 2009; Chiaroni *et al.*, 2010; Bogers, 2011). It has encouraged managers to shift their attention from the importance of controlling technologies for innovation purposes to developing the know-where and absorptive capacity that allow them to promptly and cost-effectively scan, identify, and use technological knowledge already available outside organizational boundaries.

The OI concept has two dimensions, namely, outside–in and inside–out OI. Outside–in OI entails systematically accessing and using external technological knowledge to improve internal innovation processes. On the other hand, inside–out OI requires transferring internal technologies, disembodied from physical artefacts, to external partners in exchange of a monetary compensation (Gassmann and Enkel, 2004; Chesbrough and Crowther, 2006; Dahlander and Gann, 2010; Huizingh, 2011; Lichtenthaler, 2011; Usman and Vanhaverbeke, 2017).

Firms have been adopting different approaches to realize outside–in OI, such as using online OI intermediaries, opening proprietary OI platforms, working with OI service providers, entering joint development agreements, networks or consortia with universities, suppliers, clients or firms from other industries, realizing corporate venture capital investments, and working with innovative start-ups (Spithoven *et al.*, 2010; Parida *et al.*, 2012; West and Bogers, 2014; Brunswicker and Vanhaverbeke, 2015). On the other hand, inside–out OI can be pursued through selling or out-licensing patents, spinning-out new ventures, entering joint ventures or other non-equity partnerships to jointly commercialize a technology (Frishammar *et al.*, 2012; Inauen and Schenker-Wicki, 2012).

Research has clearly indicated that successfully implementing OI not only require firms to create organizational solutions that allow them to quickly and effectively access external knowledge, but also that it is equally important that firms modify their internal organization (including structures, processes, culture, norms, and beliefs) to allow for a proper sharing, integration, and adaptation of the externally accessed knowledge for innovation purposes (Spithoven *et al.*, 2010; Brunswicker and Vanhaverbeke, 2015; Huang *et al.*, 2015; Ahn *et al.*, 2016; Jantunen, 2005).

Most OI research has addressed OI processes and activities in large, R&D-intensive companies, which have been the enthusiastic early adopters of this strategic approach to industrial innovation.

Few studies have investigated the antecedents, processes, and consequences of OI in Small and Medium-Sized Enterprises (SMEs). Some notable exceptions are: van de Vrande *et al.* (2009), Bianchi *et al.* (2010), Lee *et al.* (2010), Parida *et al.* (2012), Vanhaverbeke (2012, 2017), Spithoven *et al.* (2013), Brunswicker *et al.* (2014), Brunswicker and Vanhaverbeke (2015) and Dufour and Son (2015).

These studies indicate that SMEs are adopting OI to overcome a numbers of challenges, including lack of necessary resources and complementary assets, increasing globalization, aggregating competition, commoditization pressure, and funding constraints. These challenges prompt SMEs to explore new avenues of innovation, among which, pursuing OI (Van de Vrande *et al.*, 2009; Crema *et al.*, 2013; Dahlander and Gann, 2010). In particular, the common lack of internal R&D capability drives SMEs toward the search for external sources for technology exploitation (Spithoven *et al.*, 2010). Moreover, SMEs have flexible organizational structures and are more adoptable to change. This flexibility and adoptability can allow SMEs to benefit from OI more than their larger counterparts (Parida *et al.*, 2012; Dufour and Son, 2015). Gassmann (2006) suggests that suppliers' integration can serve as a boost for innovation in SMEs. Overall, OI appears to be a useful tool to overcome the liability of smallness that characterizes SMEs (Gassmann *et al.*, 2010), as research pointing to the positive effect of OI on SMEs' innovation performance suggests (Bianchi *et al.*, 2010, Suh and Kim, 2012; Minguela-Rata *et al.*, 2014; Bjerke and Johansson, 2015; Vanhaverbeke, 2017).

The limited understanding of OI in SMEs is surprising for a number of reasons. On one hand, the impact of SMEs in industrialized and developing economies is unquestionable. For instance, in the European Union, SMEs account for 99% of businesses, provide 67% of the jobs, and generate 57% of every euro of value-added (Muller *et al.*, 2015). In the UK, there are 5.5 million SMEs accounting for over 99% of the private business sector and covering 60% jobs (Rhodes, 2006). Similar data characterize the economy of the United States (U.S. Census Bureau Update 2002–2010). In the US, SMEs were responsible for 24% of all R&D expenses in 2005, compared with 4% in 1981 (National Science Foundation, 2006). As a result of this, encouraging innovation in SMEs has been acknowledged as a priority for policymakers at the local, regional, and national levels (Jones and Tilley, 2003).

Moreover, research has shown that innovation in SMEs is subject to idiosyncratic challenges and follows peculiar processes (Acs and Audretsch, 1990; Cobbenhagen, 2000). A common finding of this research is that SMEs have the advantage of flexibility and adaptability in comparison with larger companies, which makes them especially good at producing inventions and new ideas and increases their R&D productivity (Audretsch and Vivarelli, 1996; Laursen and Salter, 2004). However, SMEs often do not control the complementary assets (such as manufacturing, distribution, and marketing capabilities) that play a critical role in transforming inventions into new products and processes and in capturing value from innovation (Freel, 2000; Narula, 2004).

Furthermore, SMEs usually do not have the required breadth and diversity of in-house knowledge and technological resources needed to keep them abreast of the most recent technological developments and to promptly respond to changing market and competitive conditions (Lambrechts *et al.*, 2017). Opening the boundaries of their innovation process and systematically using externally generated knowledge are therefore of paramount importance for SMEs to successfully innovate over time, as noted above. Indeed, there are preliminary data and anecdotal evidence suggesting that SMEs are increasingly using OI to overcome these liabilities (van de Vrande *et al.*, 2009; Bianchi *et al.*, 2010; Spithoven *et al.*, 2013; Theyel, 2013; Vanhaverbeke, 2017).

This notwithstanding, the above-mentioned lack of systematic research on OI in SMEs leaves practitioners with little or no guidance as to how this approach can and should be adopted in smaller firms and with no understanding of the barriers that should be overcome to successfully use OI to foster innovation. This explains the recent calls for more research on OI in SMEs, where the antecedents, dynamics, and performance consequences of OI may substantially differ from what is known in existing research.

This edited book aims to offer a response to these calls, by collecting the most recent theoretical and empirical research on the challenges, peculiarities, and approaches of OI in SMEs. It covers a number of important aspects that advance theoretical and empirical research on OI in SMEs and provides managers and policymakers with valuable insights about how OI could be encouraged and adopted in these organizations.

We divide the book into five parts:

Part I: "State of the Art on Open Innovation in SMEs" (Chapters 1 and 2)
Part II: "Adopting Open Innovation in SMEs" (Chapters 3–6)
Part III: "Networks and Innovation Ecosystems" (Chapters 7 and 8)
Part IV: "Sectoral Patterns of Open Innovation in SMEs" (Chapters 9–12)
Part V: "Measuring, Evaluating and Stimulating Open Innovation in SMEs" (Chapters 13–15).

In the first part of the book, an updated and detailed overview of the current knowledge on OI in SMEs is provided, with one chapter offering a systematic literature review of the topic and another chapter addressing the challenges that adopting OI in SMEs entails.

In the second part, the focus of the volume shifts to the implementation of OI in SMEs, by illustrating a number of approaches, tools, and mechanisms that influence the capability of SMEs to successfully engage in OI processes. One chapter introduces an innovative model, which summarizes the key factors motivating and hindering the adoption of OI in SMEs. Another chapter shows how the concept of frugal innovation can be used to understand the impact that OI in SMEs can have on societal and inclusive development. Another topic covered in this section deals with the role of entrepreneurial CEOs in the implementation of OI in SMEs.

In the third part, the volume places OI in SMEs into an innovation ecosystems perspective, by focusing on the relationships that SMEs establish with external actors in networked innovation processes. One chapter studies how SMEs embedded in a regional innovation ecosystem perceive and manage OI processes by interacting with external stakeholders such as suppliers and competitors. The other chapter carefully examines the breadth and depth of SMEs' collaborations for innovation. In this chapter, the authors explore the nature of innovation within four SMEs in low- and medium-tech industries and the extent to which these firms are leveraging OI modes to support their innovation activity in terms of the breadth and depth of the collaborations engaged with.

The fourth part then explores the application of OI in SMEs working in different sectors, in the attempt to highlight the idiosyncratic behaviors due to different industry belonging. This represents an important contribution to the development of a generalizable and eternally valid theory of OI in SMEs. One chapter in this part of the volume looks into the OI practices of clustered SMEs in a French Technopole, another focuses on the collaborative practices between SMEs and external designers in the Italian furniture industry. The next chapter unravels how the personality of the entrepreneurial orchestrator reflects upon the SME ecosystem forming around him/her. The findings of this chapter suggest that studying the psychology of entrepreneurial orchestrators can reveal a great deal about the networks and ecosystems they help shape. The last chapter in this part looks into crowd funding as an emerging approach to OI. In this chapter, the authors show how SMEs can leverage their social relations with extant customers and innovative users to raise equity capital.

Finally, in the last part, the volume closes by discussing the issues linked with measuring OI in SMEs, by presenting the concept of the OI project canvas for SMEs and by illustrating how policy intervention can foster OI in small firms.

By covering this breadth of topics, this edited volume contributes to pushing the frontier of academic research on OI in SMEs forward, and it will hopefully open a lot of avenues for future studies. Scholars will find plenty of suggestions about how to continue the academic debate around OI in SMEs, which will also strongly benefit OI research in general. Indeed, by investigating the peculiar antecedents, processes, and consequences of OI in SMEs, the volume contributes to increase the external validity and generalizability of the findings of OI research to an empirical setting, which is different from the OI research has developed so far. Of course, the volume also represents a valuable source of insights for managers working in SMEs, who will find useful material to identify the benefits that OI may bring in their organizations, the challenges and barriers to be overcome to successfully adopt OI, the peculiar approaches that its implementation requires, and the relationships that SMEs can establish with the broader ecosystem in which they are embedded for innovation purposes.

Innovating SMEs are fragile since they do not have extensive funding and in-house competencies to adopt new technologies and apply them in their new products or services. The volume shows in different chapters — and in the last chapter in particular — that OI in SMEs can and should be stimulated by governments at a regional, national, and international level. SMEs innovate in a specific way and face particular problems during the innovation process: policy-makers have to examine how to adapt policy instruments as the traditional measures apt for large companies and high-tech start-ups might not be appropriate for the bulk of the innovating SMEs.

References

Acs, Z. J. and D. B. Audretsch (1990). *Innovation and Small Firms*, MIT Press.

Ahn, J. M., Y. Ju, T. H. Moon, T. Minshall, D. Probert, S. Y. Sohn and L. Mortara (2016). Beyond absorptive capacity in open innovation process: The relationships between openness, capacities and firm performance, *Technology Analysis & Strategic Management*, 28(9), pp. 1009–1028.

Audretsch, D. B. and M. Vivarelli (1996). Firms size and R&D spillovers: Evidence from Italy, *Small Business Economics*, 8(3), pp. 249–258.

Bianchi, M., S. Campodall'Orto, F. Frattini and P. Vercesi (2010). Enabling open innovation in small- and medium-sized enterprises: How to find alternative applications for your technologies, *R&D Management*, 40(4), pp. 414–431.

Bjerke, L. and S. Johansson (2015). Patterns of innovation and collaboration in small and large firms. *The Annals of Regional Science* 55(1), pp. 221–247.

Bogers, M. (2011). The open innovation paradox: Knowledge sharing and protection in R&D collaborations, *European Journal of Innovation Management*, 14(1), pp. 93–117.

Brunswicker, S. and W. Vanhaverbeke (2015). Open innovation in small and medium-sized enterprises (SMEs): External knowledge sourcing strategies and internal organizational facilitators, *Journal of Small Business Management*, **53**(4), pp. 1241–1263.

Chesbrough, H. (2003). Open *Innovation: The New Imperative for Creating and Profiting from Technology*, Harvard Business Press.

Chesbrough, H. (2004). Managing open innovation, *Research-Technology Management*, **47**(1) pp. 23–26.

Chesbrough, H. (2006). Open innovation: A new paradigm for understanding industrial innovation, In: *Open Innovation: Researching A New Paradigm*. H. Chesbrough, W. Vanhaverbeke and J. West, pp. 1–19.

Chesbrough, H. and A. K. Crowther (2006). Beyond high tech: Early adopters of open innovation in other industries, *R&D Management*, **36**(3), pp. 229–236.

Chiaroni, D., V. Chiesa and F. Frattini (2010). Unravelling the process from closed to open innovation: Evidence from mature, asset-intensive industries, *R&D Management*, **40**(3), pp. 222–245.

Cobbenhagen, J. (2000). *Successful Innovation: Towards a New Theory for the Management of Small and Medium Sized Enterprises*, Edward Elgar Publishing.

Dahlander, L. and D. M. Gann (2010). How open is innovation? *Research Policy*, **39**(6), pp. 699–709.

Durst, S. and P. Ståhle (2013). Success factors of open innovation — A literature review,*International Journal of Business Research and Management*, **4**(4), pp. 111–131.

Feller, J., P. Finnegan, J. Hayes and P. O'Reilly (2009). Institutionalising information asymmetry: Governance structures for open innovation, *Information Technology & People* **22**(4), pp. 297–316.

Freel, M. S. (2000). Do small innovating firms outperform non-innovators? *Small Business Economics*, **14**(3), pp. 195–210.

Frishammar, J., U. Lichtenthaler and J. Rundquist (2012). Identifying technology commercialization opportunities: The importance of integrating product development knowledge, *Journal of Product Innovation Management*, **29**(4), pp. 573–589.

Gassmann, O. and E. Enkel (2004). Towards a theory of open innovation: Three core process archetypes. Proceedings of The R&D Management Conference, Lisbon, Portugal, July 6–9.

Huang, H. C., M. C. Lai and W. W. Huang (2015). Resource complementarity, transformative capacity, and inbound open innovation, *Journal of Business & Industrial Marketing*, **30**(7), pp. 842–854.

Huizingh, E. K. (2011). Open innovation: State of the art and future perspectives, *Technovation*, 31(1), pp. 2–9.

Inauen, M. and A. Schenker-Wicki (2012). Fostering radical innovations with open innovation, *European Journal of Innovation Management*, **15**(2), pp. 212–231.

Jantunen, A. (2005). Knowledge-processing capabilities and innovative performance: An empirical study, *European Journal of Innovation Management*, **8**(3), pp. 336–349.

Jones, O. and F. Tilley (2003). *Competitive Advantage in SMEs: Organising for Innovation and Entrepreneurship*, Chichester: Wiley.

Lambrechts, F., W. Voordeckers, N. Roijakkers and W. Vanhaverbeke (2017). Exploring open innovation in entrepreneurial private family firms in low- and medium-technology industries, *Organizational Dynamics*. Published online.

Laursen, K. and A. Salter (2004). Searching high and low: what types of firms use universities as a source of innovation? *Research Policy*, **33**(8), pp. 1201–1215.

Lichtenthaler, U. (2011). Open innovation: Past research, current debates, and future directions, *The Academy of Management Perspectives*, **25**(1), pp. 75–93.

Minguela-Rata, B., J. Fernández-Menéndez and M. Fossas-Olalla (2014). Cooperation with suppliers, firm size and product innovation, *Industrial Management & Data Systems*, **114**(3), pp. 438–455.

Muller, P., C. Caliandro, V. Peycheva, D. Gagliardi, C. Marzocchi, R. Ramlogan and D. Cox (2015). Annual Report on European SMEs 2014/2015. *European Commission*.

Narula, R. (2004). R&D collaboration by SMEs: New opportunities and limitations in the face of globalisation,*Technovation*, **24**(2), pp. 153–161.

National Science Foundation (2016). *Survey of Industrial Research and Development*, Division of Science.

Parida, V., M. Westerberg and J. Frishammar (2012). Inbound open innovation activities in high-tech SMEs: The impact on innovation performance, *Journal of Small Business Management*, **50**(2), pp. 283–309.

Rhodes, C. (2016). Business Statistics. House of Common Library, pp. 1–14.

Spithoven, A., B. Clarysse and M. Knockaert (2010). Building absorptive capacity to organise inbound open innovation in traditional industries, *Technovation*, **30**(1), pp. 130–141.

Spithoven, A., W. Vanhaverbeke and N. Roijakkers (2013). Open innovation practices in SMEs and large enterprises, *Small Business Economics*, **41**(3), pp. 537–562.

Suh, Y. and M. S. Kim (2012). Effects of SME collaboration on R&D in the service sector in open innovation, *Innovation-Management Policy & Practice*, **14**(3), pp. 349–362.

Theyel, N. (2013). Extending open innovation throughout the value chain by small and medium-sized manufacturers, *International Small Business Journal*, **31**(3), pp. 256–274.

U.S. Census Bureau, S., CPS; *International Trade Administration; Bureau of Labor Statistics, BED; Advocacy-Funded Research* (Update 2002–2010). Small Business GDP.

Usman, M. and W. Vanhaverbeke (2017). How start-ups successfully organize and manage open innovation with large companies, *European Journal of Innovation Management*, **20**(1), pp. 171–186.

van de Vrande, V., J. P. J. de Jong, W. Vanhaverbeke and M. de Rochemont (2009). Open innovation in SMEs: Trends, motives and management challenges, *Technovation*, **29**(6–7), pp. 423–437.

Vanhaverbeke, W. (2017). *Managing Open Innovation in SMEs*, Cambridge University Press, Cambridge: UK.

West, J. and M. Bogers (2014). Leveraging external sources of innovation: A review of research on open innovation, *Journal of Product Innovation Management*, **31**(4), pp. 814–831.

Part I

State of the Art on Open Innovation in SMEs

Chapter 1

A Systematic Review of the Literature on Open Innovation in SMEs

Muhammad Usman, Nadine Roijakkers, Wim Vanhaverbeke and Federico Frattini*

**muhammad_usman@uhasselt.be*

Abstract

This chapter provides a systematic review of the open innovation (OI) research carried out within the context of small and medium-sized enterprises (SMEs). The chapter should provide an accurate understanding of current literature about OI in SMEs, and it elaborates many future research avenues. The recent increase in the number of publications in this field shows that it has gained the attention of the academicians. The existing research shows that SMEs organize and manage OI in an entirely different way from large companies. SMEs get involved in OI based on their own strategic needs, and OI mechanisms have to be designed differently for SMEs.

* Corresponding author.

The chapter maps the research through Web of Science — Core Collection database. We investigated the "OI in SMEs" literature from January 2003 till June 2017. A total of 118 articles, eight book chapters, and six books are analyzed. The discussion is mainly categorized into six broad themes: adoption of OI, the benefits of OI, challenges, role of networking, sectoral patterns, and the role of policymaking. The chapter concludes with several avenues for future research in the field.

1.1. Introduction

This chapter provides a systematic review of the open innovation (OI) research carried out within the context of small and medium-sized enterprises (SMEs). As such, the chapter builds on and extends the work carried out by Brunswicker and Van de Vrande (2014) who have equally conducted a literature search and have identified several interesting themes for further research, such as the important role of SMEs as coordinators/orchestrators of their OI relations and how to optimally fulfill this role. Furthermore, they have concluded that, while the literature on OI has grown exponentially ever since Henry Chesbrough published his groundbreaking work in 2003 (Chesbrough, 2003), SMEs have received only little attention. Most researchers have studied large multinationals and their OI practices, and the lessons drawn from these cases and studies cannot be easily transferred to SMEs; SMEs represent unique contexts in terms of their resource endowments, skill sets, the tight connection between the entrepreneur and the (OI) strategy of the company, etc. (Van de Vrande *et al.*, 2009; Vanhaverbeke, 2012). An explicit focus on the specifics of the SME context when studying these companies is thus warranted. Hence, for the purpose of this book, this chapter extends the efforts of Brunswicker and Van de Vrande (2014) and Hossain and Kauranen (2016) with respect to mapping the OI in SMEs research field, albeit with a focus on the themes covered in subsequent chapters of this edited book.

SMEs are crucial catalysts of both developed and developing economies, accounting for over 99% of all businesses and more than

60% of all jobs created, as well as important sources of innovation (Audretsch, 1995; Muller *et al.*, 2015). As such, governments are searching for ways to increase the productivity of SMEs. Several researchers have identified OI as an important strategy to overcome typical SME weaknesses such as resource (time, money, etc.) constraints and skill gaps (Bougrain and Haudeville, 2002; Dahlander and Gann, 2010; Edwards *et al.*, 2005; Lee *et al.*, 2010; Rahman and Ramos, 2010; Wynarczyk *et al.*, 2013). In order to optimally benefit from OI, however, companies need to develop completely new skill sets (e.g., orchestration capabilities) and create the right atmosphere (e.g., open culture) within their firms. While researchers have generated best practices and lessons on how to bring these changes about within the context of large companies, the insights on SMEs are still relatively thin. This chapter intends to remedy this gap, and one important first step is to map the existing work on OI in SMEs, on the basis of which valuable new insights can be generated. The aims of this chapter are thus twofold:

- mapping the existing literature in the OI in SMEs field so as to provide a basis for subsequent chapters;
- identifying promising areas for future research that subsequent chapters connect to.

In terms of methodology, we conducted a thorough review and analysis of papers published on Web of Science (Core Collection) during the period starting January 1, 2003 till June 19, 2017, following established approaches for systematic literature reviews by many authors (Pittaway *et al.*, 2004; Rousseau *et al.*, 2008; Spender *et al.*, 2017). Our review is concentrated not only on the field of innovation management but also includes related fields (e.g., strategic management) where appropriate. The following methodological steps were taken:

(1) The identification of keywords based on authors' prior experience, an initial assessment of the literature, and brainstorming

sessions. Identified keywords include "open innovation," "SMEs," "collaborative innovation," "strategic alliance," "innovation," "small businesses," etc. These keywords were then organized into search strings, for example: [*open innovation* OR *collaborative innovation* AND *small and medium-sized enterprises* OR *SMEs* OR *small businesses*].

(2) The carrying out of a preliminary search on Google Scholar using the basic search strings. This initial search was used for identifying additional keywords for the main search. Additional keywords include, for example, "inbound innovation," "outbound innovation," "coupled innovation," "entrepreneurial ventures," etc.

(3) The use of the basic search string *open innovation* and *SMEs* in the Web of Science — Core Collection (search engine) database to identify the key citation indexes for review. The selection was made based on the volume of citations relevant to the basic search string. The citation database, Web of Science, was then reviewed using the search strings identified in steps 1 and 2.

(4) The review of identified citations according to the inclusion and exclusion criteria summarized in Table 1.1. Two stages were undertaken to reduce the number of citations. In the first stage, we analyzed the titles of articles according to the exclusion criteria; in the second stage, we analyzed the abstracts according to the inclusion criteria. We then cross-checked the reference sections of the included articles to assess the search strategy.

(5) The review of 126 citations that met the search criteria. These publications were organized into different categories according to frequently addressed issues. As such, a list of themes was compiled that is relevant to the remainder of this book. The themes and relevant publications under each theme were then summarized and discussed.

The remainder of this chapter is structured as follows. In the Section 1.2, we present the results of our descriptive and content

Table 1.1. Inclusion and exclusion criteria.

No.	Criteria	Reasons
Inclusion criteria		
1	Studies focusing on innovative SMEs considering collaboration/strategic alliances/partnerships as a tool or source for innovation	Since the OI phenomenon is new, studies on collaborative innovation, strategic alliances, and partnerships where it leads to innovation for SMEs were also considered
2	Studies focusing on OI that consider SMEs as their point of analysis or vice versa	To keep in line with the overall theme of this book
3	Theoretical, conceptual, empirical, qualitative, quantitative methodologies, literature reviews	The aim was to include almost all relevant academic publications in this study
4	Articles/book chapters published/ indexed in Web of Science — Core Collection along with other books published on OI in SMEs	All the main journals and other outlets are indexed in the Web of Science — Core Collection database
Exclusion criteria		
1	Studies published on related area before year 2003	OI concept was coined in 2003; so, studies only after 2003 are taken into consideration
2	Simple collaboration arrangements	Only studies on collaborations that lead to innovation were considered
3	Conference papers, business magazine articles, editorials, or similar publications were not considered	Only journal articles, books, and book chapters were taken into account
4	Linguistic expression	Articles published in English were only considered; due to linguistic constraints, publications in other languages were not considered

analyses. Next, we describe the main research themes identified. Finally, we present interesting avenues for further research.

1.2. Descriptive and Content Analysis

In this section, we present the results of our review of the 126 publications considered in this study. Out of these 126 publications, 118 are published in various journals, while 8 publications are published as book chapters. In addition to these 126 publications, six books that are published on OI in SMEs, to date, are also included in our analysis and are also separately discussed in Section 1.4. Below is a descriptive and content analysis of our sample research.

1.2.1. *Descriptive analysis*

Figure 1.1 illustrates the number of relevant publications that have appeared in the field of OI in SMEs since 2003.

The figure shows that the study of OI in SMEs is a fairly recent phenomenon. It is particularly worth noting that the first publication, which is an empirical study discussing university and industry collaborations for innovating SMEs (Fontana *et al.*, 2006), appeared

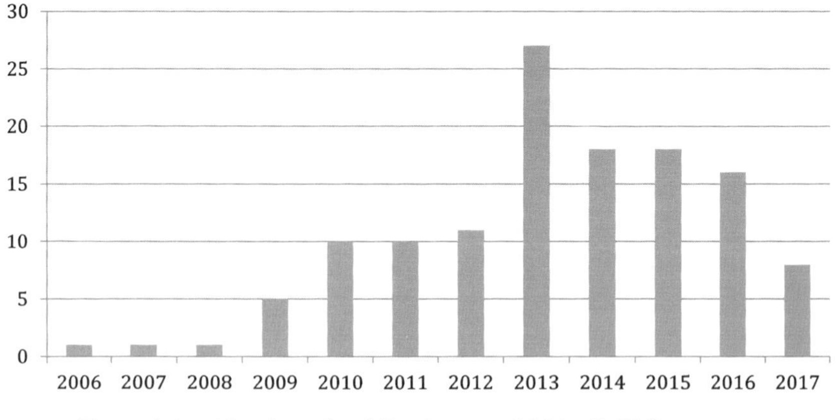

Figure 1.1. Number of publications on "OI in SMEs" per year.

in 2006. The first articles actually mentioning OI in SMEs have only started to appear from 2009 onward. Batterink (2009), for example, discusses the adoption of OI in SMEs as well as in larger companies. Lecocq and Demil (2006) study OI in a low-technology setting and discuss OI strategy from the viewpoint of new entrants to the industry in relation to incumbents that are practicing openness. As important early publications have shown that SMEs adopt OI in a unique way and in relation to different strategic needs than large firms, academics have started to pay more attention to the specific SME context as is evidenced by larger numbers of publications and an increasing citation count.

Figure 1.2 presents the citations per year to these 126 publications. The citations are growing exponentially over the years. This is the result of a combination of a growing number of publications over time and an increase of citations per publication as time progresses.

Figure 1.3 shows the impact of the 126 publications for the period 2009–2014 by showing the number of citations per publication in the

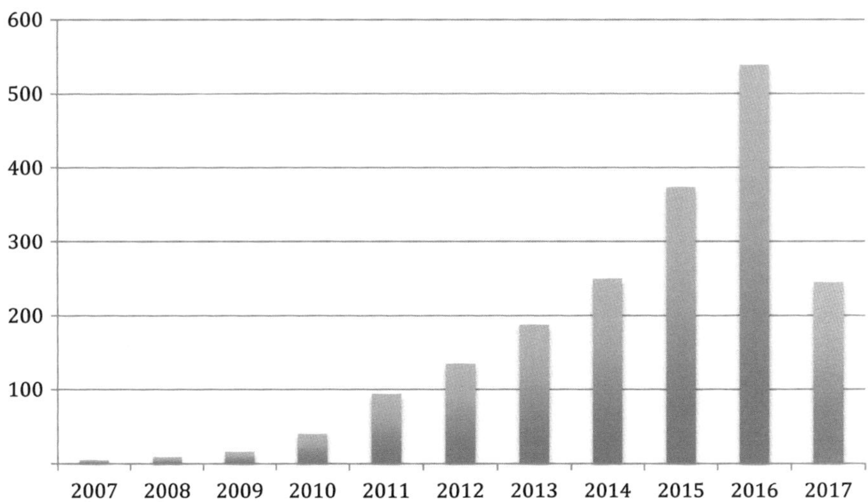

Figure 1.2. Number of citations per year.

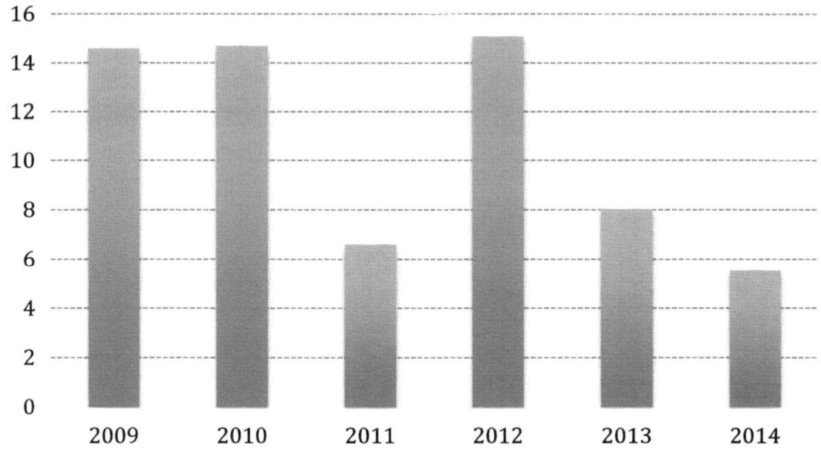

Figure 1.3. Citations per publication for the 126 publications under study (till $t + 3$).

three years following the publications published in a particular year. The articles published after 2014 are excluded since they do not have a window of three years for the citations. The figure, in general, shows the importance of some early publications in this area in 2009 and 2010. We see a higher average citation rate for publications in 2009, 2010, and 2012. This is the result of some highly cited publications — there is highly skewed distribution of the citations across the publications. Therefore, we highlight the most cited publications in Table 1.2.

Table 1.2 shows the topmost cited publications from our sample. The most cited article is from Van de Vrande *et al.* (2009) on the motives and challenges faced by SMEs while adopting various OI activities.

When it comes to publication outlets of these studies, our descriptive analysis shows that the literature on OI in SMEs is scattered across several academic journals. The majority of the articles under review in this chapter are published in *Technovation*, *International Journal of Technology Management*, *International Small Business Management*, *Technology Analysis and Strategic Management*, *Asian Journal of Technology Innovation*, *R&D Management*, and *Research Policy*. Figure 1.4 features the top journal outlets for our

Table 1.2. Top cited publications.

S. No.	Title	Author(s)	Journal	Year	Citations
1	Open Innovation in SMEs: Trends, Motives, and Management Challenges	van de Vrande, V., de Jong, J., Vanhaverbeke, W. and De Rochemont, M.	*Technovation*	2009	327
2	Open Innovation in SMEs — an Intermediated Network Model	Lee, S., Park, G., Yoon, B. and Park, J.	*Research Policy*	2010	208
3	Factors Affecting University — Industry R&D Projects: the Importance of Searching, Screening, and Signaling	Fontana, R., Geuna, A. and Matt, M.	*Research Policy*	2006	172
4	Open PHACTS: Semantic Interoperability for Drug Discovery	Williams, A., Harland, L., Groth, P., Pettifer, S., Chichester, C., Willighagen, E., Evelo, C., Blomberg, N., Ecker, G. and Goble, C.	*Drug Discovery Today*	2012	112
5	Inbound Open Innovation Activities in High-Tech SMEs: the Impact on Innovation Performance	Parida, V., Westerberg, M. and Frishammar, J.	*Journal of Small Business Management*	2012	89

(*Continued*)

Table 1.2. *(Continued)*

S. No.	Title	Author(s)	Journal	Year	Citations
6	Building Absorptive Capacity to Organize Inbound Open Innovation in Traditional Industries	Spithoven, A., Clarysse, B. and Knockaert, M.	*Technovation*	2010	89
7	Strategic Entrepreneurship, Collaborative Innovation, and Wealth Creation	Ketchen, D., Ireland, D. and Snow, C.	*Strategic Entrepreneurship Journal*	2007	63
8	Enabling Open Innovation in Small- and Medium-Sized Enterprises: How to Find Alternative Applications for Your Technologies	Bianchi, M., Campodall'Orto, S., Frattini, F. and Vercesi, P.	*R&D Management*	2010	53

Source: Web of Science.

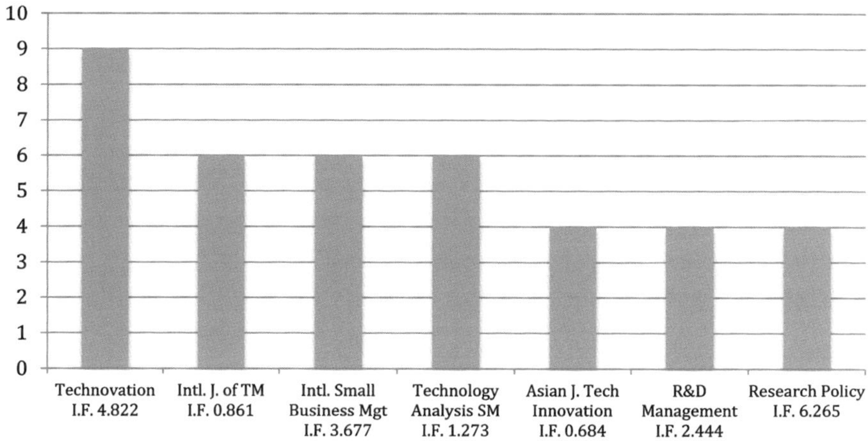

Figure 1.4. Top research outlets for publications on OI in SMEs with their five-year impact factor.

Note: To calculate the five-year impact factor, citations are counted in 2016 to the previous five years and divided by the source items published in the previous five years.

set of publications on OI in SMEs. A total of 81 articles were published in journals, where at least one more article on OI in SMEs was also published; 47 articles were stand-alone publications in their respective journals.

1.2.2. *Content analysis*

With respect to the method used, our content analysis of the 126 publications shows that the authors of 36 articles collected their data on OI in SMEs through survey research, while 32 articles were based on a combination of data from various databases. A total of 34 publications were based on case study research, and in nine publications authors based their conclusion on mixed methods or a combination of qualitative case study research and larger-scale quantitative data analysis (Table 1.3). Theoretical analyses and simulation methods were found in 19 articles. The majority of articles are exploratory in nature where authors conduct preliminary analyses (qualitative analyses in 27% of all articles considered; quantitative

Table 1.3. Methodologies used in publications on OI in SMEs.

Study methodology	No.	%
Case study	34	27.0
Multiple	*22*	17.5
Single	*12*	9.5
Surveys	36	28.6
Database	32	25.4
Mixed	9	7.1
Others (conceptual, reviews, simulation studies, etc.)	19	15.1

analyses in 54% of publications under review) of the OI practices used within the SME context with the purpose of developing new theories that can shed light on this particular type of firms and the way they engage in strategies that require them to be open to their environment and to open up their internal organization.

With respect to the topics addressed in the 126 articles that we analyzed, Figure 1.5 shows the broad categorization into themes such as the adoption of OI, the benefits and challenges of OI, sectoral patterns, etc. These themes are not mutually exclusive in the sense that publications can fall into several themes based on the set of topics they address. The majority of articles study the role of networking and being part of an OI ecosystem (25%) in the performance of SMEs and the manner in which OI practices are adopted or implemented within these small companies (23%). The benefits (16%) and challenges (10%) of OI that are typical for the SME context, such as the way in which engaging in OI helps SMEs remedy some of their (resource) shortcomings and related challenges such as protecting crucial assets such as IP, have also received noteworthy attention in the literature. Few authors (9% of all publications considered) have focused on specific industrial contexts for situating their OI studies. Even fewer studies (3%) have shed light on the importance of effective policies for stimulating SMEs to engage in OI and to thus more effectively fulfill their catalyst role in economic development. In the next section, we describe the most important insights for each of the themes summarized in Figure 1.5.

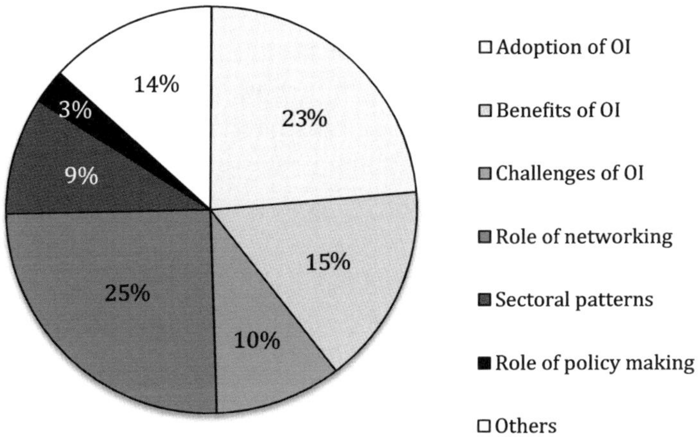

Figure 1.5. Categorization of articles based on topic.

1.3. An Overview of the Identified Research Themes

1.3.1. *Adoption of OI*

SMEs are facing a number of challenges that are typical for the SME context and are thus not typically burdening large companies. Several authors have identified these challenges pertaining to the liability of smallness, for example, a lack of several types of resources and complementary assets, commodity pressure, etc. (Bougrain and Haudeville, 2002; Dahlander and Gann, 2010; Edwards *et al.*, 2005; Rahman and Ramos, 2010). These challenges can be overcome by practicing OI effectively (Crema *et al.*, 2013; Van de Vrande *et al.*, 2009). Many SMEs are thus often facing a typical "chicken and egg" problem. In this context, Teirlinck and Spithoven (2013) point out that the likelihood of SMEs effectively engaging in OI increases with the availability and level of training (PhD) of R&D managers/experts in research collaboration. While the adoption of OI can help SMEs overcome resource challenges, they are in need of resources in order to practice OI well. On a more positive note, other researchers have pointed out that the SME context is uniquely suited for cultivating successful OI practices (much more so than the rigid large firm environment), thus increasing the likelihood of effectively adopting OI. Parida *et al.* (2012), Dufour and Son (2015), and González-Benito *et al.* (2016), for example, point to the flexible organizational structure and the easiness with which SMEs can change their strategic paths as important organizational drivers of OI that are typical for the small firm category. The advantages of the SME context thus need to be leveraged well (through effective management) in order to realize the benefits of OI.

Verbano *et al.* (2015) point out that the specific manner in which SMEs adopt OI differs according to various types of factors that can be categorized as either exogenous (e.g., industry effects, technology diffusion effects) or endogenous (e.g., the effects of product strategies, organizational culture, R&D propensity, innovation approach) to the firm. Despite these differences in adoption modes depending on firm and context peculiarities, most SMEs are found to practice inbound OI such as involving partners, employees, and customers in innovation; out-licensing of proprietary technologies and other outbound OI practices continue to be used to a very limited extent (Van de Vrande *et al.*, 2009). Few publications have focused on the stage of the innovation process where SMEs adopt open approaches, interacting with suppliers and customers in different phases to jointly generate new ideas (Bocken *et al.*, 2014). Presenza *et al.* (2016) have emphasized that SMEs' absorptive capacity to utilize external knowledge greatly increases their innovative ability. Krause and Schutte (2016) propose various OI design propositions depending on the OI life cycle framework. Several publications regarding the adoption of OI practices have studied the ways in which the implementation of OI by SMEs can be facilitated. Most notably, authors have focused on the importance of well-developed intellectual property regimes (Bianchi *et al.*, 2010) and digitalization (e.g., collaborative innovation management tools) (Chaparro-Peláez *et al.*, 2014; Garcia and Barcena, 2010; Saetta *et al.*, 2013; Schuurman *et al.*, 2016; Williams *et al.*, 2012). Gagliardi (2013) and Garcia and Barcena (2010), for instance, discuss the value of the use of Web 2.0, which is a user-friendly web-based application, that enables SMEs to effectively manage collaborative projects. Caetano and Amaral (2011) examine the usefulness of a method for technology road mapping that helps SMEs in applying technology push strategies within an OI context. Hronszky and Kovacs (2013) elaborate on the role of living labs in effectively adopting OI practices. These interactive tools integrate SMEs as partners in the technology development process, making collaborative innovation more easily accessible for these companies.

1.3.2. *Benefits of OI*

A fairly large number of studies in our sample of 126 publications have analyzed the benefits of OI adoption accruing to SMEs. In different kinds of industries, SMEs have been found to generate value from OI (Chesbrough and Crowther, 2006; Van de Vrande *et al.*, 2009) and, even more so, they are often more effective at benefiting from their openness than large companies (Spithoven *et al.*, 2013). SMEs that are successful at OI initiatives have a longer history of intensely experimenting with and adopting OI practices both at the firm level and at the project level than their unsuccessful counterparts (Yoon *et al.*, 2016). Furthermore, successful SMEs make more effective use of the supporting tools (e.g., living labs) that are available to them (Schuurman *et al.*, 2016). Studies have also pointed out that organizational factors help in creating OI culture and increasing innovation performance of SMEs (Popa *et al.*, 2017; Pustovrh *et al.*, 2017). The SMEs involved in OI are mostly first in introducing innovation in the market rather than being a follower (Hochleitner *et al.*, 2017).

Various authors have focused on studying the effects of different kinds of inbound OI approaches on innovation performance within samples of SMEs (Oke and Kach, 2012; Tranekjer and Sondergaard, 2013). Parida *et al.* (2012), for example, analyze the innovation performance of over 250 high-tech SMEs in relation to four inbound OI practices, that is, technology sourcing, horizontal technology collaboration, vertical technology collaboration, and technology scouting. Technology sourcing and vertical technology collaboration have a positive influence on radical innovation; horizontal technology collaboration and technology scouting positively affect incremental innovation within SMEs. Bjerke and Johansson (2015) find that the innovation performance of SMEs is particularly stimulated by collaborating with dissimilar technology partners that are geographically positioned at larger distances from the focal firms. Pustovrh *et al.* (2017) show that OI activities involving collaboration and knowledge exchange with the partners significantly influence the innovation performance of the SMEs. Theyel (2013) studied the effects on the innovation performance of 293 US-based SMEs of various OI practices along the value chain. This study concluded that SMEs are most likely to collaborate with OI partners in the commercialization stage of new product/service development and that collaboration in this phase mostly results in effective process innovations. Though SMEs prefer OI relationships at commercialization stage, the extent and mode of involvement in OI activities at this stage depends on the firm's capabilities and strategy for OI (Henttonen and Lehtimäki, 2017).

A study by Minguela-Rata *et al.* (2014) within the context of Spanish SMEs shows that collaborative relations with suppliers result in higher innovation performance. The authors also find that the larger companies in their sample of SMEs are more likely to benefit from OI than their smaller counterparts due to higher absorptive capacity and stronger management skills. In addition, larger companies within the SME category are more likely to generate radical innovations based on their OI practices, while the smallest of SMEs tend to realize mostly incremental innovations.

1.3.3. *Challenges of OI*

While research shows that SMEs can substantially benefit from effectively adopting OI practices, realizing these advantages requires managing and overcoming a number of OI challenges (Ullrich and Vladova, 2016). Bigliardi and Galati (2016), for example, identify four challenges in relation to OI in their sample of 157 Italian SMEs, for example, challenges related to knowledge and collaboration.

Vanhaverbeke (2017) points to the short-term outlook in most SMEs that is a challenge to successful OI, as collaborative practices typically take long to result in fruitful outcomes. Related to this challenge is the observation by Huggins and Johnston (2009) and Bianchi *et al.* (2010) that SMEs are too focused on their core business and typically lack a systematic innovation process leaving them challenged with respect to grasping and exploiting new technological developments accessed through OI relations. Qin *et al.* (2016) explore the challenges related to adoption of crowdsourcing activity. Deschamps *et al.* (2013) analyze various challenges faced by SMEs in managing collaborations with the universities and subsequent technology transfer issues. Other authors have singled out the challenges related to OI partner selection as important barriers to successful OI in SMEs (Guertler and Lindemann, 2016; Vrgovic *et al.*, 2012).

Besides identifying important challenges, researchers have also enumerated a number of ways in which SMEs can deal with some of these hurdles and manage them in such a way that they enable them to benefit from their OI practices. Spithoven *et al.* (2010), Colombo *et al.* (2012), and Bocquet and Mothe (2015) mention that it is crucial for SMEs to develop their absorptive capacity in capturing value from OI. Similarly, Kim and Park (2010) put forward the strengthening of R&D competencies and the development of a scanning function, where SMEs can diligently monitor external developments, as important drivers of OI success. Other authors (Bouncken and Kraus, 2013) have emphasized the importance of adequate knowledge protection regimes (e.g., IP, NDAs), effective knowledge management, and controlled knowledge sharing in SMEs in dealing with the OI challenge related to the loss of proprietary know-how that many SMEs are struggling with. A number of authors have examined softer coping mechanisms in relation to OI challenges such as the establishment of multidisciplinary teams within SMEs, the stimulation of an open culture and open communication, etc. (Bocken *et al.*, 2014). Finally, researchers have stressed that it is important for SMEs to develop orchestration skills and for orchestrators to take SMEs' interests within OI networks into account as these small firms tend to occupy weaker network positions leaving them dependent on partners' strategies rather than in control of the direction their OI efforts take (Dodourova and Bevis, 2014).

1.3.4. *Role of networking*

A large number of studies in our sample of articles analyzed have focused attention on the role of OI networking in SMEs' performance. Several studies, in various geographical settings and network settings featuring different types of OI partners, have pointed out that it is crucial for SMEs to ensure valuable network positions among their OI partners and to build up skills with respect to the effective orchestration of OI partners and relations (Bjerke and Johansson, 2015; Brunswicker and Vanhaverbeke, 2014; Cosh and Hughes, 2010; Dukic *et al.*, 2015; Egbetokun, 2015; Fontana *et al.*, 2006; Heger and Boman, 2015; Ketchen *et al.*, 2007; Lambrechts *et al.*, 2017; Lee *et al.*, 2010; Leyden and Link, 2013; Pullen *et al.*, 2012; Mitze *et al.*, 2015; Roper and Hewitt-Dundas, 2013; Spriggs *et al.*, 2013; Suh and Kim, 2012; Thorgren *et al.*, 2012; Tranekjer and Knudsen, 2012; Tranekjer and Sondergaard, 2013; Trippl, 2011; van Hemert *et al.*, 2013; Xie *et al.*, 2013). Mitze *et al.* (2015), in their sample of German SMEs, for example, show that SMEs that have secured a favorable position among their OI partners (in this case, educational research institutes) outperform their counterparts in less valuable spots. Indeed, social network analysis has for long evidenced the benefits of occupying central network positions where the focal firm is at the crossroad of knowledge and information flows and can thus optimally access the network's common knowledge base (Heger and Boman, 2015). Securing beneficial network positions is not only up to the networking skills developed by SMEs but is also the responsibility of the (large) network orchestrator at the helm of the network, where this leader is in a position to stimulate value creation at the level of the network as a whole by putting SMEs in strategic positions (Lecocq and Demil, 2006). Similarly, Bek *et al.* (2013) and Omta and Fortuin (2013) have highlighted the important role of effective network management in SMEs realizing their OI potential. Nordman and Tolstoy (2016) emphasize on the role of collaborations with foreign partners to be innovative and capitalize on the home and international market opportunities.

Despite these important insights, different studies continue to find that SMEs seem to be ineffective at building up the necessary networking skills and orchestration capabilities for ensuring optimal positions within OI networks and ecosystems (Dooley *et al.*, 2016). This seems to be particularly the case in transitioning economies as evidenced by studies of Serbian SMEs (Dukic *et al.*, 2015) and Hungarian SMEs (Torok and Toth, 2013). Bek *et al.* (2013), in their analysis of Russian SMEs and their positions in OI clusters, conclude that there is an important role for the government in disseminating best practices regarding entrepreneurs that have managed to secure valuable OI network positions based on their skills, in providing training in networking to SME leaders, as well as in carrying out/ensuring network management that safeguards SMEs' positions and roles.

1.3.5. *Sectoral patterns*

Several articles that we investigate in this chapter have positioned analyses within specific industrial contexts (Cagno *et al.*, 2015; Oughton *et al.*, 2013). It is important to study OI within SMEs operating in various sectoral settings as research shows that differences in generated insights are often due to the unique characteristics of specific types of industries, for example, high-tech (Colombo *et al.*, 2014; Rolandsson *et al.*, 2011) versus low-tech and medium-tech sectors (Dodourova and Bevis, 2014), service-dominated industries versus manufacturing-based sectors (Hosseini and Narayanan, 2014), R&D-intensive in life sciences sector (Marangos and Warren, 2017), etc.

Studies of high-tech sectors, such as medical devices, typically point at the complexity of technologies and the new product development process that call for unique configurations of large numbers of dissimilar partners that are united through common goals (Pullen *et al.*, 2012). Investigations of low-tech industries, such as the food industry (Ferto *et al.*, 2016), have identified a typical lack of openness characterizing firms in this particular sector that poses specific challenges for SMEs seeking horizontal collaborations to boost innovation. Saguy and Sirotinskaya (2014) explain the challenges SMEs are facing in the food industry and propose a number of solutions to overcome these challenges. Openness in this sector also seems to increase to the extent that foreign partners are part of OI networks (Dries *et al.*, 2014; Pettenella and Maso, 2011).

More so than in the analyses of manufacturing industries, the role of digital tools supporting OI and stimulating interconnectedness among partners is particularly stressed by researchers investigating service industries, such as tourism (Carlisle *et al.*, 2013; Chaparro-Peláez *et al.*, 2014).

Vanhaverbeke (2012, 2017) has been studying OI in low-tech and medium-tech industries, such as the bicycle parts industry, the textile industry, quilts and pillows, radiators, etc. He argues that SMEs in those industries can benefit substantially by applying OI, but how OI networks are managed by SMEs is hardly comparable to the management of OI in large companies (Chesbrough, 2003, 2006). OI in low-tech and medium-tech industries relies on personal relations between SME managers (not on OI teams), and OI is a direct consequence of business model innovations initiated by the entrepreneur(s). OI in SMEs can be studied in isolation of SMEs' strategy (business model changes) and the essential role of the entrepreneurs. Managing the network of partners is the most difficult part, and failing to manage the networks in a proper way explains why so many SMEs quit OI initiatives prematurely. There is an urgency among entrepreneurs to discover good OI practices in SME networks and to disseminate these good practices through effective guidelines (see also Chapter 15).

1.3.6. *Role of policymaking*

Studies on policymaking in our sample of publications have nearly always emphasized the important role of governments in stimulating OI in SMEs through, for example, financial support, legal support, the provision of training, the setting up of research centers and incubators, etc. (Aggestam and Weiss, 2011; Bocken *et al.*, 2014; Cosh and Hughes, 2010; Kang *et al.*, 2013; Padilla-Melendez *et al.*, 2013; Pervan *et al.*, 2015; Spithoven *et al.*, 2013; Suh and Kim, 2012; van Hemert *et al.*, 2013). More specifically, governments can help SMEs overcome some of the challenges associated with OI and more effectively realize the potential benefits.

A number of studies have highlighted the significant governmental task of helping SMEs close the skill and knowledge gaps with academia and other industrial partners, which will enable them to take on more valuable roles within OI networks and ecosystems (Pervan *et al.*, 2015; Reschwamm and Wolf, 2008). In this context, authors stress the crucial function of publicly funded research centers and universities that are particularly suited for assisting SMEs in bridging research gaps by diffusing key basic knowledge (Flores *et al.*, 2009). As such, by providing SMEs with access to core knowledge relevant to their fields of business, by helping them to digest this knowledge, and by educating them with respect to how to integrate this know-how into their innovation processes, governments help small firms build up their absorptive capacity, which is crucial for successful OI (Ferto *et al.*, 2016). Others have emphasized the intermediating role of governments where they can help SMEs identify and connect to the right partners, such as end-users, through, for example, setting up innovation hubs (Vrgovic *et al.*, 2012). Further intermediating tasks of governments relate to helping SMEs develop necessary OI skills at different phases of their development where they set up extensive SME support networks focusing both on content (e.g., which technologies to access through OI) and management (e.g., how to obtain valuable network positions) (McAdam *et al.*, 2014; Xiaoyuan and Yanning, 2011). Finally, studies in our sample point at the important role of governments as OI network orchestrators or as parties that ensure that ecosystem development and orchestration occur in a way that is beneficial for SMEs (van Hemert *et al.*, 2013). When left to large commercial orchestrators, network coordination may lead SMEs to leave the network prematurely or to SMEs' inability to generate and capture value from their participation

(Roper and Hewitt-Dundas, 2013). Governments are thus also stimulated to collaborate with large industrial partners in ensuring that SMEs can play their role effectively within OI networks (Tsekouras and Kompis, 2014).

Besides highlighting the way in which governments can make a positive difference in the OI performance of SMEs, one stream of research within this theme has also emphasized the difficulties SMEs often experience when trying to access public support. Particularly, the administrative and bureaucratic burden on SMEs in need of governmental aid is mentioned as a factor that is preventing many SMEs from making optimal use of the available facilities (Aggestam and Weiss, 2011).

1.4. Books about OI in SMEs

To our knowledge, only a few publications exist on this topic. Most of these books are expensive, and those that are less expensive are only recently published. As a result, most OI scholars do not know these books, and that is why we introduce them briefly in this section. Two of the books we discuss below are edited volumes (Rahman and Ramos, 2012a, 2012b). The other books are monographs: one uses a highly academic approach to OI in SMEs (Wynarcyk, 2014) and the others — the more recent ones — focus on management issues of OI in SMEs (Pickert, 2015; Kasende, 2017; Vanhaverbeke, 2017). The first two focus on a specific SME setting, while the last provides guidelines and a general framework for managing OI in small firms.

1. *SMEs and Open Innovation: Global Cases and Initiatives* by Hakikur Rahman and Isabel Ramos (2012a). This book is an edited volume with 16 chapters, each focusing on individual themes and cases. The authors discuss diverse policy, economic, and cultural issues, including numerous opportunities and challenges surrounding OI strategies, risks and risk management, and evolution pattern of SMEs on adopting OI strategies through measurable criteria. It also intends to assist practitioners in

designing action plans to empower SMEs. The book is divided into four sections: (1) concepts, researches, and practices; (2) innovation marketing, communication, and growth management; (3) entrepreneurship, strategies, and crisis management; and (4) creativity, collaboration, and cocreation. The 16 chapters focus on different business developments in SMEs covering many aspects of research and practices of OI in smaller companies.

2. *Cases on SMEs and Open Innovation: Applications and Investigations* is a second volume edited by Rahman and Ramos (2012b). This book reviews applications of OI concepts and strategies for SMEs development by accommodating theoretical perspectives and case studies. It covers diverse aspects of OI in terms of policy, politics, economy, and culture, making it a useful reference for researchers, practitioners, and academics. The book incorporates twelve case studies and is divided into three sections: (1) government-sponsored programs and projects, (2) public–private partnerships in innovation, and (3) OI approaches in innovation.
3. *The Dynamics of Open Innovation in SMEs* by Pooran Wynarczyk (2014) (Routledge, Routledge Studies in Innovation, Organizations, and Technology). This book is written in an academic style and targets academic audience. Increasingly, research and policy attention has come to focus on the SME sector as a key source of new product development, innovation, and suppliers of new technologies. Increasingly, firms are moving away from the "closed innovation paradigm" to a more open way of working in which firms actively collaborate with other companies and institutions. It has been claimed that this new form of collaboration provides access to technologies and facilities that would otherwise take years and major investment to acquire in-house. This "OI" is of great benefit to any company, but is incredibly useful for SMEs in particular. Although OI is receiving more attention in academic research, the existing literature is largely qualitative and focused on the practices of high-technology multinational firms and generally ignores the work being carried out in this area by SMEs. Using a combination of

theory and cutting-edge empirical (quantitative) investigation, the author strives to address this major gap in current literature.

4. *Open Innovation in Family Firms* by Pickert Michael (2015). This is an interesting book focusing on the particular topic of how two different and simultaneously active generations influence the OI activities in family businesses. The book provides a literature review on the topics of family firms, generations in family firms, and OP, and it combines these topics as far as possible. The empirical work is based on semistructured interviews with family members and nonfamily members, who are responsible for the OI activities in different companies. The interviews generate new insights about the influence of multiple generations in family firms on OI, and practical insights are compared to the theoretical assumptions of the literature review. The main result is that OI is fostered by two simultaneously active generations in family firms. Typical restrictions for undertaking OI activities are the financial restrictions and the fact that managers in family firms do not understand the potential of OI.
5. *Open Innovation for Manufacturing in Small and Medium Enterprises: Open Innovation in SMEs* by Christelle Kasende (2017). This book describes how manufacturing SMEs without R&D department in South Africa can innovate with less resources. It also describes how they can overcome problems that are affecting their long-term success and survival. One of the solutions proposed is the adoption of open innovation strategy (OIS). The book shows that an OIS in SMEs with fewer resources can be effective to help SMEs achieve their goals and create sustainable competitive advantage in the market. OIS also implies internal and external resources sharing with employees, other SMEs, large firms, and other stakeholders such as customers and suppliers. This strategy requires changes in the company's organizational structure and encourages employees' empowerment, motivating factors, and developing program to reward innovative employees. From SMEs to corporation managers, this book

provides management guidance on becoming innovative with fewer resources. It can also be very useful for startups as it streamlined the way such structures can tap their most important resources, employees, to innovate and survive as novice in a competitive market.

6. *Managing Open Innovation in SMEs* by Wim Vanhaverbeke (2017, Cambridge University Press). In line with the previous book, this provides a practical guideline for entrepreneurs and SME managers on how they can innovate better and faster with less resources through OI in networks of partners. The challenge is to manage the partners in such networks.

   Most small firms face a harsh business environment through increased global competition. Various factors including changing market conditions or new regulations force them to reinvent their business through new technologies or novel value propositions. As small companies have insufficient financial resources and technical competencies to develop technology in-house, they have to innovate in collaboration with external partners — suppliers, customers, research labs and universities, large companies, or networks of other SMEs. However, OI is not yet a common practice in small firms, and although innovation is imperative for most European SMEs, few understand how to benefit from innovating through partnerships. There is an urgent need to understand how entrepreneurs can organize OI.

   This book studies how OI can be managed and implemented in SMEs, including a framework to implement OI strategies successfully. Managing OI in small companies is actually quite specific, and we have to reinvent OI to make it useful for entrepreneurs in small firms; therefore, the book pays a lot of attention to the role of the entrepreneur, and the integration of strategy, business model changes, and OI. The book provides an in-depth analysis of OI practices in small enterprises based on rich case studies from successful European firms. The result is a set of practical guidelines for entrepreneurs at the end of each chapter.

Innovation is usually associated with high-tech industries and top-notch technology. In contrast, most of the companies studied in this book are examples of low-tech or medium-tech industries — bicycle parts, quilts and pillows, chemical treatment of textiles, barometers, radiators, etc. They are representatives for the majority of the European SMEs. Innovation and OI in these industries are not about inventing new technologies or pushing the technological frontier, but it boils down to find interesting applications of existing technologies. All cases show value chain partners are important for OI, but OI connects SMEs with new partners from completely different industries. Take for instance Quilts of Denmark that succeeded in developing the first functional quilt with technology originally developed by NASA. Applying technologies in a new industry context requires a lot of applied technology development, which becomes over time a valuable asset for the innovating SMEs, and allows them to transform the company into a complete new business.

The book focuses in three main topics: the first half is about the OI networks among SMEs in low-tech and medium-tech industries; the second part explains how small firms (high-tech) can team up with large companies; and, finally, attention is paid to the potential role of intermediaries that have customized their services for innovating SMEs to facilitate OI activities in the latter.

1.5. Future Research Avenues

The purpose of this chapter has been to provide a systematic review of the literature on OI in SMEs with a particular focus on describing the most prevalent research themes in this field and setting the stage for subsequent chapters in this book. Our analysis shows that authors' attention has turned toward the OI efforts of SMEs (besides large companies) most notably in the last five years. Since then we have witnessed a growing number of publications in journals such as *Technovation* and *IJTM* as well as an increasing number of citations to these works. Based on both qualitative and quantitative research methodologies, the authors have particularly explored how SMEs adopt OI practices and how these ways differ from those of large firms as well as the important role of effective networking in SMEs' OI performance. Our analysis suggests several themes for further investigation.

With respect to research methods and settings, our review shows that most studies so far have been of an exploratory nature, making it difficult to generalize results across larger populations of SMEs. While these existing studies and the insights they generate are highly valuable, future (multidisciplinary) investigations will need to quantitatively test hypotheses in larger, longitudinal datasets on the OI practices of SMEs. Besides successful OI implementation cases, we will also need to study failure cases for useful lessons to be retrieved. For example, cases of entrepreneurial OI ecosystems where SMEs successfully function as orchestrators and manage to create self-sustaining networks need to be compared and contrasted to ecosystems that disintegrate once the SME orchestrator exits the ecosystem. Which OI skills are found in SMEs being part of the first type of ecosystem versus the latter? Furthermore, our review shows that OI in SMEs is mostly studied within the context of developed (Western European) economies. Governments of developing nations in, for example, Asia and Africa are also in need of insights with respect to how best to stimulate OI in local SMEs and how to develop effective facilitating policies.

With respect to research topics addressed, this chapter shows that, while the role of networking has received significant attention so far, there is still much need for additional insights within this theme. Specifically, research needs to shed light on how SMEs can secure valuable OI network positions and how they can become effective orchestrators of their OI relationships. Relationships with large OI partners seem to present unique challenges to SMEs (most notably related to IP management), requiring knowledge on the types of management skills SMEs would need to coordinate these partners well. Finally, considering the lack of financial resources within most (nonfamily-owned) SMEs, research is needed on the role of venture capitalists, business angels, as well as public support funds in stimulating the OI performance of SMEs.

References

Aggestam, F. and Weiss, G. (2011). Innovation in EU forestries: A science–policy dialogue. *Innovation In Forestry: Territorial and Value Chain Relationships*, p. 294.

Audretsch, D. B. (1995). Innovation, growth and survival. *International Journal of Industrial Organization* **13**(4), pp. 441–457.

Batterink, M. (2009). The adoption of open innovation. A longitudinal study of large firms and SMEs, in high-, medium-, and low-tech industries in the Netherlands. Profiting from External Knowledge: How Firms Use Different Acquisition Strategies to improve Their Innovation Performance, **3**, pp. 25–45.

Bek, M. A., Bek, N. N., Sheresheva, M. Y. and Johnston, W. J. (2013). Perspectives of SME innovation clusters development in Russia. *Journal of Business & Industrial Marketing* **28**(3), pp. 240–259.

Bianchi, M., Campodall'Orto, S., Frattini, F. and Vercesi, P. (2010). Enabling open innovation in small- and medium-sized enterprises: how to find alternative applications for your technologies. *R&D Management* **40**(4), pp. 414–431.

Bigliardi, B. and Galati, F. (2016). Which factors hinder the adoption of open innovation in SMEs? *Technology Analysis & Strategic Management* **28**(8), pp. 869–885.

Bocken, N. M., Farracho, P., M. Bosworth, R. and Kemp, R. (2014). The front-end of eco-innovation for eco-innovative small and medium sized companies. *Journal of Engineering and Technology Management* **31**, pp. 43–57.

Bougrain, F. and Haudeville, B. (2002). Innovation, collaboration and SMEs internal research capacities. *Research Policy* **31**(5), pp. 735–747.

Bouncken, R. B. and Kraus, S. (2013). Innovation in knowledge-intensive industries: the double-edged sword of coopetition. *Journal of Business Research* **66**(10), pp. 2060–2070.

Brunswicker, S. and Van de Vrande, V. (2014). *Exploring Open Innovation in Small and Medium-Sized Enterprises*. Oxford: Oxford University Press.

Caetano, M. and Amaral, D. C. (2011). Roadmapping for technology push and partnership: a contribution for open innovation environments. *Technovation* **31**(7), pp. 320–335.

Chaparro-Peláez, J., Pereira-Rama, A. and Pascual-Miguel, F. J. (2014). Inter-organizational information systems adoption for service innovation in building sector. *Journal of Business Research* **67**(5), pp. 673–679.

Chesbrough, H. W. (2003). *Open Innovation: The New Imperative for Creating and Profiting from Technology*. Harvard Business Press, Boston.

Chesbrough, H. and Crowther, A. K. (2006). Beyond high tech: early adopters of open innovation in other industries. *R&D Management* **36**(3), pp. 229–236.

Cosh, A. and Hughes, A. (2010). Never mind the quality feel the width: university–industry links and government financial support for innovation in small high-technology businesses in the UK and the USA. *The Journal of Technology Transfer* **35**(1), pp. 66–91.

Crema, M., Verbano, C. and Venturini, K. (2013). Open innovation profiles in italian manufacturing companies. *Entrepreneurial Knowledge, Technology and the Transformation of Regions* **68**, p. 68.

Dahlander, L. and Gann, D. M. (2010). How open is innovation? *Research Policy* **39**(6), pp. 699–709.

Deschamps, I., Macedo, M. G. and Eve-Levesque, C. (2013). University–SME collaboration and open innovation: intellectual-property management tools and the roles of intermediaries. *Technology Innovation Management Review* **3**(3), p. 33.

Dodourova, M. and Bevis, K. (2014). Networking innovation in the European car industry: does the open innovation model fit? *Transportation Research Part A — Policy and Practice* **69**, pp. 252–271.

Dufour, J. and Son, P.-E. (2015). Open innovation in SMEs — towards formalization of openness. *Journal of Innovation Management* **3**(3), pp. 90–117.

Edwards, T., Delbridge, R. and Munday, M. (2005). Understanding innovation in small and medium-sized enterprises: a process manifest. *Technovation* **25**(10), pp. 1119–1127.

Ferto, I., Molnar, A. and Toth, J. (2016). Borderless ideas — open innovation in the Hungarian food chain. *British Food Journal* **118**(6), pp. 1494–1515.

Flores, M., Boër, C., Huber, C., Plüss, A., Schoch, R. and Pouly, M. (2009). Universities as key enablers to develop new collaborative environments for innovation: successful experiences from Switzerland and India. *International Journal of Production Research* **47**(17), pp. 4935–4953.

Fontana, R., Geuna, A. and Matt, M. (2006). Factors affecting university–industry R&D projects: the importance of searching, screening and signalling. *Research Policy* **35**(2), pp. 309–323.

Gagliardi, D. (2013). Next generation entrepreneur: innovation strategy through Web 2.0 technologies in SMEs. *Technology Analysis & Strategic Management* **25**(8), pp. 891–904.

Garcia, M. C. J. and Barcena, N. S. (2010). Web 2.0 applications by SMEs as tool for innovation and improvement. *DYNA* **85**(8), pp. 662–666.

González-Benito, Ó., Muñoz-Gallego, P. A. and García-Zamora, E. (2016). Role of collaboration in innovation success: differences for large and small businesses. *Journal of Business Economics and Management* **17**(4), pp. 645–662.

Guertler, M. R. and Lindemann, U. (2016). Identifying open innovation partners: a methodology for strategic partner selection. *International Journal of Innovation Management* **20**(05), pp. 1640011.

Heger, T. and Boman, M. (2015). Networked foresight — the case of EIT ICT labs. *Technological Forecasting and Social Change* **101**, pp. 147–164.

Henttonen, K. and Lehtimäki, H. (2017). Open innovation in SMEs: collaboration modes and strategies for commercialization in technology-intensive companies in forestry industry. *European Journal of Innovation Management* **20**(2), pp. 329–347.

Hochleitner, F. P., Arbussà, A. and Coenders, G. (2017). Inbound open innovation in SMEs: indicators, non-financial outcomes and entry-timing. *Technology Analysis & Strategic Management* **29**(2), pp. 204–218.

Hossain, M. and Kauranen, I. (2016). Open innovation in SMEs: a systematic literature review. *Journal of Strategy and Management* **9**(1), pp. 58–73.

Hosseini, S. M. P. and Narayanan, S. (2014). Adoption, adaptive innovation, and creative innovation among SMEs in Malaysian manufacturing. *Asian Economic Papers* **13**(2), pp. 32–58.

Hronszky, I. and Kovacs, K. (2013). Interactive value production through living labs. *Acta Polytechnica Hungarica* **10**(2), pp. 89–108.

Huggins, R. and Johnston, A. (2009). Knowledge networks in an uncompetitive region: SME innovation and growth. *Growth and Change* **40**(2), pp. 227–259.

Krause, W. and Schutte, C. (2016). Developing design propositions for an open innovation approach for SMEs. *South African Journal of Industrial Engineering* **27**(3), pp. 37–49.

Lambrechts, F., Voordeckers, W., Roijakkers, N. and Vanhaverbeke, W. (2017). Exploring open innovation in entrepreneurial private family firms in low- and medium-technology industries. *Organizational Dynamics*, ISSN 0090-2616, https://doi.org/10.1016/j.orgdyn.2017.05.001. (article in press).

Lecocq, X. and Demil, B. (2006). Strategizing industry structure: the case of open systems in a low-tech industry. *Strategic Management Journal* **27**(9), pp. 891–898.

Lee, S., Park, G., Yoon, B. and Park, J. (2010). Open innovation in SMEs — an intermediated network model. *Research Policy* **39**(2), pp. 290–300.

Marangos, S. and Warren, L. (2017). A mapping for managers: open innovation for R&D intensive SMEs in the life sciences sector. *European Journal of Innovation Management* **20**(2), pp. 210–229.

Minguela-Rata, B., Femández-Menéndez J., Fossas-Olalla, M. (2014). "Cooperation with suppliers, firm size and product innovation." *Industrial Management & Data System* **114**(3), pp. 438–455.

Mitze, T., Alecke, B., Reinkowski, J. and Untiedt, G. (2015). Linking collaborative R&D strategies with the research and innovation performance of SMEs in peripheral regions: do spatial and organizational choices make a difference? *The Annals of Regional Science* **55**(2–3), pp. 555–596.

Muller, P., Caliandro, C., Peycheva, V., Gagliardi, D., Marzocchi, C., Ramlogan, R. and Cox, D. (2015). Annual report on European SMEs. *European Commission.*

Nordman, E. R. and Tolstoy, D. (2016). The impact of opportunity connectedness on innovation in SMEs' foreign-market relationships. *Technovation* **57**, pp. 47–57.

Oke, A. and Kach, A. (2012). Linking sourcing and collaborative strategies to financial performance: the role of operational innovation. *Journal of Purchasing and Supply Management* **18**(1), pp. 46–59.

Oughton, D., Mortara, L. and Minshall, THW. (2013). Managing asymmetric relationships in open innovation: lessons from multinational companies and SMEs. In: *Open Innovation in the Food and Beverage Industry*. M. G. Martinez, ed., Woodhead Publishing, pp. 276–293.

Parida, V., Westerberg, M. and Frishammar, J. (2012). Inbound open innovation activities in high-tech SMEs: the impact on innovation performance. *Journal of Small Business Management* **50**(2), pp. 283–309.

Pittaway, L., Robertson, M., Munir, K., Denyer, D. and Neely, A. (2004). Networking and innovation: a systematic review of the evidence. *International Journal of Management Reviews* **5**(3–4), pp. 137–168.

Popa, S., Soto-Acosta, P. and Martinez-Conesa, I. (2017). Antecedents, moderators, and outcomes of innovation climate and open innovation: an empirical study in SMEs. *Technological Forecasting and Social Change* **118**, pp. 134–142.

Presenza, A., Abbate, T., Meleddu, M. and Cesaroni, F. (2016). Small- and medium-scale Italian winemaking companies facing the open innovation challenge. *International Small Business Journal: Researching Entrepreneurship,* DOI: https://doi.org/10.1177/0266242616664798.

Pullen, A. J. J., de Weerd-Nederhof, P. C., Groen, A. J. and Fisscher, O. A. M. (2012). Open innovation in practice: goal complementarity and closed NPD networks to explain differences in innovation performance for SMEs in the medical devices sector. *Journal of Product Innovation Management* **29**(6), pp. 917–934.

Pustovrh, A., Jaklič, M., Martin, S. A. and Rašković, M. (2017). Antecedents and determinants of high-tech SMEs' commercialisation enablers: opening the black box of open innovation practices. *Economic Research-Ekonomska Istraživanja* **30**(1), pp. 1033–1056.

Qin, S., Van Der Velde, D., Chatzakis, E., McStea, T. and Smith, N. (2016). Exploring barriers and opportunities in adopting crowdsourcing based new product development in manufacturing SMEs. *Chinese Journal of Mechanical Engineering* **29**(6), pp. 1052–1066.

Rahman, H. and Ramos, I. (2010). Open innovation in SMEs: from closed boundaries to networked paradigm. *Issues in Informing Science and Information Technology* **7**, pp. 471–487.

Roper, S. and Hewitt-Dundas, N. (2013). Catalysing open innovation through publicly-funded R&D: a comparison of university and company-based research centres. *International Small Business Journal* **31**(3), pp. 275–295.

Rousseau, D. M., Manning, J. and Denyer, D. (2008). 11 evidence in management and organizational science: assembling the field's full weight of scientific knowledge through syntheses. *The Academy of Management Annals* **2**(1), pp. 475–515.

Saetta, S., Tiacci, L. and Cagnazzo, L. (2013). The innovative model of the Virtual Development Office for collaborative networked enterprises: the GPT network case study. *International Journal of Computer Integrated Manufacturing* **26**(1–2), pp. 41–54.

Saguy, I. S. and Sirotinskaya, V. (2014). Challenges in exploiting open innovation's full potential in the food industry with a focus on small and medium enterprises (SMEs) (vol 38, pg 136–148, 2014). *Trends in Food Science & Technology* **40**(1), pp. 136–148.

Schuurman, D., De Marez, L. and Ballon, P. (2016). The impact of living lab methodology on open innovation contributions and outcomes. *Technology Innovation Management Review* pp. 7–16.

Spender, J.-C., Corvello, V., Grimaldi, M. and Rippa, P. (2017). Startups and open innovation: a review of the literature. *European Journal of Innovation Management* **20**(1), pp. 4–30.

Spithoven, A., Vanhaverbeke, W. and Roijakkers, N. (2013). Open innovation practices in SMEs and large enterprises. *Small Business Economics* **41**(3), pp. 537–562.

Teirlinck, P. and Spithoven, A. (2013). Research collaboration and R&D outsourcing: different R&D personnel requirements in SMEs. *Technovation* **33**(4–5), pp. 142–153.

Theyel, N. (2013). Extending open innovation throughout the value chain by small and medium-sized manufacturers. *International Small Business Journal* **31**(3), pp. 256–274.

Torok, A. and Toth, J. (2013). Open characters of innovation management in the Hungarian wine industry. *Agricultural Economics-Zemedelska Ekonomika* **59**(9), pp. 430–439.

Tranekjer, T. L. and Sondergaard, H. A. (2013). Sources of innovation, their combinations and strengths — benefits at the NPD project level. *International Journal of Technology Management* **61**(3–4), pp. 205–236.

Trippl, M. (2011). Regional innovation systems and knowledge-sourcing activities in traditional industries — evidence from the Vienna food sector. *Environment and Planning A* **43**(7), pp. 1599–1616.

Tsekouras, G. and Kompis, C. (2014). Public–Private innovation strategic alliances For SMEs: An emerging model. In: *Managing Public-Private Strategic Alliances. Reserach in Strategic Alliances*. IAP, USA, pp. 27–84.

Ullrich, A. and Vladova, G. (2016). Weighing the pros and cons of engaging in open innovation. *Technology Innovation Management Review*, pp. 34–40.

Van de Vrande, V., De Jong, J. P., Vanhaverbeke, W. and De Rochemont, M. (2009). Open innovation in SMEs: trends, motives and management challenges. *Technovation* **29**(6), pp. 423–437.

Vanhaverbeke, W. (2012). Open innovation in SMEs: How can small companies and start-ups benefit from open innovation strategies? Leuven, Belgium: Flanders District of Creativity.

van Hemert, P., Nijkamp, P. and Masurel, E. (2013). From innovation to commercialization through networks and agglomerations: analysis of sources of innovation, innovation capabilities and performance of Dutch SMEs. *Annals of Regional Science* **50**(2), pp. 425–452.

Verbano, C., Crema, M. and Venturini, K. (2015). The identification and characterization of open innovation profiles in Italian small and medium-sized enterprises. *Journal of Small Business Management* **53**(4), pp. 1052–1075.

Vrgovic, P., Vidicki, P., Glassman, B. and Walton, A. (2012). Open innovation for SMEs in developing countries — an intermediated communication network model for collaboration beyond obstacles. *Innovation-Management Policy & Practice* **14**(3), pp. 290–302.

Williams, A. J., Harland, L., Groth, P., Pettifer, S., Chichester, C., Willighagen, E. L., Evelo, C. T., Blomberg, N., Ecker, G. and Goble, C. (2012). Open PHACTS: semantic interoperability for drug discovery. *Drug Discovery Today* **17**(21), pp. 1188–1198.

Wynarczyk, P., Piperopoulos, P. and McAdam, M. (2013). Open innovation in small and medium-sized enterprises: an overview. *International Small Business Journal* **31**(3), pp. 240–255.

Yoon, B., Shin, J. and Lee, S. (2016). Open innovation projects in SMEs as an engine for sustainable growth. *Sustainability* **8**(2), p. 146.

Chapter 2

The "Hidden Costs" of Open Innovation in SMEs: From Theory to Practice*

Cristina Marullo[†]*, Alberto Di Minin,*
Chiara De Eleonora Marco and Andrea Piccaluga

[†] *c.marullo@sssup.it*

Abstract

This study explores the process of open innovation (OI) implementation in small and medium-sized enterprises (SMEs), with the aim of shedding light on the factors that may either enhance or undermine effectiveness in different contexts. The hypothesis of different paths of OI implementation is shaped across three theoretical dimensions: (1) the combination of open innovation approaches

---

*This chapter is based on a research project conducted with the support of the EU Commission Institute for Perspective Technological Studies (IPTS JRC, Seville, ES) within the EURIPIDIS (European Innovation Policies for the Digital Shift) program. Initial findings have been published in: Di Minin *et al.* (2016). Case studies on Open Innovation in ICT. JRC100823 Institute for Prospective Technological Studies, Joint Research Centre. Luxembourg: EU Union Publications Office. (DOI: 10.2791/433370).

[†] Corresponding author.

(the choice of the governance mode); (2) the type of knowledge exchanged (scientific knowledge, technological knowledge, product/market knowledge); and (3) the sectoral innovation system to which the company belongs (characterized by different technological opportunities, innovation sources, and appropriability regimes). A multiple case study design has been adopted in order to investigate how and why the process of OI implementation in SMEs may be bounded by "hidden costs" hampering effectiveness. The study delivers four different paths of OI implementation (science-oriented, technology-driven, application-oriented, and community-driven) and demonstrates that SMEs in each path differently seek opportunities and face different risks.

2.1. Introduction

The field of research on open innovation (OI) in small and medium-sized enterprises (SMEs) has been expanding in the last decade. Relevant studies have unveiled the incentives for the adoption of OI, the "creative" governance styles characterizing the content of OI strategies, and their influence on the expected outcomes. At the same time, prior empirical work has been mostly oriented to the effectiveness argument, and still few contributions have considered the risks SMEs face in the shift from a closed innovation setting to an open approach (Brunswicker and van de Vrande, 2014; Hossain, 2015).

Due to the fact the OI paradigm has been developed and mainly discussed in the context of large enterprises and multinational companies (Chesbrough, 2003), evidence on the mechanisms underlying the process of OI implementation in SMEs is still fragmented (Vanhaverbeke *et al.*, 2008; Spithoven *et al.*, 2013; West and Bogers, forthcoming).

Differently from what happens in large enterprises, the implementation of OI in "one-project-one-technology" companies underlines a radical change in the firm's strategic objectives. The deployment of an open business model enabling collaboration mechanisms allows SMEs to realign the innovation strategy through extensive network collaborations, to leverage complementary resources, and to seize new business opportunities (van de Vrande *et al.*, 2009; Vanhaverbeke *et al.*, 2012; West and Gallagher, 2006).

Therefore, the process through which SMEs integrate technologies, partnerships, and complementary knowledge in the shift from a closed to an open approach strongly impacts the function of value creation and entails new mechanisms of appropriating a portion of that value. During this shift, context-related and unpredicted risks may come into play in the process of OI implementation in SMEs, influencing value creation and the appropriation opportunities.

This chapter focuses on the overlooked aspects of running OI in SMEs. Drawing on in-depth interviews with owners and managers of eight European SMEs, we explore the peculiar challenges of implementing OI in different settings and the underlying mechanisms bounding OI effectiveness. We combine and analyze evidence from eight case studies of OI implementation in SMEs according to three dimensions: the type of OI activities, the nature of knowledge exchanged, and the sectoral innovation system in which the company operates (the set of external context factors influencing OI implementation). We provide a detailed description of the contingent challenges (the "hidden costs") of OI emerging in different implementation paths, defined at the intersection of the three dimensions.

2.2. The Peculiarities of OI in SMEs

Throughout its evolution in the last decade, empirical literature has been increasingly interested in the multidimensional nature of OI in SMEs, analyzing both the variety of OI practices and the difference between different "types of OI implementation" (Brunswicker and van de Vrande, 2014; Lee *et al.*, 2010).

Overall, the extant literature emphasizes that the *content* of SMEs' OI strategies is shaped by multiple knowledge domains and search directions (van de Vrande *et al.*, 2009) aimed at "innovation problem-solving" activities (Felin and Zenger, 2013). At the same time, the dynamics of OI implementation in SMEs is strongly *context-dependent*. The mechanisms through which different approaches affect the emergence of opportunities and/or risks cannot be appreciated outside of the framework in which this process takes place (Edwards *et al.*, 2005).

We introduce three key traits describing the peculiarities of OI in SMEs. From these elements, we start developing our analysis. First, OI strategies in SMEs are qualified by multiple and dynamic knowledge sourcing modes, rather than by stable bilateral collaborations. Small companies benefit from relative advantages in pursuing collaborative innovation with respect to large companies (Vossen, 1998). Rapid decision-making, high risk taking, fast reaction to market requirements, and high adaptability of routines are among the behavioral characteristics of SMEs' "entrepreneurial" innovation model.

Second, from a process perspective, the alignment between the OI strategy and the company's business model allows SMEs to unlock the economic potential of the technology under development (Chesbrough, 2006). In this process, knowledge networks assume a strategic (rather than an operational) relevance (Julien, 1995). SMEs pursuing OI are therefore strongly committed to establishing long-term relationships with research partners, bringing the external network at the core of the new (open) business model.

Third, motivated management plays a central role in the implementation of OI strategies. The overlap between ownership and management in SMEs leads to shared responsibilities in setting the R&D priorities, developing a network of external alliances, and orchestrating joint value creation and value capture through the development of the business model.

Considerable research has been investigating the challenges of OI in SMEs in a wide range of activities: venturing, customer integration, user involvement, and participation in external research networks (Enkel *et al.*, 2005; Goduscheit and Knudsen, 2015; Henkel, 2006; Lee *et al.*, 2010; van de Vrande *et al.*, 2009; West and Gallagher, 2006).

Regardless of the type of strategic approach, the most significant *internal barriers* to the adoption of OI in SMEs are related to financial costs of IP protection, legal actions, bureaucracy, and to the possible emergence of organizational challenges and cognitive problems (van de Vrande *et al.*, 2009). Inexperience, lack of a strategic vision, or inadequate managerial capabilities to integrate relevant external knowledge in the R&D process may create difficulties in successfully pursuing OI. The rise of knowledge search costs, communication problems, and information asymmetries may further raise the risk of opportunistic behavior in interfirm alliances. More in general, literature on OI in SMEs acknowledges the cost-increasing effects of business model innovation — setting up and managing the external network of partners (Vanhaverbeke *et al.*, 2012). Unpredicted risks may arise, for example, from unfocused search strategies driving to an increased number of partnerships and/or collaboratively held patents (Belderbos *et al.*, 2004; Brunswicker and Vanhaverbeke, 2015; Laursen and Salter, 2006).

Regarding the external context, unpredicted risks of implementing OI may come from the emergence of knowledge gaps, information asymmetries among partners, and ineffective legal protection of intellectual property rights. In this context, knowledge leakage and unbalanced market power may strongly reduce the opportunities of innovation success. Literature on high-technology entrepreneurship adds that in high-technology small firms (HTSFs) whose new engineering knowledge (the technology under development) is the major strategic asset OI can represent a source of potential benefits, and, at the same time, a potential threat. In most cases, HTSFs do not decide to cooperate and pursue fully closed innovation models to increase the monetary value of their technology (Oakey, 2013).

Overall, research indicates that OI in SMEs is pursued when the "benefits" overcome the "costs." Some of these costs might be underestimated when new projects start. We qualify as "hidden costs" of OI the indirect strategic risks that come into play within specific contexts, undermining the effective implementation of OI in SMEs.

2.3. Research Methodology: Classifying OI in SMEs

For the purpose of reconciling different OI contexts with the related challenges, we conduct a multiple case analysis using in-depth interviews with the key representatives (CEOs, COOs, and CTOs) of eight European SMEs reporting evidence of ongoing OI projects (Di Minin *et al.*, 2016).

As analogous empirical studies demonstrate (Vanhaverbeke *et al.*, 2012), the different elements shaping OI benefits (and challenges) are specific to the framework of the company's strategy.

Case selection was therefore performed with the aim of maximizing the diversity of contexts in which OI is implemented (type of OI activities, industries, countries of operation, companies' size, and level of R&D intensity; see Table 2.1).

According to Yin's scope of research efficacy (Yin, 2003), case selection was also based on companies' willingness to provide data and detailed information on the crucial innovation project (through which the main strategic change occurred).[1] This choice allowed us to collect and compare grounded empirical information on the challenges of OI implementation at the subfirm level (Du *et al.*, 2014; Vanhaverbeke and Chesbrough, 2014).

Based on prior work (Di Minin *et al.*, 2016; Marullo *et al.*, 2015), we combine evidence from case studies according to three dimensions: the type of OI activities, the nature of knowledge exchanged, and the sectoral innovation system in which the company operates.

The first two dimensions (the type of OI approach and the nature of knowledge exchanged) are both elements of the internal context of OI implementation (OI strategy characteristics and direction); the sectoral innovation system can rather be considered as a

[1] In the first part of the interviews, the companies' representatives were guided through the company history, offering, and value proposition, and through an "innovation roadmap" including: (i) original idea; (ii) early technology development; (iii) advanced technology development; (iv) product commercialization; and (v) evolution of the business model (Di Minin *et al.*, 2016). Each interview further proceeded through in-depth unstructured questions about the most relevant elements shaping the process OI implementation nature of internal knowledge base, the type and variety of partners, and the type of appropriability mechanisms.

Table 2.1. Characteristics of companies in the sample.

Company	Country	Main activity	No. countries of operation	No. employees	R&D intensity (%)	No. patents
Graphenea SA	ES	R&D in natural sciences and engineering	3	8	30[a]	4
Guger Technologies	AT	R&D in medical engineering	3	40	40[b]	2
Celoxica Ltd.	UK	ICT (hardware)	4	32	38[a]	24
BCB Informática y Control	ES	ICT (software)	2	10	6[a]	1
Entranet Ltd.	GR	Construction	3	8	90[b]	3
Primo 1D	FR	Microelectronics/Textiles	1	8	NA	1
Zappar Ltd.	UK	IT	1	21	6[a]	1
Supponor OY	FI	Advertising and multimedia	1	19	100[b]	1

Notes: [a]Share of R&D expenditure over net sales.
[b]Share of R&D employees over total employment.

proxy of the set of external context factors influencing OI effectiveness (Edwards, *et al.*, 2005; Huizingh, 2011).

2.3.1. *The types of OI activities*

The different sets of actions adopted to combine and integrate external knowledge in the innovation process epitomize the content of OI strategies. We characterize the types of OI activities based on the definitions of Chesbrough (2003), Enkel *et al.* (2009), van de Vrande *et al.* (2009), and Dahlander and Gann (2010).

Inbound OI (sourcing/acquiring): R&D alliances; collaboration with universities and public research centers (PRCs); customer and user involvement; IP in-licensing; technology acquisition.

Outbound OI (revealing/selling): Nonequity alliances for external technology development; cooperation agreements with customers and clients (codevelopment); IP out-licensing; technology selling.

Empirical work conceptualizing OI in SMEs reports a broad range of approaches to external technological collaborations and a huge variety of partners involved (Brunswicker and Vanhaverbeke, 2015; Nieto and Santamarìa, 2010; Parida *et al.*, 2012).

Innovating SMEs extensively rely on external networking to access new relevant knowledge and missing innovation assets, while customer involvement is the most common OI practice for external technology exploitation (van de Vrande *et al.*, 2009). Among these activities, the pursuit of external collaborations for the development of technological opportunities (outbound OI) is more challenging for SMEs than for large enterprises (Spithoven *et al.*, 2013). Increased competition in the end markets and opportunistic behavior in interfirm alliances may arise from business-critical knowledge loss (Ritala *et al.*, 2015). As we see in our sample, the pursuit of outbound OI activities (e.g., for revealing purposes) in SMEs is bounded by the development of suitable appropriation mechanisms (Di Minin and Faems, 2013; Laursen and Salter, 2014; West, 2006).

2.3.2. The nature of knowledge exchanged

In our framework, we also take into account the role of different knowledge domain(s) in which external search and learning take place (Li *et al.*, 2008).

Scientific knowledge: Science search aimed at explorative learning (broadening the internal R&D knowledge and developing the ability to discover new innovation opportunities).

Technological knowledge: Technology search aimed at deepening existing technological knowledge and extending internal product innovation knowledge (exploitative learning).

Product/Market knowledge: Market search aimed at exploitative learning from the demand side. It is a complementary knowledge source with respect to the others, aimed at leveraging complementary competences in manufacturing, marketing, product customization, and distribution.

The selection of a search path and the exploitation of search mechanisms represent crucial managerial tasks associated with the successful development of collaborative innovation projects (Lopez-Vega *et al.*, 2016). In this perspective, OI partnerships in SMEs can be interpreted as a set of communication channels enabling distant search and internal/external knowledge recombination, therefore representing solutions to an innovation problem (Felin and Zenger, 2013; Spithoven *et al.*, 2013). However, different managerial capabilities are required for OI projects' success according to the nature of knowledge exchanged. With respect to science-based partnerships, a high degree of control (project formalization) is required, for example, to successfully pursue exploitative learning from the market side (Du *et al.*, 2014).

2.3.3. The sectoral innovation system in which the company operates

Sectoral innovation systems characterize the position of firms operating in different industries in terms of technological opportunities, innovation sources, and appropriability regimes (Archibugi, 2001; Dosi, 1982; Malerba, 2005; Pavitt *et al.*, 1989).

From an industrial dynamics perspective, Christensen *et al.* (2005) show that the specific modes of OI implementation in SMEs with regard to an emerging technology (consumer electronics) reflect their differential position within the innovation system. Castellacci (2008) connects technological collaboration and knowledge exchange dynamics across manufacturing and services companies in Europe to sectoral patterns of innovation. In this framework, "small and medium" is the dominant size of firms operating in advanced knowledge providers, specialized suppliers, and supplier-dominated patterns.

We analyze the sectoral innovation systems in which SMEs operate as different sets of external context characteristics (technological opportunities, type of innovation activities, sources of value creation, and appropriability mechanisms) influencing the shift from a closed approach to an OI strategy.

Science-based: Technological opportunities come from internal R&D aimed at the exploitation of scientific knowledge on the market. Advances in science and R&D, product innovation, and diversification in new markets are the main sources of value creation.

Supplier-dominated: Technological opportunities come from the incorporation of new product components, materials, software, and equipment based on the integration of knowledge from innovation partners (mainly suppliers). The development of tacit knowledge (learning by doing, learning by using) is the main source of value creation.

Specialized supplier: Technological opportunities come from user–producer cocreation. Innovation is based on performance improvement, reliability, and customization of specialized inputs (machines, instrumentation, materials, and software). The development of product-market knowledge and design skills support new value creation.

Information-intensive: Technological opportunities come from the emergence of IT-based multitechnology platform services (the supporting knowledge base to the ICT paradigm). Innovation is based on internal R&D and the role of external communities (customers, users, suppliers, and partners) as sources of value creation.

Table 2.2 reports the classification of the eight SMEs in the sample according to the three elements in our interpretative framework (types of OI activities, knowledge domains, and sectoral innovation systems).

A first analysis of the three dimensions and their relations across cases show the following:

(1) A high heterogeneity in the types of OI activities. The eight SMEs in the sample pursue inbound (sourcing) and outbound (revealing) OI activities in "isolation." In none of the cases did we observe a coupled approach, as defined by Enkel *et al.* (2009).
(2) An expected association between the type of OI activities and the domains of knowledge sourced/revealed (i.e., scientific and technological knowledge in the inbound dimension, and technological and product-market knowledge in the outbound dimension (Cesaroni *et al.*, 2005; Li *et al.*, 2008).
(3) Some recurring features at the intersection between the combination of OI approaches, the type of knowledge exchanged, and the sectoral innovation system. Science-based and supplier-dominated SMEs mainly pursue inbound OI activities aimed at scientific and technological knowledge sourcing. Conversely, specialized suppliers and information-intensive SMEs are more oriented to outbound OI aimed at integrating technological knowledge and product-market knowledge for external technology exploitation.

2.4. Case Studies

In this section, we analyze eight case studies of OI in European SMEs featuring different types of OI activities, nature (domains) of knowledge

Table 2.2. Cases classification.

	Type of OI activities				Nature of knowledge exchanged			Sectoral innovation systems			
Company	Sourcing	Acquiring	Revealing	Selling	Scientific	Technological	Product/market	Science-based	Specialized supplier	Supplier-dominated	Information-intensive
Graphenea SA	X		X		X	X		X			
Guger Technologies	X				X	X		X			
Celoxica Ltd.			X			X	X		X		
BCB Inform ática y Control			X	X		X			X		
Entranet Ltd.	X	X				X	X			X	
Primo 1D		X				X	X			X	
Zappar Ltd.			X				X				X
Supponor OY			X	X			X				X

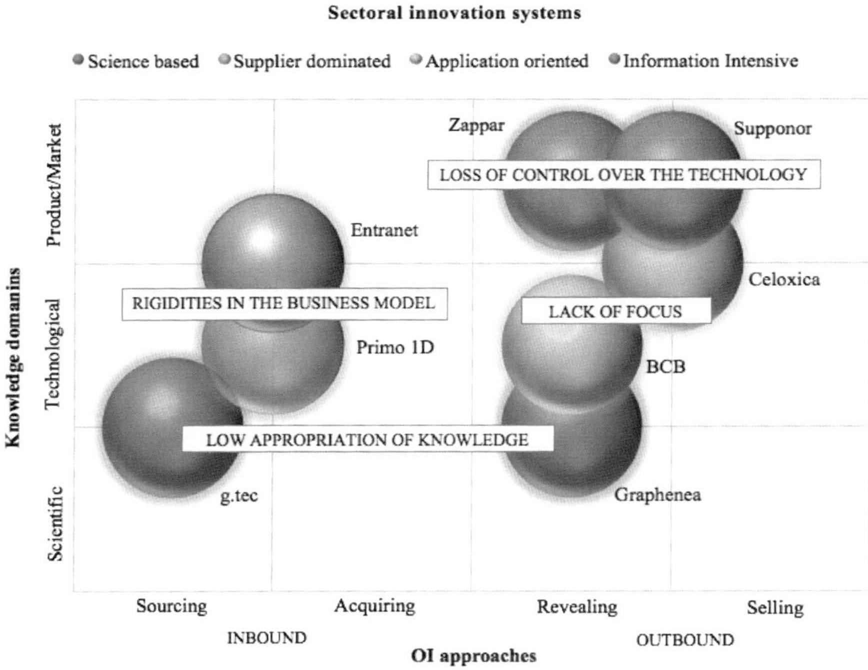

Figure 2.1. The "hidden costs" of open innovation in SMEs according to different implementation paths.

exchanged, and external context characteristics, defined by the sectoral innovation systems in which the companies operate (Figure 2.1).

In the light of this framework, we describe four different categories of "hidden costs" of OI in SMEs as they emerge in the different contexts of implementation.

2.4.1. *Managerial complexity and appropriability problems: The "hidden costs" of bringing science to the market through open innovation*

The two companies examined in this section arose from a pure research setting (R&D in natural science and medical engineering) and engaged in OI projects with industrial partners to commercialize the results of scientific research developed in academic institutions. Innovation paths aimed at bringing science to the market build on the integration of scientific knowledge (advances in science and R&D) and technological/market knowledge through open business models targeting the industrial (B2B) or the final use (B2C).

The shift from a closed to an open business model in this path implies a transition from science-based cooperative research (universities, R&D consortia) to market-based partnerships (industrial companies) aimed at pursuing diversification strategies to enter new markets. However, when science gets close to the market, important strategic risks may emerge.

The two following cases in Sections 2.4.1.1 and 2.4.1.2 show that, for SMEs operating in science-based settings, the shift toward open business models focused on the search and integration of external knowledge raises managerial complexities (difficulties in the alignment of partners' incentives; risks of contractual opportunism) and appropriability problems (scientific knowledge revealing).

2.4.1.1. *Guger Technologies (g.tec)*

Guger Technologies (g.tec)[2] is a medical engineering company founded for the development of brain–computer interface (BCI) systems.[3] The idea behind g.tec's technology was born from the development of an operating system serving real-time applications process data without buffering delays. The first phase of the technology development strongly benefited from scientific knowledge inflows originating from the company's ecosystem. Joining large R&D consortia facilitated R&D priority setting and the development

[2] The company was founded in 2009 as a spin-off of Gratz University of Technology. It employs eight people and sells its products in more than 60 countries worldwide.

[3] BCI systems are direct communication pathways between the brain and an external device assisting, augmenting, or repairing human cognitive or sensory–motor functions.

of the company's business model. g.tec's OI strategy was motivated by the need of pushing the technology exploitation phase in cooperation with international market-based partners. It is qualified by product codesign projects (mainly devoted to technological knowledge sharing, in order to offer state-of-the-art equipment to the B2B market) and product customization projects, benefiting from external product-market knowledge. According to this strategy, the company evolved its business model from targeting the public research market to a focus on B2B and B2C segments. This strategic move was intended to create opportunities for value capture from different sources and accelerate the commercialization of the innovation.[4] Still, unpredicted managerial complexities emerged in the implementation phase, due to substantial misalignments in the partners' objectives. Switching from science-based cooperative research to industrial partnerships increased the need of formalization (Du *et al.*, 2014). The shift toward an open business model was undermined by increased coordination costs of monitoring and control. Due to the high levels of confidentiality required, the company was forced to center its appropriation strategy on trade secrets, nondisclosure agreements, confidentiality, and retention of human resources. Still, under this tension between the development of an open business model and the emergence of the "hidden costs" of knowledge revealing (driving the company in the direction of closed innovation), only some forms of external collaboration take place. g.tec attentively manages these collaborations, designing them through a cautious combination of HR retention strategies and enforcing with its customers standardized contracts that allow an accurately controlled revealing of the scientific knowledge needed in the customization. Only through this strictly controlled strategy, g.tec is able to pursue an open and collaborative business model.

[4]The IntendiX BCI system for home use market was therefore developed, and in 2014 the company started developing the first end-user product for patients.

2.4.1.2. *Graphenea SA*

Graphenea is a small Spanish company founded around the idea of synthesizing graphene products for industrial applications.[5] Graphene is a material that does not exist in nature: it was isolated and characterized in 2007, but the current method of adoption was still not valid for industrial applications, remaining enclosed in universities and public R&D labs. The idea behind Graphenea was to find a way to integrate graphene materials in different industrial applications (batteries, composites, polymers, metals). In the early phase of the technology development, Graphenea pioneered research and advanced the state-of-the-art graphene production. The company deploys an open business model since the beginning of its activity: as a science-based start-up, cooperation was the essential driver of product innovation success.[6] In a first phase, Graphenea's OI strategy was based on scientific and technological knowledge sourcing from small companies, universities, and big players in the S&T ecosystem, aimed at the development of a niche product based on scientific research. At the end of 2013, Repsol, a petrochemical multinational company, entered the equity of Graphenea with 10% of shares. The industrial partnership was aimed at the advanced technological development and commercialization of the polymer composite through a shift toward an open business model. The high levels of confidentiality required by scientific knowledge revealing prevent the opportunities of advanced technology development with industrial partners. Therefore, Graphenea benefits from its nonmarket interactions with public partners and clients: universities and PRCs represent the first customers for graphene products. Working in close collaboration with them, Graphenea is able to conduct collaborative research projects that result not only in new opportunities of technology application, but also in a reduction of R&D costs for the company.

---

[5] Founded in 2010, Graphenea has been established in a large nanotechnology center (CIC nanoGUNE), Spain. The company employs 12 people and currently exports graphene materials to 40 countries.

[6] Graphenea participated in five European projects, and it is a partner of the Graphene Flagship initiative, one of the biggest EU investments in R&D for the development of graphene applications.

2.4.2. *Loss of focus and commercial viability risks: The "hidden costs" of application-oriented open innovation*

OI represents a strong opportunity for SMEs developing application-oriented technology solutions across multiple sectors. Specialized suppliers are indeed strongly dependent on demand-side technological inputs (cooperation with clients) in order to focus on performance improvement, reliability, and customization of their technologies. However, pursuing OI in an application-oriented path can result in distraction of resources, lack of focus, and failed attempts to find synergies between the new project and the core business of the company.

The two cases reported in this section show how losing focus in the implementation of OI could represent a serious and unpredicted risk in application-oriented paths, potentially compromising the commercial viability of the company's business model. Both companies are specialized suppliers of transversal knowledge solutions in the B2B market (Machine Vision techniques and FPGA applications, respectively).

Openness to external technology codevelopment is therefore motivated by the necessity of integrating technical specifications and product/market knowledge in the engineering, prototyping, and manufacturing phases. However, in this context, OI success can be undermined by rising search costs, low return on R&D investments, and the insurgence of managerial challenges.

2.4.2.1. *Celoxica Ltd.*

Celoxica is an ICT company developing hardware acceleration solutions targeted to the global financial services industry.[7] The company was born from an OI strategy aimed at exploiting academic R&D on the HANDELC language in multiple markets. In a first phase, the company was oriented at extending the domains of application for the technology. The searching strategy was characterized by multiple

---

[7]The company was established in 2000 by incorporating Embedded Solutions Ltd. (ESL), a technology spin-out from Oxford University. Today, it employs 32 people in three different countries.

research alliances for the exploitation of FPGA system architectures, aimed at integrating technical specifications and expertise from different partner industries. Accordingly, the business model was not being focused on a specific value proposition. The lack of focalization on a specific market path raised serious managerial challenges in R&D priority setting. In the first years of implementation of such an OI strategy, the company faced increased search costs and R&D expenditure, leading to almost negative operating margins from the single products. This unpredicted risk seriously challenged the commercial viability of such an open business model.

The evolution of Celoxica's OI strategy was a direct consequence of this strategic risk. With the management team renewal, the composition of the company's internal competences radically changed; the strong technological competences arising from 15-year R&D on FPGA technologies were balanced with a new management team coming from a financial market environment. Celoxica's business model shifted from products to specific services (ultra-low latency solutions) for the electronic trading community, based on the FPGA technology. A deep integration between technical background and trading expertise was the key to successfully develop the new business.[8]

2.4.2.2. *BCB Informática y Control*

BCB Informática y Control (BCB) is a small ICT engineering company (computer systems design and data processing).[9] BCB's core business is based on system integration and quality control services through the development of Machine Vision applications. The company offers "control and management solutions" to its clients' industrial processes based on hardware integration and software codevelopment, usually selling the prototype to the customer at the

[8] A focused strategy based on a single FPGA application for the financial industry characterized the new round, facilitated by partnerships with banks and Corporate VC investments.

[9] Based in Spain, the company employs 10 people in two different countries.

end of each collaboration project. The main benefits of this OI approach came from a business model aimed at exploiting transversal knowledge solution through user–producer codevelopment and customization projects. External networking and research partnerships within the market for technology were the drivers of a huge expansion of BCB's R&D activity in the industrial automation, automotive aeronautics, thermosolar, optical, and ICT fields.[10] On one side, BCB's embeddedness in large research consortia played a fundamental role in expanding the company's internal knowledge base in different industries. On the other side, an unfocused R&D strategy created huge barriers to priority setting for go-to-market. The "hidden cost" emerging from this path was the risk of not being able to pursue a successful exploitation strategy. SMEs involved in multiple R&D projects may be exposed to the risk of failing to set the market to exploit their core technologies and, therefore, missing the integration of important assets for innovation development. In the words of the owner: "Only large enterprises can do this."

2.4.3. *Rigidity and knowledge lock-in: The "hidden costs" of open innovation in low-tech paths*

The successful implementation of OI in low-technology industries is strongly dependent on the availability of external technological inputs (components, materials, software, or equipment), and on the ability of the entrepreneur to effectively search, acquire, assimilate, and integrate technological knowledge from other industries (Vanhaverbeke, 2017). OI approaches in such a technology-driven path are therefore aimed at exploitative learning (technological capabilities deepening). Companies establish strategic alliances and cooperation to integrate, refine, and extend product innovation

[10] The company developed several prototypes of Machine Vision industrial applications: 2D measurement equipment, pharma-packaging control system, optical inspection equipment, remote tele-control, automatic number-plate recognition, and optical devices.

knowledge to new markets (Cesaroni *et al.*, 2005). In this section, we report two case studies showing that, although the implementation of OI in supplier-dominated SMEs can be triggered by successful sourcing strategies, an excessive focalization of the business model can drive rigidities. The evolution of business models aimed at integrating technologies and pursuing collaborative product design and development in niche markets may indeed constrain the enrichment of the company's internal knowledge base. The evolution of alliances inside a focused business model exposes supplier-dominated SMEs to the risk of remaining locked-in to the development of specific technological knowledge and not being able to move away from a nonscalable business model. Unfortunately, the estimation of the risks of an excessive focalization is limited when new projects start, and this risk increases if the availability of new external technological knowledge is absent or scarce (e.g., for companies located in poor ecosystems).

2.4.3.1. *Entranet Ltd.*

Entranet Ltd. is a small construction company developing innovative embedded systems enhanced by Automated Speech Recognition and Speech Synthesis Technologies.[11] In the words of the founder, "The core business of the company is filling the usability gap between buildings and technologies." Entranet's OI path was developed through external search, technology acquisition (IP in-licensing), and exploitation of highly reliable ASR and SST technologies on top of the company's applications (noise reduction algorithms) through four years of internal R&D. In the commercialization phase, Entranet was strongly oriented at the identification of potential applications of the new system of use in a growing market niche. The development of an open business model was therefore centered on product codevelopment with customers and users in order to establish new standards for the assisted living market (elderly individuals)

[11] Based in Thessaloniki (GR), the company employs eight people and has two commercial branches, in London and in the US.

and possibly other new markets (blind people assistance). On one side, the development of an open business model through the integration of manufacturers, technological partners, and lead users allowed the successful integration of ICTs into a non-ICT knowledge base (construction industry). On another side, the process of implementation of such an OI path was not relevant for the evolution of the company's core competences: it simply allowed the identification of the specific business case/applications for a technology that had already been developed. From a technical perspective, it provided a "customization" of the single solution with respect to specific uses (market applications). Such a setting raised the strategic risk to remain locked in a nonscalable business model: the result of an excessive focalization raised the risks of losing the adaptability to new market segments, limiting the evolution of the internal knowledge base to the specific technology. This risk was aggravated by the lack of access to qualified/cutting-edge resources, since the company was not located in a vibrant innovation ecosystem.

2.4.3.2. *Primo1D*

Primo1D is a French company founded in 2013. It offers solutions for embedding electronics into objects and materials using E-Thread, a 3D micropackaging technology. The company was born out of an OI strategy (technology spin-out) resting on collaboration with its local ecosystem.[12] The company's founder actively attempted to build an industrial project through in-depth market analysis and external technology scouting in the textile sector. As a result, while the idea of application initially considered was the "traceability" of products, the project moved to other potential applications in the fields of

[12] The E-Thread technology was first developed in the CEA-Leti nanotechnology center labs (Grenoble high-technology cluster), also through the participation in the European PASTA Project consortium, funded by the 7th European Framework Program. CEA filed 18 families of patents on the E-Thread technology (i.e., technology, its improvements, its applications). Primo1D does not own the technology: the company has an exclusive license on it, paid through a mix of fixed fees and royalties on future turnover.

"anticounterfeiting or recycling" of luxury goods. The OI project combined research on electronic packaging and interconnection technology with textile research to realize an innovative approach of smart textile.[13] The development of the company's business model was therefore focused on meeting the needs of potential customers in the identified market of application. Small R&D projects for collaborative product design and development — smart yarns — enabled the company to grow inside its supportive ecosystem but, at the same time, limited the advanced technology development to a specific market path. Therefore, the evolution of the company's internal competences can be locked in by the knowledge available in the ecosystem.

2.4.4. *Open but controlled: The hidden costs of OI in community-driven OI*

The shift from a closed to an open business model comes along with increasing complexity and increases the costs of coordination of internal and external resources. Achieving the right balance between external knowledge sourcing and technology control could represent a severe managerial challenge to the implementation of OI in SMEs. With the rise of the Web 2.0, information-intensive SMEs are increasingly developing open business models whose fast scalability is allowed by network effects.

IT-based platforms allow SMEs to engage collaborative communities of innovators (partners, suppliers, customers, and users) as external sources of value creation. The implementation of OI in a community-driven path is the result of different steps, characterizing an evolution of the business model into more complex and ambitious partnerships and requiring the redefinition of new avenues of collaborations.

The cases described in this section highlight two serious challenges for SMEs operating in information-intensive path. A first managerial challenge lies in the evolution of the business model from a closed setting toward the inclusion of collaborative communities. Finding the

[13]The E-Thread traceability can be used to fight theft and counterfeiting in retail and to control textile production. It provides information along the whole product life cycle.

right balance between technological knowledge revealing and resources sharing represents a critical task for successful appropriation strategies. A second challenge comes from the alignment of partners' incentives while retaining technology control. Where the companies implemented OI 2.0 strategies, they claimed they had built communities and platforms starting from the assets that they could best control.

We conclude that the creation of a collaborative marketplace leading the convergence between different industries increases the risk of knowledge leakage and, therefore, the costs of technology control.

2.4.4.1. *Zappar Ltd.*

Zappar Ltd. is a British company operating in the IT industry.[14] The Zappar app was launched in 2013 as an evolution of the QR code, allowing users to create their own AR contents independently. The business was initially set according to a closed, consulting-based model based on one-to-one contracts aimed at fostering the technology diffusion among users (become the technology standard for a community of content creators). This approach changed when Zappar expanded its business model to involve customers from a variety of sectors in the development of the products. The new open business model allowed the company to expand from the entertainment industry to a wider spectrum of possible markets. This evolution was enabled by the Zapcode Creator platform, an online tool offering a simple, detailed, and cost-effective AR creation instrument to the B2B and B2C markets (big corporations, small businesses, and individuals). As a result, the sources of revenue streams evolved from one-to-one contracts with single users to revenue-sharing contracts with its partners. The ultimate company's strategic goal is to "become the central engine of an ecosystem in a democratized AR." Up to this point, a further evolution toward a 2.0 business model enabled the aggregation and crowd-sourced AR development around Zappar's technology across the world. The evolution of the company's business model was slowed down by the necessity of finding

[14] Born in 2011 as a spin-off from Cambridge University (UK), Zappar develops augmented reality (AR) applications for smartphones, tablets, and wearables.

a right balance between the increasing need of technological knowledge revealing (aimed at scaling up the platform) and the necessity of retaining control over the technology. Zappar still works closely with the communities of customers to codevelop AR contents and applications, but the technology is still being exploited internally.

2.4.4.2. *Supponor OY*

Supponor OY is a Finnish sports media and technology company whose technology replaces traditional perimeter advertising in sports broadcasts with digital content.[15] The company's operational teams work with local partners and specialists at live sporting events across the globe. Supponor's technology is the DBRLive (Digital Billboard Replacement Technology). This proprietary technology allows to replace the existing perimeter systems (LED billboards) with new billboards, keeping their traditional look for the on-site audience but whose appearance on the broadcast feed can be changed through digitally generated graphics. The shift from a closed setting to an open business model was a result of the need to exploit the potential of the technology. Supponor explicitly said that working openly and collaborating with the other stakeholders operating in its market sector were essential in developing its business; nevertheless, the company realized that control mechanisms to ensure a proper return on investment in innovation were needed. The evolution of Supponor's business model was characterized by the shift from the exploitation of digital billboards replacement solutions based on traditional approaches to the market (technology selling) to a service-based value proposition driven from the disintermediation of the traditional advertising distribution chain for digital

[15] The company was started in 2000, and in the first phase of its evolution (the early development of the technology) went through different rounds of VC investments. In 2006, Conor Venture Partners invested in VAS to start Supponor. In 2008, Northzone, a European technology investment partnership, further invested in Supponor. Today, the company employs 19 people, moved its headquarters to the UK, but maintained two branches located in Finland, a commercial office, an operational center, and its R&D department.

content.[16] Through the implementation of OI, Supponor made the active market converge around its technology. However, in order to pursue technology collaborations with its partners, Supponor first patented its innovation. The identification of a scalable business model based on the application of a potential disruptive technology dangerously rose the costs of technology control. Patenting the DBRLive allowed Supponor to protect and keep under control the results of its research. This case emphasizes the relevance of IP protection and technology control for information-intensive SMEs developing collaborations downstream with key partners.

2.5. Conclusions

Small firms have more opportunities of successful exploitation of new technological trajectories in comparison with large companies, due to their high flexibility in their R&D processes allowing the swift use of external resources and competences (Acs and Audretsch, 1987).

Nonetheless, the implementation of an OI strategy in SMEs can be a source of mistakes and managerial problems (Vanhaverbeke, 2017).

In spite of the surge of relevant empirical research on the determinants of OI effectiveness, little attention has been paid to *how* and *why* SMEs' managerial approaches in implementing OI may give rise to unpredicted risks.

This chapter has focused on the interpretation of the "hidden costs" of implementing OI in SMEs. Based on previous works (Di Minin *et al.*, 2016; Marullo *et al.*, 2015), we use an interpretative framework defining different contexts of OI implementation in SMEs through different combinations of OI activities, nature of knowledge exchanged, and sectoral innovation systems in which the companies operate. Combining evidence from multiple case studies,

[16] The open business model relied on three elements: (i) DBRLive technology, a technically innovative media platform that integrates with live broadcast feeds; (ii) value creation for sports rights holders by enhancing their perimeter billboard; and (iii) creating and delivering marketing opportunities to brands looking to engage specific sports audiences with relevant marketing.

we provide a detailed description of four categories of contingent challenges emerging when companies go from theory to practice.

The main contribution of the study is to advance the state-of-the-art of research on OI challenges in SMEs through a framework bringing back different types of OI challenges to specific internal and external contexts of implementation.

Strategic resources planning for OI is not an easy task and requires resilience against unexpected difficulties. In some cases, the process of implementation of OI requires more resources and/or time than companies originally planned. This can be the result of a mismatch of goals in alliances or may come from dimensions of transaction costs increasing beyond the control of the individual companies. Regardless of the specific set of OI practices, the rise of unpredicted risks related to the appropriation of relevant knowledge, loss of focus in R&D priority setting, rigidities in the business model, and weak technology control may compromise the implementation process, driving SMEs in the direction of closed innovation.

In this conclusive section, we provide a comprehensive description of four different categories of "hidden costs" of OI in SMEs, as related to different types of implementation contexts.

The low appropriability of scientific knowledge drives closeness in science-oriented OI paths.

SMEs are very sensitive to unintended knowledge spillovers in research ecosystems from which they successfully source scientific and preindustrial knowledge. However, the commercialization of research and technology developed in academic institutions through open business models entails a shift from science-based to market-based partnerships (cooperation agreements with clients/customers). In this context, increased managerial complexities and appropriability problems can seriously constrain the implementation of OI strategies.

We have examined two cases of science-based SMEs, developing leading-edge technologies in nanoscience and in the biomedical sector through a wide use of R&D collaborations (scientific and technological knowledge sourcing from research partners).

Both companies agreed that explorative learning through external knowledge sourcing was fundamental to exploit their potential in expanding the market of research. However, when approaching the phase technology exploitation (cooperation agreements and alliances with industrial partners), confidentiality became a big impeding factor in most research projects. Science-based SMEs often operate in very difficult situations in which they have to overcome high barriers to enter or build the market themselves.

Appropriability problems related to the nature of "inventive" (codified) knowledge lower the incentives to pursue market-based partnerships: high monitoring costs, coordination costs, and the risk of loss of relevant knowledge represent the "hidden costs" of OI in this context.

Appropriation problems often make the only chance of commercialization to expand the market by selling to scientific communities. The pursuit of external collaborations with trusted partners (universities, PRCs) might help overcoming the main barriers that SMEs face to enter science-based industries.

Lack of focus can compromise the commercial viability of application-oriented business models.

OI paths aimed at unlocking the market potential of general purpose technologies may be undermined by the "hidden costs" of an unfocused search. We have analyzed two cases of specialized suppliers engaging in outbound OI projects (integrating technological and product/market knowledge from clients to develop new applications of the technology). R&D priority setting represented, given the context, a crucial factor influencing outbound OI effectiveness.

Managing the "fuzzy front-end" of OI in application-oriented paths has the highest impact on new projects' success. In this context, search strategies aimed at exploiting new technological solutions in a variety of markets and/or the integration of multiple partners in the early phases of the technology development might determine increasing search costs and R&D over commitment.

Unfocused search exposes SMEs to the risk of failing to set the relevant market to exploit their core technologies and the related R&D priorities (e.g., failing the integration of important assets for innovation development).

Opening up to external ideas has the undoubted benefit of driving new opportunities to enrich the internal knowledge base. However, overfocalized OI paths in low-technology SMEs can drive rigidities and technological lock-in.

The implementation of OI in supplier-dominated (low-technology) SMEs is characterized by strong learning opportunities arising from the integration of external technological knowledge and internal R&D through business model innovation. The process of extracting and appropriating value from the newly developed technology is a critical success factor in this path (Vanhaverbeke and Chesbrough, 2014).

We analyzed two cases of companies successfully pursuing inbound OI aimed at technological knowledge sourcing/acquiring. For companies in this path, the development of an open business model complementing the internal technology base with externally developed knowledge represented the key managerial challenge in implementing OI.

The development of overfocalized business models targeted at specific market niches (supplier-dominated paths) may constrain the evolution of SMEs' core competences and rise the strategic risks of remaining locked-in to the available technological knowledge. The risk of losing the adaptability of the business model to different technology domains increases as soon as the availability of companies is located in poor ecosystems.

The risk of losing control over the technology slows down the implementation of community-driven open business models.

The rise of unpredicted risks of knowledge leakage might limit the scalability of OI 2.0 business models. The "hidden costs" of OI for the two information-intensive SMEs in the sample clearly emerged in the potential loss of business-critical technological knowledge shared through platforms. We find evidence that companies tend to build alliances and collaborate when they can secure control of their own technologies and knowledge. Succeeding in the journey from a closed setting (based on new proprietary technologies) toward distributed ecosystems implies a lot of trial and error, but most of all, adequate capabilities to balance competitive markets and collaborative communities (Boudreau and Lakhani, 2009; West and Lakhani, 2008).

In this chapter, we reported the first results of an explorative research project on OI in SMEs. Based on prior work (Di Minin *et al.*, 2016; Marullo *et al.*, 2015), we classify and analyze the challenges of implementing OI in eight case studies of European SMEs through an original interpretative framework and provide evidence of how different context-related elements may drive specific "hidden costs" when companies go from theory to practice.

The definition of the three dimensions in our framework of OI implementation in SMEs helped to constrain variability in empirical results and to clarify the relation between different implementation paths. Nonetheless, further research is needed to shape the external validity of our findings and achieve a more in-depth understanding of the phenomenon of OI in its complexity.

Three elements clearly emerge from this first exploratory study as key managerial tasks to successfully prevent strategic risks in the OI implementation.

The first element is the importance of adopting a balanced approach to R&D investments. The shift from a closed setting to an open business model in one-project-one-technology companies requires a clear balance between external knowledge sourcing and the development of internal competences. A constant evolution in the company's key competences helps constrain the risks of rigidities in the business model for companies operating with a limited number of high-risk R&D projects.

Second, the main benefits of outbound OI in SMEs arise from the access to very specific product and market knowledge. In our cases, successful technology exploitation came from bottom-up approaches to external technology development. R&D priority setting and focused search strategies allow for achieving proper resources orientation toward this objective. Rising search costs are among the most diffused causes of failing OI implementation in SMEs. Focusing on a specific value proposition allows SMEs to set priorities to go to the market and to build scalable business models.

Finally, clear appropriation strategies unlock the potential of outbound OI in SMEs. Across the different paths, we have observed specific combinations of IP management, selective revealing, and collaborative R&D. Managers need to develop the flexibility to balance the risk of revealing business-critical knowledge with the opportunities to access relevant external assets for technology exploitation.

References

Acs, Z. J. and Audretsch, D. B. (1987). Innovation, market structure and firm size, *The Review of Economics and Statistics*, 71, pp. 567–574.

Archibugi, D. (2001). Pavitt's taxonomy sixteen years on: a review article, *Economics of Innovation and New Technology*, 10, pp. 415–425.

Belderbos, R., Carree, M. and Lokshin, B. (2004). Cooperative R&D and firm performance, *Research Policy*, 33, pp. 1477–1492.

Boudreau, K. and Lakhani, K. (2009). How to manage outside innovation, *MIT Sloan Management Review*, 50, pp. 69–76.

Brunswicker, S. and van de Vrande, V. (2014) Exploring open innovation in small and medium-sized enterprises, eds. Chesbrough, H., Vanhaverbeke, W. and West, J., *New Frontiers in Open Innovation* (Oxford University Press, Oxford), pp. 135–156.

Brunswicker, S. and Vanhaverbeke, W. (2015). Open innovation in small and medium-sized enterprises (SMEs): external knowledge sourcing strategies and internal organizational facilitators, *Journal of Small Business Management*, 53, pp. 1241–1263.

Castellacci, F. (2008). Technological paradigms, regimes and trajectories: manufacturing and service industries in a new taxonomy of sectoral patterns of innovation, *Research Policy*, 37, pp. 978–994.

Cesaroni, F., Di Minin, A. and Piccaluga, A. (2005). Exploration and exploitation strategies in industrial R&D, *Creativity and Innovation Management*, 14, pp. 222–232.

Chesbrough, H. W. (2003). *Open Innovation: The New Imperative for Creating and Profiting from Technology* (Harvard Business School Press, Boston, MA).

Chesbrough, H. (2006). *Open Business Models: How to Thrive in the New Innovation Landscape* (Harvard Business School Press, Boston, MA).

Christensen, J. F., Olesen, M. H. and Kjær, J. S. (2005). The industrial dynamics of open innovation — evidence from the transformation of consumer electronics, *Research Policy*, 34, pp. 1533–1549.

Dahlander, L. and Gann, D. M. (2010). How open is innovation? *Research Policy*, 39, pp. 699–709.

Di Minin, A., De Marco, C. E., Marullo, C., Piccaluga, A., Casprini, E., Mahdad, M. and Paraboschi, A. (2016). *Case Studies on Open Innovation in ICT*. Institute for Prospective Technological Studies (IPTS JRC) Science for Policy Report EUR 27911 EN.

Di Minin, A. and Faems, D. (2013). Building appropriation advantage: An introduction to the special issues on intellectual property management. *California Management Review*, 55(4), pp. 7–14.

Dosi, G. (1982). Technological paradigms and technological trajectories, *Research Policy*, 11, pp. 147–162.

Du, J., Leten, B. and Vanhaverbeke, W. (2014). Managing open innovation projects with science-based and market-based partners, *Research Policy*, 43, pp. 828–840.

Edwards, T., Delbridge, R. and Munday, M. (2005). Understanding innovation in small and medium-sized enterprises: a process manifest, *Technovation*, 25, pp. 1119–1127.

Eisenhardt, K. M. (1989). Building theories from case study research, *Academy of Management Review*, 14, pp. 532–550.

Enkel, E., Gassmann, O. and Chesbrough, H. (2009). Open R&D and open innovation: exploring the phenomenon, *R&D Management*, 39, pp. 311–316.

Enkel, E., Kausch, C. and Gassmann, O. (2005). Managing the risk of customer integration, *European Management Journal*, 23, pp. 203–213.

Felin, T. and Zenger, T. R. (2013). Closed or open innovation? Problem solving and the governance choice, *Research Policy*, 43, pp. 914–925.

Goduscheit, R. C. and Knudsen, M. P. (2015). How barriers to collaboration prevent progress in demand for knowledge: a dyadic study of small and medium-sized firms, research and technology organizations and universities, *Creativity and Innovation Management*, 24, pp. 29–54.

Henkel, J. (2006). Selective revealing in open innovation processes: the case of embedded Linux, *Research Policy*, 35, pp. 953–969.

Hossain, M. (2015). A review of literature on open innovation in small and medium-sized enterprises, *Journal of Global Entrepreneurship Research*, 5, pp. 1–12.

Huizingh, E. K. R. E. (2011). Open innovation: state of the art and future perspectives, *Technovation*, 31, pp. 2–9.

Julien, P. (1995). New technologies and technological information in small businesses, *Journal of Business Venturing*, 10, pp. 459–475.

Laursen, K. and Salter, A. (2006). Open for innovation: the role of openness in explaining innovation performance among UK manufacturing firms, *Strategic Management Journal*, 27, pp. 131–150.

Laursen, K. and Salter, A. J. (2014). The paradox of openness: appropriability, external search and collaboration, *Research Policy*, 43, 867–878.

Lee, S., Park, G., Yoon, B. and Park, J. (2010). Open innovation in SMEs, an intermediated network model, *Research Policy*, 39, pp. 290–300.

Li, Y., Vanhaverbeke, W. and Schoenmakers, W. (2008), Exploration and exploitation in innovation: reframing the interpretation, *Creativity and Innovation Management*, 17, pp. 107–126.

Lopez-Vega, H., Tell, F. and Vanhaverbeke, W. (2016). Where and how to search? Search paths in open innovation, *Research Policy*, 45, pp. 125–136.

Malerba, F., (2005). Sectoral systems: How and why innovation differs across sectors, eds. Fagerberg, J., Mowery, D. C. and Nelson, R. R. *The Oxford Handbook of Innovation* (Oxford University Press, Oxford), pp. 380–406.

Marullo, C., De Marco, C. E., Di Minin, A. and Piccaluga, A. (2015). The hidden costs of OI in SMEs, *Proceedings of 2nd World Open Innovation Conference (WOIC 2015)*, Santa Clara (CA).

Nieto, J. and Santamaría, L. (2010). Technological collaboration: bridging the innovation gap between small and large firms, *Journal of Small Business Management*, 48, pp. 44–69.

Oakey, R. P. (2013). Open innovation and its relevance to industrial research and development: the case of high-technology small firms, *International Small Business Journal*, 31, pp. 319–336.

Parida, V., Westerberg, M. and Frishammar, J. (2012). Inbound open innovation activities in high-tech SMEs: the impact on innovation performance, *Journal of Small Business Management*, 50, pp. 283–309.

Pavitt, K., Robson, M. and Townsend, J. (1989). Technological accumulation, diversification, and organisation in UK companies, 1945–1983, *Management Science*, 35, pp. 81–99.

Ritala, P., Olander, H., Michailova, S. and Husted, K. (2015). Knowledge sharing, knowledge leaking and relative innovation performance: an empirical study, *Technovation*, 35, pp. 22–31.

Roberts, E. B. (1988). What we've learned: managing invention and innovation, *Research-Technology Management*, 31, pp. 11–29.

Spithoven, A., Vanhaverbeke, W. and Roijakkers, N. (2013). Open innovation practices in SMEs and large enterprises, *Small Business Economics*, 41, pp. 537–562.

van de Vrande, V., de Jong, J., Vanhaverbeke, W. and De Rochemont, M. (2009). Open innovation in SMEs: trends, motives and management challenges, *Technovation*, 29, pp. 423–437.

Vanhaverbeke, W. (2017). *Managing Open Innovation in SMEs* (Cambridge University Press, Cambridge, UK).

Vanhaverbeke, W. and Chesbrough, H. (2014). A classification of open innovation and open business models, eds. Chesbrough, H., Vanhaverbeke, W. and West, J., *New Frontiers in Open Innovation* (Oxford University Press, Oxford), pp. 50–68.

Vanhaverbeke, W., Cloodt, M. and van de Vrande, V. (2008). Connecting absorptive capacity and open innovation. Available at SSRN 1091265.

Vanhaverbeke, W., Vermeersch, I. and De Zutter, S. (2012). Open innovation in SMEs: how can small companies and start-ups benefit from open innovation strategies? Issues in Informing Science & Information Technology, Brussels.

Vossen, R. (1998). Relative strengths and weaknesses of small firms in innovation, *International Small Business Journal*, 16, pp. 88–94.

West, J. (2006). Does appropriability enable or retard open innovation? eds. Chesbrough, H., Vanhaverbeke, W. and West, J., *Open Innovation: Researching a New Paradigm* (Oxford University Press, Oxford), pp. 1–48.

West, J. and Bogers, M. (2017). Open innovation: Current status and research opportunities, *Innovation: Organization and Management*, **19**(1), pp. 43–50.

West, J. and Gallagher, S. (2006). Open innovation: The paradox of firm investment in open source software, *R&D Management*, 36, pp. 319–331.

West, J. and Lakhani, K. R. (2008). Getting clear about communities in open innovation, *Industry & Innovation*, 15, pp. 223–231.

Yin, R. K. (2003). *Case Study Research: Design and Methods* (Sage Publications, Thousand Oaks, CA).

Part II

Adopting Open Innovation in SMEs

Chapter 3

An Open Innovation Model for SMEs

Barbara Bigliardi and Francesco Galati*

Department of Engineering and Architecture
University of Parma, V. le delle Scienze 181/A, 43124 Parma, Italy
**barbara.bigliardi@unipr.it*

Abstract

Notwithstanding the paradigm of open innovation that was introduced and studied within large companies, in the last decade, studies on this concept in small and medium-sized enterprises (SMEs) have experienced a rapid growth (e.g., van de Vrande *et al.*, 2009; Spithoven *et al.*, 2013). However, several issues remain unexplored. Theories and models for the management of open innovation in this context are not well established in the literature, but are necessary for SMEs' managers. The aim of this chapter is to understand, through a systematic review of the literature, if and how the specific characteristics and behaviors of SMEs affect their implementation of the open innovation paradigm. Specifically, a "4W open innovation model" including motives, facilitators and hindering factors (*Why/why not*), type of exchanged knowledge

* Corresponding author.

(*What*), practices and activities (*hoW*), and actors (*With whom*) is proposed. The results allow us to point out a future research agenda.

3.1. Introduction

Innovation has always been of paramount importance in business. Researchers agree in recognizing innovation as a crucial factor for the success and long-term survival of companies. For decades, companies followed a closed innovation perspective, but in today's context characterized by increasing global competition and rising innovation costs, they are no longer able to survive with their own internal resources. Consequently, they must look for new and more open modes of innovation (OECD, 2008). Empirical evidence shows that companies are creating and adopting an ever wider range of explorative and exploitative ideas, concepts, expertise, and methods. In other words, companies have revised their innovation strategies and are moving from the traditional (closed) model to a new (open) model of innovation, where the mindset can be reformulated as "If we make the best use of both internal and external ideas, we will win" (Bigliardi and Galati, 2016a). Chesbrough (2003) formalized the open innovation (OI) paradigm as "the use of purposive inflows and outflows of knowledge to accelerate internal innovation, and to expand the markets for external use of innovation, respectively." According to this paradigm, companies can and should use both external and internal ideas, and thus they also benefit from external knowledge (De Jong *et al.*, 2008).

Since Chesbrough published his book on OI in 2003, there has been a proliferation of studies regarding this issue. Most of the examples of OI implementation are from high-tech industries and large companies. Specifically, we can cite the cases of open-source software (Henkel, 2006; Remneland-Wikhamn *et al.*, 2011), information and communication technology (Bigliardi *et al.*, 2012), biotech (Chiaroni *et al.*, 2011), the pharmaceutical industry (Chiaroni *et al.*, 2009; Galati and Bigliardi, 2016), and so on. However, the literature also investigates the adoption of OI in low-tech or mature industries such as the food industry (e.g., Bigliardi and Galati, 2013).

As for the size of the companies adopting OI, this paradigm has mainly been analyzed in large or multinational enterprises (e.g., Chesbrough, 2003; Kirschbaum, 2005). Indeed, pioneering studies focused mostly on the implementation of OI in large high-tech firms such as IBM (Chesbrough, 2003) and Procter & Gamble (Dodgson *et al.*, 2006). While scholars agree that innovation also plays an important role in SMEs, these types of companies have often been excluded from the discussion on OI. However, in recent years, research on OI in smaller companies has experienced rapid growth (e.g., Lee *et al.*, 2010; van de Vrande *et al.*, 2009), which demonstrates that a move from a closed to an open approach to innovation is also emerging within SMEs.

Before delving into the main arguments of the chapter, we formally define the distinction between large companies and SMEs. According to EU recommendation 2003/361, SMEs are defined by their number of employees (<250), turnover (≤50 € million), and balance sheet total (≤43 € million).

The analysis of the literature in the context of SMEs highlights that OI is a broader concept that encompasses diverse dimensions. In discussing these dimensions, we follow the distinction between content, context, and process made by Pettigrew *et al.* (1990) in his discussion of organization change research, later adopted in the OI context by Huizingh (2011), and reconsidered by Bigliardi and Galati (2016b). Following their perspectives, we structure our systematic review in a framework comprised of four dimensions: *Why/why not* (reasons for implementing or not implementing the paradigm), *What* (the content of OI), *hoW* (the process of OI), and *With whom* (the actors involved in the process). The first dimension (*Why/why not*) refers to the reasons why SMEs may decide to rely or not to rely on external knowledge to support their own innovation process. It also considers the factors that facilitate or impede the adoption of OI. In addition, for an effective implementation of the OI paradigm, proper management of knowledge is required. In fact, as previously stated, OI implies the systematic exploration of a wide range of knowledge sources (West *et al.*, 2006), which could reside both inside and outside the companies' boundaries. Thus, the second dimension is the type of exchanged knowledge, and in the following, we will refer to this dimension as *What*. An analysis of the examples of OI implementation shows that this paradigm can be adopted by firms in three different ways: companies can engage in inbound OI activities, outbound OI activities, or they can adopt a coupled OI process by combining inbound and outbound OI (Gassmann and Enkel, 2004). Thus, the practices and activities adopted must also be investigated. We will refer to this dimension as *hoW*. Finally, OI implies collaboration with different actors, which may include suppliers, customers, universities, etc. We will label this dimension *With whom*.

Based on these premises, the aim of this chapter is to understand if and how the idiosyncratic characteristics and behaviors of SMEs affect their implementation of the OI paradigm. Specifically, the framework depicted in Figure 3.1, including the four abovementioned dimensions, will be proposed as a guide for future research in the SMEs' context.

With this purpose in mind, the chapter is structured as follows: first, we describe the methodology utilized for conducting this

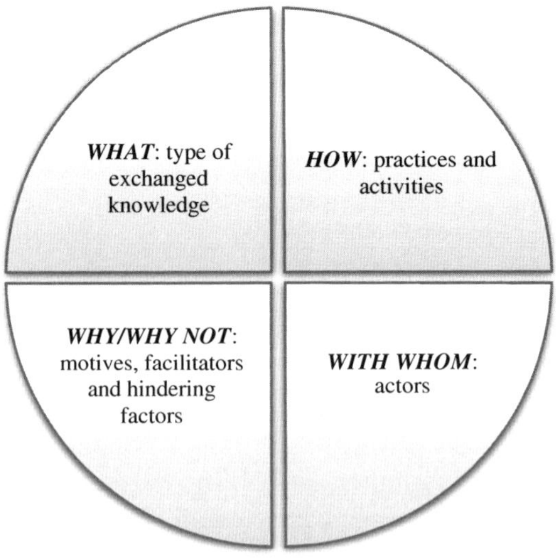

Figure 3.1. The OI model used as a guide for the research.

systematic review, and then we review the most recent contributions on the topic of OI in SMEs, thus highlighting the peculiarities that distinguish SMEs from larger companies in terms of OI; next, we analyze each of the dimensions included in the model; finally, we conclude the chapter by discussing the 4W model and describing a future research agenda.

3.2. Methodology

This study consists of a systematic review of articles published in scholarly papers on the topic of OI implementation in SMEs. In particular, we sum up and systematize the progress of research on the relationship between the specific characteristics and behaviors of SMEs and their implementation of the OI paradigm, as discussed in the extant scientific literature.

In general, the aim of a systematic review is to identify key scientific contributions to a field, and its results are often descriptively presented and analyzed. Applying the principles of systematic review also helps to ensure more reliable results, upon which conclusions may be drawn and decisions made. The reason we chose to perform a systematic rather than a narrative review is that the latter has been widely criticized for its lack of relevance, related to the use of a personal and usually subjective methodology (Hart, 1998). Transfield *et al.* (2003) proposed applying the principles of the systematic review methodology, which is usually implemented in medical sciences research, in the management context, using a rigorous, replicable, scientific, and transparent process.

Table 3.1 schematically represents the review process we adopted, together with the review flow diagram (depicted in Figure 3.2). More specifically, we carried out an extensive review of the literature by adopting a four-phase methodology to identify studies dealing with both OI and SMEs. First, we conducted a search for journal articles within the ABI/INFORM, JSTOR, JCR, SCOPUS, EMERALD, and EBSCO databases (Phase 1). We used a combination of the following keywords (Phase 2): Open Innovation, SME*,

Table 3.1. Overview of the systematic review process adopted.

Phase 1. Selecting the databases
• ABI/INFORM, JSTOR, JCR, SCOPUS, EMERALD, EBSCO (academic search elite and business source elite).
Phase 2. Search words used
• Open Innovation, SME*, Innovation, Absorptive Capacity, Collaboration, Small and Medium-sized Enterprise, Small and Medium Enterprise, Small Business, External Knowledge, Networking, Co-creation, Knowledge Acquisition, Knowledge Sharing, Partner*.
Phase 3. Articles filtered
• We read each one of the 163 articles in detail to make sure that they suited the requirements of the research. At this stage, we excluded 97 articles which did not fit the research aim.
Phase 4. Analysis
• Each article was analyzed by both authors with reference to the four dimensions in the scheme depicted in Figure 3.1.

Note: For the sake of clarity, SME* and Partner* stand for all the keywords starting with SME and Partner respectively.

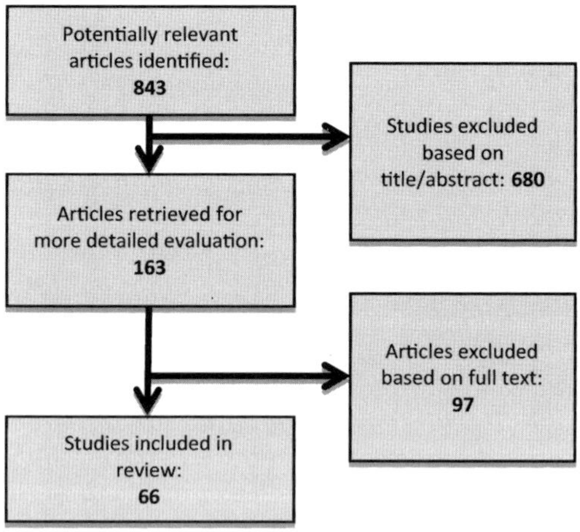

Figure 3.2. Systematic review flow diagram.

Innovation, Absorptive Capacity, Collaboration, Small and Medium-sized Enterprise, Small and Medium Enterprise, Small Business, External Knowledge, Networking, Co-creation, Knowledge Acquisition, Knowledge Sharing, Partner*. At this stage, we did not restrict the area of the research, as we were aware of the numerous terms that could be related to the topic of the chapter. Thus, we decided to consider all the potential contributions to the research and to defer the selection to a later stage.

We identified 843 potential articles for our systematic review, which were then subjected to a double screening. A first sorting of the articles' titles and abstracts allowed us to exclude 680 papers that did not meet the inclusion criteria. This left us with 163 potential articles for thorough analysis. The second screening extended beyond the titles and abstracts to the main text of the articles. Hence, we read each one of these 163 articles in detail to ensure that they suited the requirements of the research (Phase 3). At the end of this stage, we excluded 97 articles that did not fit the research aim. This left a total of 66 studies that matched the inclusion criteria and could be further analyzed. Finally, we analyzed each article, with reference to the four dimensions included in the framework presented in Figure 3.1 (Phase 4).

3.3. Specific Characteristics and Behaviors of SMEs with Regard to OI

Before discussing the four dimensions mentioned above, this section presents the main characteristics of SMEs. For the sake of clarity, and since our focus is on OI, we highlight the main factors that are most relevant concerning the adoption of the OI paradigm in the SME context. This means that other factors distinguishing SMEs and larger companies will not be presented in our discussion if they are not relevant for the opening of their innovation processes.

SMEs comprise, especially in Europe, the majority of all enterprises, reaching almost 99% of the total number of businesses (Klewitz and Hansen, 2014). Consequently, they are recognized as playing a central role in the European economy. Indeed, SMEs are the main source of entrepreneurial skills, employment, and innovation.

SMEs and larger companies differ obviously in terms of size, but they also differ with regard to policymaking procedures, structures, and utilizations of resources (Dufour and Son, 2015). Consequently, they are accustomed to conducting their businesses in different ways. This is also true for the adoption of the OI paradigm (Dufour and Son, 2015; Xiaobao *et al.*, 2013). Understanding the characteristics of SMEs is thus important both for SME scholars and managers interested in OI adoption in this context.

By nature, SMEs are more inclined to take risks and are more result-oriented (Dufour and Son, 2015; Parida *et al.*, 2012; Rocheska *et al.*, 2014) than large firms. SMEs are often entrepreneurial firms, where their founders have a desire to maintain control. Obviously, this entails an unwillingness to disclose information about their innovation processes and, consequently, an unwillingness to accept openness (Dufour and Son, 2015; Hernandez-Mogollon *et al.*, 2010).

SMEs also differ from larger companies in terms of organizational structure. Indeed, SMEs have a structure that is more flexible and organic than that of larger companies (Brunswicker and Vanhaverbeke, 2015; Dufour and Son, 2015; Rocheska *et al.*, 2014). Because of their flexibility, they react faster to changing market demands (Parida *et al.*, 2012). As they are organic structures, SMEs are characterized by a lower level of specialization, standardization, and formalization (Dufour and Son, 2015). Moreover, in today's changing environment, even though SMEs may face more difficulties in recognizing the need for change, their structure promotes innovativeness and allows them to adapt to the environment better than their larger counterparts (Brunswicker and Vanhaverbeke, 2015; Dufour and Son, 2015). In terms of OI, it is relatively easy to adjust the organizational structure of an SME, which is required to open the innovation process.

SMEs are less formalized and generally able to adjust their processes (Brunswicker and Vanhaverbeke, 2015; Dufour and Son, 2015; Lee *et al.*, 2010; van de Vrande *et al.*, 2009). The decision-making process is usually centralized at the level of the manager, who is responsible for many facets of the enterprise and for most management decisions (Dufour and Son, 2015; Zahra and Filatotchev, 2004). As for the communication process, it is more efficient and less complex to manage and organize than within larger companies

(Dufour and Son, 2015; Zahra and Filatotchev, 2004). Similarly, the planning process is not formal (Dufour and Son, 2015).

Because of their size, the number of people employed in SMEs is lower, and therefore loose and informal working relationships prevail (Dufour and Son, 2015). In a similar way, SMEs have a limited customer and supplier base. Their smallness also encourages individual creativity. As stressed by van de Vrande *et al.* (2009), SMEs often rely on the initiatives of their employees when looking for new methods of innovation.

SMEs have an obvious and fundamental role in innovation. However, while some authors have shown that SMEs tend to have a higher R&D productivity than larger firms (Lee *et al.*, 2010), they may lack the resources and capabilities required in industrial processes, such as manufacturing, distribution, and marketing, which are fundamental in developing an innovation. Further, the innovation models of SMEs differ from those of their larger counterparts: innovation in SMEs usually has an external focus and is often strongly based on social and personalties (Baum *et al.*, 2000; Brunswicker and Vanhaverbeke, 2015; Ceci and Iubatti, 2012; Edwards *et al.*, 2005). Due to their size, SMEs cannot cover all the innovation activities required to successfully realize an innovation (Brunswicker and Vanhaverbeke, 2015; Lee *et al.*, 2010). Consequently, innovation in SMEs is almost always characterized by an interorganizational and boundary-spanning component and is based on collaboration with other firms (Brunswicker and Vanhaverbeke, 2015; Edwards *et al.*, 2005; Lee *et al.*, 2010). Moreover, SMEs have less bureaucracy, a greater willingness to take risks, and the ability to react and adapt their structures more quickly to changing environments (Lee *et al.*, 2010).

SMEs are also characterized by other peculiarities, which are often presented as challenges they should face. Among others, the extant literature often proposes their limited resources as the main challenge: SMEs suffer from a lack of human capital, financial resources, managerial and technical skills, as well as know-how and technological assets (Bigliardi and Galati, 2016b; Brunswicker and Vanhaverbeke, 2015; Caloghirou *et al.*, 2004; Dahlander and Gann,

2010; Dufour and Son, 2015; Lee *et al.*, 2010; Oakey, 2013; Rahman and Ramos, 2010; Spithoven *et al.*, 2013; van de Vrande *et al.*, 2009; Wynarczyk *et al.*, 2013). In addition to limited resources, the cultural mindset of SMEs is seen as one of the main elements challenging SMEs (Dufour and Son, 2015; van de Vrande *et al.*, 2009). The OI paradigm may help SMEs in overcoming these challenges (Gassmann *et al.*, 2010; Hossain, 2015). For example, drawing on the resource-based view, as supported by Barney and Clark (2007) and Bigliardi and Galati (2016b), as well as on the transaction cost theory (Brouthers and Nakos, 2004), this lack of resources may lead SMEs to adopt OI. Collaborations, partnerships, and strategic and multiactor alliances are key factors in helping SMEs access the resources they lack (Baum *et al.*, 2000; Lee *et al.*, 2010; van de Vrande *et al.*, 2009).

The main characteristics of SMEs that may affect (both positively or negatively) the adoption of OI and that distinguish them from larger companies are summarized in Table 3.2, along with the main references.

3.3.1. *Motives, facilitators, and hindering factors: The Why/why not dimension*

We can identify a first stream of literature focusing on the reasons why SMEs adopt OI as well as the factors that facilitate or hinder its adoption. This refers to the *Why/why not* dimension of our model.

As far as the *Why* dimension is concerned, the literature includes several studies indicating that the adoption of OI by SMEs may provide the firms with a series of benefits and identifies several reasons to adopt the paradigm. Specifically, the motives can be classified into two main groups: entrepreneurial and financial reasons on the one hand and innovation reasons on the other.

With regard to the first group, different studies (e.g., Chesbrough and Crowther, 2006; Yun and Mohan, 2012) found that the two most common reasons to adopt OI are the need to maintain growth and the need to generate revenues. In terms of growth, Csath (2012) argued that SMEs typically embrace OI to grow internationally.

Table 3.2. The idiosyncratic characteristics and behaviors of SMEs with reference to OI.

Main characteristics of SMEs	Main references
More inclined to take risks and result-oriented	Dufour and Son (2015), Parida *et al.* (2012), Rocheska *et al.* (2014)
Limited customer and supplier base, more focus on their needs, higher bargaining power over enterprises	Dufour and Son (2015)
Entrepreneurial firms	Dufour and Son (2015), Hernandez-Mogollon *et al.* (2010)
Structure more flexible and organic, less bureaucracy	Brunswicker and Vanhaverbeke (2015), Dufour and Son (2015), Rocheska *et al.* (2014)
Structure able to react faster to changing market demands	Parida *et al.* (2012)
Lower level of specialization, standardization, and formalization	Dufour and Son (2015)
Easier to adjust the organizational structure required to open up the innovation process	Dufour and Son (2015)
Fluid culture	Dufour and Son (2015), van de Vrande *et al.* (2009)
External focus and often strongly based on social and personalties	Baum *et al.* (2000), Brunswicker and Vanhaverbeke (2015), Ceci and Iubatti (2012), Edwards *et al.* (2005)
Characterized by an interorganizational and boundary-spanning component, and based on the collaboration with other firms	Brunswicker and Vanhaverbeke (2015), Edwards *et al.* (2005), Lee *et al.* (2010)

(*Continued*)

Table 3.2. (*Continued*)

Main characteristics of SMEs	Main references
Less formalized and likely to be able to adjust their processes	Brunswicker and Vanhaverbeke (2015), Dufour and Son (2015), Lee *et al.* (2010), van de Vrande *et al.* (2009)
Decision-making process used to be centralized to the manager	Dufour and Son (2015), Zahra and Filatotchev (2004)
Communication process more efficient and less complex to be managed and organized	Dufour and Son (2015), Zahra and Filatotchev (2004)
Lower number of people employed and on which their culture relies; loose and informal working relationships prevail	Dufour and Son (2015)
Individual creativity encouraged, rely on initiatives of their employees when looking for new ways for innovation	van de Vrande *et al.* (2009)
Lack of human capital, financial resources, managerial and technical skills, as well as know-how and technological assets	Bigliardi and Galati (2016b), Brunswicker and Vanhaverbeke (2015), Caloghirou *et al.* (2004), Dahlander and Gann (2010), Dufour and Son (2015), Lee *et al.* (2010), Oakey (2013), Rahman and Ramos (2010), Spithoven *et al.* (2013), van de Vrande *et al.* (2009), Wynarczyk *et al.* (2013)

Chesbrough (2003) also stressed that SMEs start adopting OI to adapt their organizations to changes in the environment. Moreover, SMEs adopt OI to reduce costs, achieve economies of scale, reduce risks, increase operational flexibility, and market their products effectively (Hoffman *et al.*, 2001; Lee *et al.*, 2010; Rocheska *et al.*, 2014; Spithoven *et al.*, 2013; van de Vrande *et al.*, 2009). As far as the second group of reasons is concerned (i.e., the innovation reasons), SMEs may embrace OI to attain strong innovation performance and to successfully develop and commercialize new products (Pullen *et al.*, 2008). Moreover, OI may provide SMEs with R&D services required in the innovation process (Hossain, 2015; Segarra-Oña *et al.*, 2011). Other studies (e.g., Hoffman *et al.*, 2001; van de Vrande *et al.*, 2009) proposed reasons that refer to the innovation collaborations established by SMEs: to acquire missing knowledge and complementary resources and to enlarge their social networks (Pullen *et al.*, 2012). The results of the study by Gans and Stern (2003) indicated that SMEs adopt OI with the intention to serve their customers effectively or when opening new markets. Similarly, van de Vrande *et al.* (2009) proposed the overcoming of difficulties in the commercialization of new products and the effectiveness of the innovation process as main motivations. Finally, a recent and less explored reason is the development of eco-innovation: Bocken *et al.* (2014) claimed that eco-innovation is increasingly important for SMEs, and hence they may adopt the OI paradigm to support such innovation.

As for the variables that facilitate an SME in opening its innovation process, we identified three types of enabling factors in the literature that are suitable for both small and medium-sized firms and for large companies. These factors are technologies and financial and networking policies. With respect to the former, the diffusion of new technologies such as information and communication technologies (ICT) allows external partners and SMEs' employees to exchange required knowledge (e.g., Awazu *et al.*, 2009; Chesbrough, 2003; Galati *et al.*, 2016; Gassmann, 2006; Huston and Sakkab, 2006; Lichtenthaler, 2008; Torkkeli *et al.*, 2007). In addition, Rahman and Ramos (2013) highlighted that Web 2.0 and information systems (IS) play a key role in the assimilation of knowledge and are thus critical to the adoption of OI. As for the financial factors, the existence of a corporate venturing unit (Chesbrough, 2003; Lichtenthaler, 2008; van de Vrande *et al.*, 2009; Vanhaverbeke *et al.*, 2008) as well as the availability of financial resources (van de Vrande *et al.*, 2009) may facilitate OI adoption. Finally, with regard to networking policies, there are different examples in the literature (e.g., Torkkeli *et al.*, 2007; van Hemert *et al.*, 2013) demonstrating how policies can help SMEs in adopting the OI paradigm. For example, van Hemert *et al.* (2013) presented a subsidy program similar to the Dutch "innovation vouchers," which stimulated SMEs to explore their (technology) opportunities. The benefits of this kind of policy, often based on university–SME interaction, are supported by studies that show that, in the early stages of innovation, firms do indeed benefit from external networks with universities (Kaufmann and Tödtling 2001; Tödtling *et al.*, 2009).

Alternatively, instead of focusing on *Why* SMEs engage in open innovation, one could also consider reasons why SMEs do not adopt OI and the barriers that impede its adoption (*Why not*). This second dimension is less explored in SMEs (Bigliardi and Galati, 2016b). Following the same approach adopted for analyzing the motives to adopt OI, it is possible to classify the reasons that hinder SMEs' adoption of OI into two main groups, namely, economic and financial reasons and internal reasons. The former are probably the most obvious ones. Indeed, OI (with regard to alliances) is a costly process (Bigliardi and Galati, 2016b; Teirlinck and Spithoven, 2013; van de Vrande *et al.*, 2009; Verbano *et al.*, 2015) that makes product development projects costlier (Knudsen and Mortensen, 2011). Consequently, SMEs often prefer not to open their innovation processes, developing new products following a closed approach instead. Internal reasons refer to the internal and intrinsic characteristics of SMEs. While introducing OI in the SME context, we stressed that SMEs must face numerous challenges, some of which are related to internal factors. The lack of managerial competencies (Bigliardi and Galati, 2016b; McAdam *et al.*, 2014; Teirlinck and Spithoven, 2013; van de Vrande *et al.*, 2009; Verbano *et al.*, 2015) as well as the high complexity of managing the OI process (Bigliardi and Galati, 2016b; van de Vrande *et al.*, 2009; Verbano *et al.*, 2015) may cause SMEs to not adopt OI. Similarly, due to the entrepreneurial behavior of SMEs, they often encounter cultural resistance that forces them to choose a closed innovation model instead of an OI model (Bigliardi and Galati, 2016b; Verbano *et al.*, 2015). A final internal challenge is the "not invented here" (NIH) syndrome (Chesbrough and Crowther, 2006; Scozzi *et al.*, 2005; van de Vrande *et al.*, 2009).

The extant literature refers to the factors hindering the adoption of OI as obstacles and barriers. These factors have recently been classified by Bigliardi and Galati (2016b) into four main types: knowledge, organizational, collaboration, and financial-strategic factors. Knowledge barriers refer to the loss of know-how or to the risk of imitation by competitors (Bigliardi and Galati, 2016b; Teirlinck and Spithoven, 2013; Ussman *et al.*, 2001; Verbano *et al.*, 2015) as well as to the extent to which relevant knowledge is not available (Breschi and Malerba, 2005). A knowledge-poor environment can inhibit a firm's ability to adopt OI (Bigliardi and Galati, 2016b; van de Vrande *et al.*, 2009; Verbano *et al.*, 2015). Organizational barriers are related to a lack of the managerial skills required to establish an effective collaboration with external partners (Bigliardi and Galati, 2016b; McAdam *et al.*, 2014; Teirlinck and Spithoven, 2013; van de Vrande *et al.*, 2009) as well as to resistance to change (Dufour and Son, 2015; Hernández-Mogollon *et al.*, 2010; Verbano *et al.*, 2015). Collaboration barriers refer to the collaboration mechanisms behind OI, specifically to the loss of know-how (Bigliardi and Galati, 2016b; Teirlinck and Spithoven, 2013), difficulties in finding the right partners (Bigliardi and Galati, 2016b; van de Vrande *et al.*, 2009; Verbano *et al.*, 2015), the risk of opportunistic behavior on the part of partners (Bigliardi and Galati, 2016b; McAdam *et al.*, 2014), and cultural differences with partners (Bigliardi and Galati, 2016b; Teirlinck and Spithoven, 2013; van de Vrande *et al.*, 2009). Finally, the main — and probably the most obvious — obstacles to the opening of the innovation process are financial and strategic barriers, which refer to economic and financial issues (van de Vrande *et al.*, 2009; Verbano *et al.*, 2015) as well as a lack of strategic vision on the part of the firms in terms of innovative knowledge (van de Vrande *et al.*, 2009).

A schematic summary of the discussion reported above is provided in Table 3.3, listing all the reasons (both "why" and "why not"), facilitators, and hindering factors.

3.3.2. *The type of knowledge exchanged: The What dimension*

In adopting the OI model, SMEs must look for external sources of innovation due to their intrinsic characteristics, exploring their environments and searching for the missing knowledge, skills, and technology rather than relying only on their internal R&D (Spithoven *et al.*, 2013). A generally accepted classification of this knowledge exchange divides knowledge into internal and external knowledge.

Internal knowledge, which is derived from sources internal to an SME, is mainly represented by internal design activities, internal R&D, and the acquisition of mechanical equipment (Filippetti, 2011; Kruse, 2012; Santoro *et al.*, 2016). Given their size, SMEs cannot cover all the innovation activities required to successfully realize an innovation (Brunswicker and Vanhaverbeke, 2015; Lee *et al.*, 2010). Therefore, innovation in SMEs is almost always characterized by an interorganizational and boundary-spanning component, based on collaboration with other firms (Brunswicker and Vanhaverbeke, 2015; Lee *et al.*, 2010). Teirlinck and Spithoven (2013) argued that research cooperation and R&D outsourcing offer possibilities to complement a firm's internal research resources.

In terms of "external knowledge," SMEs typically exchange information with their customers regarding their needs, contexts, experiences, and preferences (Baldwin and von Hippel, 2011; Brunswicker and Vanhaverbeke, 2015; Dahlander and Frederiksen, 2012; Franke *et al.*, 2006; van de Vrande *et al.*, 2009) as well as new insights into new business opportunities beyond existing products and markets (Enkel *et al.*, 2005; Santoro *et al.*, 2016). SMEs tend to establish partnerships with their suppliers to learn their technical

Table 3.3. Reasons, facilitators, and hindering factors.

Facilitators, Reasons, Hindering Factors			Main references
Why	Entrepreneurial and financial reasons	To maintain growth and obtain revenues	Chesbrough and Crowter (2006), Yun and Mohan (2012)
		To grow internationally	Csath (2012)
		To adapt their organization to changes in the environment	Chesbrough (2003)
		To reduce costs, to achieve economies of scale, to reduce risks, to increase operational flexibility, and to market their products effectively	Hoffman and Schlosser (2001), Lee *et al.* (2010), Rocheska *et al.* (2014), Spithoven *et al.* (2013), van de Vrande *et al.* (2009)
	Innovation reasons	To attain strong innovation performance and successfully develop and commercialize new products	Pullen *et al.* (2008)
		To obtain the R&D services required	Hossain (2015), Segarra-Oña *et al.* (2011)
		To acquire missing knowledge and complementary resources, to enlarge their social networks	Hoffman and Schlosser (2001), Pullen *et al.* (2012), van de Vrande *et al.* (2009)
		To serve customers effectively or to open up new markets	Gans and Stern (2003), van de Vrande *et al.* (2009)
		To overcome difficulties in the commercialization of new products and to reach effectiveness of innovation process	van de Vrande *et al.* (2009)
		To develop eco-innovation	Bocken *et al.* (2014)

(*Continued*)

Table 3.3. (*Continued*)

Facilitators, Reasons, Hindering factors			Main references
Facilitators	Technologies	ICTs	Awazu *et al.* (2009), Bessant and Tidd (2009), Chesbrough (2003), Chiaroni *et al.* (2011), Galati *et al.* (2016), Gassmann (2006), Huston and Sakkab (2006), Lichtenthaler (2008), Torkkeli *et al.* (2007)
		Web 2.0 and IS	Bugshan (2015), Rahman and Ramos (2013)
	Financial	Existence of corporate venturing unit	Bessant and Tidd (2009), Chesbrough (2003), Gassmann and Enkel (2004), Lichtenthaler (2008), Vanhaverbeke *et al.* (2008), van de Vrande *et al.* (2009)
		Availability of financial resources	Dodgson *et al.* (2006), Nakagaki *et al.* (2012), van de Vrande *et al.* (2009)
	Networking policies	Regional innovation system	Torkkeli *et al.* (2007), van Hemert *et al.* (2013)

Why not	Economic and financial reasons	OI is a costly process	Bigliardi and Galati (2016b), Knudsen and Mortensen (2011), Teirlinck and Spithoven (2013), Verbano *et al.* (2015)
	Internal reasons	Lack of managerial competencies	Bigliardi and Galati (2016b), McAdam *et al.* (2014), Teirlinck and Spithoven (2013), van de Vrande *et al.* (2009), van de Vrande *et al.* (2009), Verbano *et al.* (2015)
		High complexity in managing the OI process	Bigliardi and Galati (2016b), Scozzi *et al.* (2005), van de Vrande *et al.* (2009), Verbano *et al.* (2015)
		NIH syndrome	Chesbrough and Crowther (2006), van de Vrande *et al.* (2009)
Hindering factors	Knowledge barriers	Loss of know-how and risks of imitation by competitors	Bigliardi and Galati (2016b), Teirlinck and Spithoven (2013), van de Vrande *et al.* (2009), Verbano *et al.* (2015)
		Extent to which relevant knowledge is not available	Breschi and Malerba (2005), van de Vrande *et al.* (2009)

(*Continued*)

Table 3.3. (*Continued*)

Facilitators, Reasons, Hindering factors		Main references
Organizational barriers	Lack of the managerial skills required to establish an effective collaboration with external partners	Bigliardi and Galati (2016b), McAdam *et al.* (2014), Teirlinck and Spithoven (2013), van de Vrande *et al.* (2009), Verbano *et al.* (2015)
	Resistance to change	Hernandez-Mogollon *et al.* (2010), Verbano *et al.* (2015)
Collaboration barriers	Loss of know-how	Dufour and Son (2015), Bigliardi and Galati (2016b), Teirlinck and Spithoven (2013)
	Difficulties in finding the right partners	Bigliardi and Galati (2016b), van de Vrande *et al.* (2009), Verbano *et al.* (2015)
	The risk of opportunistic behavior of partners of collaboration	Bigliardi and Galati (2016b), McAdam *et al.* (2014)
	Cultural differences with partners	Bigliardi and Galati (2016b), Teirlinck and Spithoven (2013), van de Vrande *et al.* (2009), Verbano *et al.* (2015)
Financial and strategic barriers	Economic and financial issues	Bigliardi and Galati (2016b), Teirlinck and Spithoven (2013), van de Vrande *et al.* (2009), Verbano *et al.* (2015)
	Lack of strategic vision of the firms in terms of innovative knowledge	van de Vrande *et al.* (2009)

expertise (Brunswicker and Vanhaverbeke, 2015; Chesbrough and Prencipe, 2008; Tsai, 2009). The knowledge exchanged with universities or research centers is usually represented by R&D equipment, inventive and preindustrial knowledge, and knowledge on how to appropriate technologies (Brunswicker and Vanhaverbeke, 2015; Dufour and Son, 2015; Freel and Robson, 2016; Spithoven *et al.*, 2013). In addition, SMEs may exchange technological trends, ideas, and knowledge with their intermediates (Bader, 2006; Brunswicker and Vanhaverbeke, 2015; Vega-Jurado *et al.*, 2008). Finally, as highlighted by Kruse (2012), SMEs may also need, and thus exchange, documentations in the form of patents and licenses, standards and regulations, knowledge from the Internet, magazines, and search engines, as well as knowledge and ideas about new market trends from fairs, conferences, and workshops. It is thus possible to further classify external knowledge as follows: "market knowledge" (i.e., knowledge from customers, suppliers, competitors, and other firms), "scientific knowledge" (derived from universities, laboratories, public and private research centers, etc.), and "media knowledge" (i.e., documentations in the form of patents and licenses, regulations, information from the Internet and from events, and so on).

SMEs only collaborate with science and technology partners or with value chain partners if they can easily absorb such knowledge (Spithoven *et al.*, 2013) and exploit it quickly for commercial purposes (Dufour and Son, 2015; Parida *et al.*, 2012; Rocheska *et al.*, 2014). Wynarczyk *et al.* (2013) argued that SMEs tend to collaborate for the sake of product introductions. Other studies have confirmed that collaboration for SMEs is more important in the commercialization stage than in the early stages of innovation (e.g., van de Vrande *et al.*, 2009).

All these considerations affirm that, on the one hand, *what* SMEs exchange may vary according to their collaboration partners, and on the other hand, the unique characteristics and orientations of SMEs affect the type of knowledge exchanged with each partner. In other words, collaboration with different types of partners leads SMEs to exchange different types of knowledge: market knowledge when considering a customer as a partner and scientific knowledge when considering a university as a partner. However, this is a general affirmation that is also true for large companies. SMEs are distinguished from larger companies, and the type of knowledge exchanged with each partner is affected by their unique characteristics (e.g., their limited internal resources) and orientations (e.g., their result-oriented behavior). This implies that if a large and a small firm are involved in a collaborative innovation project with universities, their knowledge exchange is quite different. Small firms tend to prefer short-term projects, characterized by the ability to rapidly develop a commercial application, given their result-oriented focus. On the contrary, larger firms may also undertake longer R&D projects, given their long-term orientation and stock of resources. Thus, smaller firms are interested in application-oriented research and applied knowledge, whereas larger firms are often interested in basic research. This consideration also derives from the fact that large firms usually possess the means (in financial, human, and facility terms) needed to internally exploit such basic knowledge.

A summary of the discussion above is reported in Table 3.4.

3.3.3. *Practices and activities of OI: The hoW dimension*

Chesbrough and Crowther (2006) identified two main dimensions of OI: inbound and outbound. The term inbound refers to a company that increases its innovativeness using in-house and external R&D provided by actors such as suppliers or customers. Conversely, the term outbound refers to companies that look for external organizations that are better suited to commercialize their knowledge or technologies. Subsequently, Gassmann and Enkel (2004) suggested a third dimension, the "coupled process," according to which companies cooperate with other enterprises in strategic networks by coupling the outside-in and inside-out processes.

In regard to inbound processes, SMEs may directly involve customers in their innovation process, for example, by developing or codeveloping products based on customers' specifications (Gassman, 2006; Spithoven *et al.*, 2013; van de Vrande *et al.*, 2009). These

Table 3.4. The type of knowledge exchanged by SMEs in the OI process.

Type of knowledge		Main references
Internal knowledge	Internal design activities, internal R&D, and the acquisition of mechanical equipment	Filippetti (2011), Kruse (2012), Santoro *et al.* (2016)
External knowledge		
Market knowledge	Customers' information about their needs, context, experiences and preferences, new insights into new business opportunities	Baldwin and von Hippel (2011), Brunswicker and Vanhaverbeke (2015), Dahlander and Frederiksen (2012), Enkel *et al.* (2005), Franke *et al.* (2006), Santoro *et al.* (2016), van de Vrande *et al.* (2009)
	Suppliers' technical expertise	Brunswicker and Vanhaverbeke (2015), Chesbrough and Prencipe (2008), Tsai (2009)
	Technological trends, ideas, and knowledge	Bader (2006), Brunswicker and Vanhaverbeke (2015), Vega-Jurado *et al.* (2008)
Scientific knowledge	R&D equipment, inventive and preindustrial knowledge, and knowledge on how to appropriate technologies	Brunswicker and Vanhaverbeke (2015), Dufour and Son (2015), Freel and Robson (2016), Ramos *et al.* (2009), Spithoven *et al.* (2013)
Media knowledge	Documentations in the form of patents and licenses, standards and regulations, knowledge from Internet, magazines, and search engines, as well as knowledge and ideas about new market trends from fairs, conferences, and workshops	Kruse (2012)

activities are often associated with a market-based short-term orientation, which is allowed by the organizational characteristics of SMEs (e.g., low formalization, organizational flexibility, and entrepreneurial orientation). Similarly, SMEs may systematically collaborate with external partners (e.g., universities, research centers, other firms) to support the innovation process (Lee *et al.*, 2010; van de Vrande *et al.*, 2009; Wynarczyk *et al.*, 2013) or to purchase R&D work from others (Wynarczyk *et al.*, 2013). Alternatively, SMEs may buy or use external IP such as patents (Gassmann, 2006; Lee *et al.*, 2010; van de Vrande *et al.*, 2009; Wynarczyk *et al.*, 2013). According to these authors, these practices are legitimized by SMEs' characteristics, such as limited resources or short-term and market-based orientation, when they realize the benefits of applying external research for developing and commercializing innovative products.

Similarly, SMEs may follow different approaches for the inside-out process. They can, for example, sell or license their patents or know-how (Lee *et al.*, 2010; van de Vrande *et al.*, 2009; Wynarczyk *et al.*, 2013), mainly to monetize their existing technologies. In this way, SMEs can enlarge their resource pool. However, the licensing of technologies requires prior financial investment, which few SMEs can afford (Bianchi *et al.*, 2010; Dufour and Son, 2015; van de Vrande *et al.*, 2009). Alternatively, SMEs can make their own unused innovations available to others for free (Lee *et al.*, 2010; Wynarczyk *et al.*, 2013) or actively participate in others' innovation (Lee *et al.*, 2010; van de Vrande *et al.*, 2009; Wynarczyk *et al.*, 2013). The inside-out practices of SMEs can be partially explained by the difficulty SMEs experience in exploiting (internally) all the possible applications of their technological discoveries, given their limited resource base.

Finally, as for the coupled process, SMEs may combine the outside-in process (aiming to gain external knowledge) with the inside-out process (aiming to bring ideas to market). In coupled OI, companies cooperate with other organizations in strategic networks (Gassmann and Enkel, 2004). Vanhaverbeke (2017) recently explored how these strategic networks are organized by innovative SMEs.

In general, when adopting OI, different researchers indicate that SMEs prefer nonmonetary activities (e.g., networking and informal knowledge sourcing) to pecuniary, resource-intensive, and complex transaction-based ones (e.g., acquisitions and in-licensing) (Dahlander and Gann, 2010; Dufour and Son, 2015).

SMEs have limited resources and will therefore rely on OI to develop new products and services. Few of them have top-notch technologies that they can license to other organizations. Accordingly, inbound OI practices are more commonly used than the outbound ones (Brunswicker and van de Vrande, 2014; Dufour and Son, 2015). For some authors (e.g., Lee *et al.*, 2010; Rocheska *et al.*, 2014), the outbound practices are also more associated with the commercialization stage. Recently, OI practices have been increasingly adopted by SMEs (Rocheska *et al.*, 2014). A categorization of the above-mentioned practices following the inbound, outbound, and coupled classification (Bigliardi and Galati, 2016b) is proposed in Table 3.5.

Table 3.5. The main OI practices in SMEs.

Types of practices		Main references
Inbound practices	Customer involvement	van de Vrande *et al.* (2009)
	External networking	Lee *et al.* (2010), van de Vrande *et al.* (2009), Wynarczyk *et al.* (2013)
	Inward IP licensing	Lee *et al.* (2010), van de Vrande *et al.* (2009), Wynarczyk *et al.* (2013)
	Internet exploration	Bigliardi and Galati (2016b), Kruse (2012)
	Know-how acquisition	Wynarczyk *et al.* (2013)
Outbound practices	Outward IP licensing	Lee *et al.* (2010), van de Vrande *et al.* (2009), Wynarczyk *et al.* (2013)
	Knowledge exploitation	Lee *et al.* (2010), Wynarczyk *et al.* (2013)
	Knowledge provision	Lee *et al.* (2010), van de Vrande *et al.* (2009), Wynarczyk *et al.* (2013)
Coupled practices	Alliances with complementary companies	Bigliardi and Galati (2016b)

3.3.4. *The actors involved in the OI process: The With whom dimension*

Finally, there is considerable literature describing the types of partners in OI. Von Hippel (1998) identified four main external sources of knowledge, namely: (i) suppliers and customers; (ii) university, government, and private laboratories; (iii) competitors; and (iv) other nations. According to his theory, most new products and services are codeveloped, or at least refined, by the end users. After the introduction of the OI paradigm in 2003, other scholars continued arguing that companies increasingly use relationships to seek out and acquire external knowledge for their innovative processes and to increase their absorptive capacity, that is, their ability to recognize and appropriate valuable external technologies (e.g., Spithoven *et al.*, 2013). The inside-out, outside-in, and coupled processes require the construction of external networks to internalize and/or externalize technologies (Dufour and Son, 2015). Despite having few direct contacts, SMEs may typically access more external sources through collaborative networks (Agüero and Sánchez, 2010). However, using too many external sources and search channels is time-consuming, laborious, and too expensive compared to the resulting benefits (Cantner *et al.*, 2009).

External knowledge sourcing may span various kinds of external innovation partners. According to authors such as Danneels and Kleinschmidt (2001), Spithoven *et al.* (2013), and more recently Santoro *et al.* (2016), the actors for collaboration may be classified into market-related and technology-related partners, depending on the knowledge that has to be exchanged. Similarly, Parida *et al.* (2012) and later Hossain (2015) distinguished between vertical and horizontal collaboration, which are relevant for radical and incremental innovation, respectively. Another classification was provided by Brunswicker and Vanhaverbeke (2015), who grouped the actors for collaboration as follows: (i) direct and indirect customers; (ii) suppliers; (iii) universities and research centers; (iv) experts on intellectual property rights; and (v) network partners. In general, it is possible to identify the following key groups of SMEs' OI partners: market partners, scientific partners, and intermediate service providers.

Market partners are involved in collaboration with the actors belonging to the value chain of the SME, that is, customers (both direct and indirect) and suppliers as well as competitors. SMEs tend to collaborate with customers to gain knowledge about their needs, tastes, and behaviors in order to satisfy their requirements (Baldwin and von Hippel, 2011; Brunswicker and Vanhaverbeke, 2015; Dahlander and Frederiksen, 2012; Franke *et al.*, 2006; van de Vrande *et al.*, 2009). As emphasized by different studies (Baldwin and von Hippel, 2011; Dahlander and Frederiksen, 2012; Franke *et al.*, 2006; Gassmann, 2006), the involvement of customers is an important alternative to internal innovation process development, as the information they have may aid internal R&D in developing new products or discovering new markets. It is possible to distinguish between direct and indirect customers (Brunswicker and Vanhaverbeke, 2015). The integration of information from direct customers is a popular practice among SMEs (van de Vrande *et al.*, 2009). The typically flat and organic structure of SMEs facilitates the early integration of users in the innovation process (Dufour and Son, 2015; Lee *et al.*, 2010). Differently from larger companies and mainly due to their limited base of customers, SMEs tend to integrate their customers by developing unstructured and informal practices that do not require massive investment (van de Vrande *et al.*, 2009). In a similar way, the involvement of indirect customers/users is useful to SMEs in searching for new business opportunities beyond existing products and markets (Enkel *et al.*, 2005; Santoro *et al.*, 2016). Finally, within the customer class, the extant literature identifies extreme users as a third segment (Santoro *et al.*, 2016). These users provide an active contribution to the development of new innovative ideas and to the commercialization of innovations (Flowers, 2008).

With respect to suppliers, SMEs typically collaborate with them to benefit from their technological expertise (Brunswicker and Vanhaverbeke, 2015; Chesbrough and Prencipe, 2008; Tsai, 2009). Specifically, SMEs usually introduce such collaborations early in the innovation process (Gassmann, 2006) by adopting unstructured and informal practices with customers (Brunswicker and Vanhaverbeke, 2015). The collaboration of SMEs with suppliers is less explored than that with customers, and there is not yet a clear consensus regarding whether customers or suppliers are the most important partners (Corral De Zubielqui *et al.*, 2016; Enkel *et al.*, 2009). Finally, contacts with competitors are also important for successful innovation performance (van Hemert *et al.*, 2013). SMEs show a preference for networking with other players because of their fear of giving away their technologies to their competitors. However, close contact with competitors allows SMEs to be at the forefront of new market developments and participate with other firms in innovation cooperation. As previously suggested by Hamel *et al.* (1989), successful companies can use alliances with competitors to build skills in areas outside the formal agreement, and thus to disseminate new knowledge throughout their organizations.

Scientific partners include universities and research organizations (Brunswicker and Vanhaverbeke, 2015), higher education institutes (HEI), and government research organizations (Corral De Zubielqui *et al.*, 2016). These kinds of partners represent a relevant source of knowledge for three reasons. First, science may significantly alter the search for inventions (Brunswicker and Vanhaverbeke, 2015; Fabrizio, 2009; Shinn and Lamy, 2006; Tsai, 2009). Second, they may offer more timely access to innovative trends (Fabrizio, 2009). Finally, universities may help SMEs that do not have sufficient resources to afford knowledge brokers and crowdsourcing platforms (Tsai, 2009). However, different studies rank these partners as a poor source of innovation for SMEs compared to market partners (Corral De Zubielqui *et al.*, 2016; Mina *et al.*, 2014; Theyel, 2013). In fact, several barriers to external knowledge sourcing in university–industry relationships exist, including cultural differences, long-term scientific research, and incompatible rewards systems (Harryson *et al.*, 2008).

Finally, Brunswicker and Vanhaverbeke (2015) observed that interaction with "intermediate service providers" is key for SMEs, and they include, for example, interaction with experts on intellectual property rights (IPR), which provides crucial information services, thus helping SMEs to bridge the gap between technological opportunities and their successful commercialization (Bader, 2006; Brunswicker and Vanhaverbeke, 2015; Vega-Jurado *et al.*, 2008). Notwithstanding, the involvement of IPR experts is slow and costly for SMEs (Hurmelinna-Laukkanen *et al.*, 2007). Moreover, it requires SMEs to deal with complex regulations and patent protection procedures (Brunswicker and Vanhaverbeke, 2015).

Other actors with whom larger companies tend to collaborate, such as knowledge brokers, are not key partners of SMEs because of their lack of financial and human resources. Moreover, the kinds of partners discussed above are not mutually exclusive and can be combined to maximize the value of OI (Brunswicker and Vanhaverbeke, 2015).

The above-mentioned actors for collaboration for SMEs are summarized in Table 3.6.

Table 3.6. The main actors of collaboration.

Types of actors		Main references
Market-related	Direct customers	Baldwin and von Hippel (2011), Brunswicker and Vanhaverbeke (2015), Dahlander and Frederiksen (2012), Dufour and Son (2015), Franke *et al.* (2006), Kim and Park (2010), Lee *et al.* (2010), van de Vrande *et al.* (2009)
	Indirect customers	Brunswicker and Vanhaverbeke (2015), Dufour and Son (2015), Enkel *et al.* (2005), Santoro *et al.* (2016)
	Extreme users	Flowers (2008), Füller *et al.* (2008), Santoro *et al.* (2016)
	Suppliers	Brunswicker and Vanhaverbeke, (2015), Chesbrough and Prencipe (2008), Gassmann (2006), Tsai (2009)
	Competitors	van Hemert *et al.* (2013)

(*Continued*)

Table 3.6. (*Continued*)

Types of actors		Main references
Scientific partners	Universities and research organizations	Brunswicker and Vanhaverbeke (2015), Fabrizio (2009), Shinn and Lamy (2006), Tsai (2009)
	HEI and government research organizations	Corral de Zubielqui *et al.* (2016)
Intermediate service providers	Experts on IPR	Bader (2006), Ramsden and Bennett (2005), Vega-Jurado *et al.* (2008)

3.4. The 4W Open Innovation Model

The present chapter aims to understand if and how the specific characteristics and behaviors of SMEs impact their implementation of OI. Their typical characteristics and behaviors legitimize the growing interest of scholars in OI in SMEs. The results obtained through our systematic review of the literature allow us to conclude that the characteristics and behaviors of SMEs heavily affect the implementation of OI, differentiating OI in SMEs from that in large firms. Among the various possibilities to innovate openly, SMEs prefer short-term and low-cost partnerships compared with larger firms. This can be easily connected to SMEs' result-oriented behaviors, their shortage of resources, and their lower absorptive capacity. However, this also implies that SMEs implement OI in a distinctive way, quite differently from large companies, and this is deserving of specific research attention.

The above-mentioned elements can now be placed into the framework proposed in the introduction section (as shown in Figure 3.3). This figure depicts the *4W open innovation model.*

In this model, we propose that the 4W dimensions are related to the other dimensions. These relationships derive from the systematic analysis of the literature and lead us to suppose the holistic nature of the model. The study by Brunswicker and Vanhaverbeke (2015)

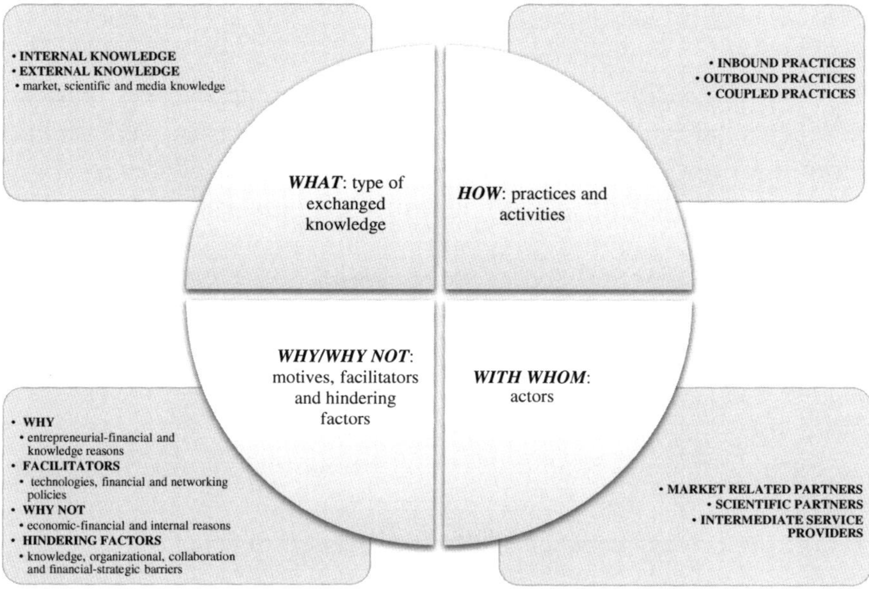

Figure 3.3. The 4W open innovation model.

inspired us to hypothesize a correlation within the four dimensions. Specifically, the authors stressed that external knowledge sourcing may span various kinds of external innovation partners, who relate to different knowledge flows and can provide access to widely different knowledge domains such as science, technology, design, customer insights, and product–market trends, which in turn implies nonpecuniary direct interactions with external actors. In addition, they affirmed that SMEs may exhibit different patterns of sourcing, given that access to each innovation source and the value expected in each case can differ significantly. Finally, they emphasized the centrality of combined innovation sources, knowledge exchange, motives for exploration activities, and OI practices for the success of a firm's sourcing strategy.

With our model, we argue that an SME may adopt OI based on the definition of one of the four dimensions (usually, the *Why* dimension). In other words, starting from various motives, it is possible to identify several collaboration modes, using various combinations of actors and practices and facing specific facilitators or barriers. Each choice made on one dimension has an impact on the other dimensions and reduces the range of alternatives that can be taken in the other dimensions. For example, if an SME should develop a radical new product (*Why*) that requires advanced scientific knowledge (*What*), it may decide to collaborate with scientific partners such as universities and research centers and not with another set of external actors (*With whom*). This OI process could force the company to adopt mainly specific inbound practices such as inward IP licensing or know-how acquisition (*hoW*) and to face barriers such as cultural differences and long-term scientific research (*Why not*).

Even though this is only one of the possible examples that can be derived from the holistic model, it could support the understanding of the connections between the four proposed dimensions.

3.5. Agenda for Future Research

Our analysis of the extant literature has led to the identification of several challenges for the implementation of OI in SMEs while providing ideas for future research. We decided to group these challenges according to the four dimensions of the model mentioned above.

3.5.1. *The Why/why not dimension*

In order to find the right partner — one of the main barriers identified — it could be important to rely more on professional actors such as innovation intermediaries, as suggested by the literature on matchmaking problems in networks (e.g., Holzmann *et al.*, 2014; Montelisciani *et al.*, 2014). This has implications for both SME managers and policymakers. SME managers could overcome this barrier in identifying the right partner (both commercial and scientific partners) through the expert advice of an innovation intermediary. Additionally, policymakers could stimulate such collaborations through *ad hoc* financial incentives.

Furthermore, SME managers should pursue an adequate and transparent IPR strategy with partners, avoiding unclear or ambiguous situations in which an established (and potentially fruitful) collaboration could be compromised (Bigliardi and Galati, 2016b). Again, the possibility to rely on experts such as the above-mentioned innovation intermediaries could help SMEs that lack the capability to solve these problems. For example, the experience of experts in innovative collaborations could guide SMEs in assessing the risks related to the loss of their know-how and in tailoring a suitable, individualized IPR strategy.

We thus suggest that future research could further explore the reasons why an SME adopts OI or chooses closed innovation, considering the role of facilitators (such as innovation intermediaries) and trust.

3.5.2. *The What dimension*

The *What* dimension is less explored in the extant literature. It thus deserves specific attention in future research. Starting from a careful analysis of the literature, we emphasized that *What* SMEs exchange varies according to the partners involved in the collaboration. Moreover, the unique characteristics (e.g., their limited internal resources) and orientations (e.g., their result-oriented behavior) of SMEs affect the type of knowledge exchanged with each partner. Therefore, it is vital to analyze the differences between OI in SMEs and in their larger counterparts. Vanhaverbeke (2017) described the specific nature of OI in SMEs in detail, but this study shows that the lessons learned from OI in large companies are not applicable in the SME context.

However, while for the other dimensions the relationships between such unique SMEs' characteristics and behaviors are explicitly defined in the literature, those for the *What* dimension have only been implicitly argued. Therefore, future research should provide a better understanding of how the unique characteristics and behaviors of SMEs affect the type of knowledge exchanged with each partner and how this can lead to a specific OI framework for SMEs.

3.5.3. *The hoW dimension*

OI should be realized in specific ways. For instance, the effective implementation of OI requires that SMEs properly manage knowledge flows with their partners. Indeed, while OI processes may enable SMEs to overcome their liabilities related to their small size, they need to be careful in terms of their IP management, and they should protect their technologies selectively. As discussed by Bigliardi and Galati (2016b), the fear of losing relevant knowledge and the risk of opportunistic behavior on the part of partners are prominent risks in the SME context. The IP management issue is a trust-related problem. Thus, further empirical evidence on the relationship between OI strategies and value appropriation by SMEs as well as on how this relationship may vary according to the dimensions of our model would be useful to our understanding of effective IP management within SMEs.

3.5.4. *The With whom dimension*

As previously stated, the implementation of OI activities and practices in SMEs depend on trust building among the actors involved in the partnerships. In this chapter, we have shown that SMEs may engage in collaboration with a broad set of partners. Thus, future research may investigate how trust can be built between SMEs and each partner and how the behavior and role of entrepreneurs are crucial in this regard. In addition, it could be of interest to investigate the same issues (trust formation and the entrepreneur's role) but from a network perspective. We believe that, contrary to a classic approach based on a one-to-one analysis of partnerships, the role played by these factors in a networked environment may be very different.

3.6. Conclusions

Openness in SMEs is an exciting research topic. Even though OI is an important issue for SMEs, as it can improve innovation performance, their intrinsic characteristics often influence OI implementation. In this chapter, we have investigated if and how the specific characteristics and behaviors of SMEs affect their implementation of the OI paradigm. Several peculiarities emerged from the analysis, which lead to a different adoption of the OI paradigm by SMEs compared with their larger counterparts. For instance, SMEs are characterized by a high inclination to take risks as well as a great focus on their needs, a flexible and organic structure, and a low level of specialization, standardization, and formalization. These traits may contribute to an easier adjustment of their organizational structures to the OI process. In addition, their lack of human capital, financial resources, managerial and technical skills, as well as their shortage of know-how and technological assets may negatively influence their ability to implement the OI paradigm. However, these liabilities may also provide a stimulus for adopting OI. Moreover, we argued that this shortage of resources, know-how, and technological assets could heavily influence the motives behind the decision to adopt OI, the selection of partners, the type of knowledge exchanged with each partner, and the identification of the appropriate OI practices to be implemented.

Based on a systematic review of previous works, we build on a new framework, the *4W open innovation model*, which could facilitate the interpretation of OI in SMEs. Specifically, we identified four dimensions: motives, facilitators of, or barriers to OI (*Why/why not*), type of exchanged knowledge (*What*), practices and activities (*hoW*), and actors (*With whom*).

The main contribution of the study is the idea that an integrated vision is needed to advance the state-of-the-art of research on OI implementation in SMEs. Previously, studies have investigated one or more of the four dimensions, but to our knowledge no single study investigated the four dimensions in an integrative way. In our model, we assume that the proposed dimensions are strategic and interrelated choices for SMEs. Through their combination, our approach should facilitate SME managers to seek sources of inspiration and expertise outside of their normal circles when constructing or reconstructing their business models — and, indeed, when implementing them.

References

Agüero, C. and Sánchez, P. (2010). Intellectual capital and knowledge management in collaborative networks. *Proceedings of the European Conference on Intellectual Capital*, Lisbon, Portugal, March, pp. 29–30.

Awazu, Y., Baloh, P., Desouza, K. C., Wecht, C. H., Kim, J. Y. and Jha, S. (2009). Information-communication technologies open up innovation. *Research Technology Management*, 52(1), pp. 51–58.

Bader, M. A. (2006). *Intellectual Property Management in R&D Collaborations: The Case of the Service Industry Sector* (Springer Science & Business Media, Germany).

Baldwin, C. and Von Hippel, E. (2011). Modeling a paradigm shift: from producer innovation to user and open collaborative innovation. *Organization Science*, 22(6), pp. 1399–1417.

Barney, J. B. and Clark, D. N. (2007). *Resource-Based Theory: Creating and Sustaining Competitive Advantage* (Oxford University Press, Oxford, UK).

Baum, J. A., Calabrese, T. and Silverman, B. S. (2000). Don't go it alone: alliance network composition and startups' performance in Canadian biotechnology. *Strategic Management Journal*, 21(3), pp. 267–294.

Bessant, J. and Tidd, J. (2009). *Inovação e empreendedorismo: administração* (Bookman Editora, Porto Alegre, Brazil).

Bianchi, M., Campodall'Orto, S., Frattini, F. and Vercesi, P. (2010). Enabling open innovation in small and medium-sized enterprises: how to find alternative applications for your technologies. *R&D Management*, 40(4), pp. 414–431.

Bigliardi, B., Dormio, A. I. and Galati, F. (2012). The adoption of open innovation within the telecommunication industry. *European Journal of Innovation Management*, 15(1), pp. 27–54.

Bigliardi, B. and Galati, F. (2013). Models of adoption of open innovation in the food industry. *Trends in Food Science & Technology*, 30(1), pp. 16–26.

Bigliardi, B. and Galati, F. (2016a). Open innovation and incorporation between academia and food industry. In: Galanakis, C. ed., *Innovation Strategies in the Food Industry: Tools for Implementation* (Elsevier Academic Press, Cambridge, MA), pp. 19–39.

Bigliardi, B. and Galati, F. (2016b). Which factors hinder the adoption of open innovation in SMEs? *Technology Analysis and Strategic Management*, 28(8), pp. 869–885.

Bocken, N. M. P., Farracho, M., Bosworth, R. and Kemp, R. (2014). The front-end of eco-innovation for eco-innovative small and medium sized companies. *Journal of Engineering and Technology Management*, 31, pp. 43–57.

Breschi, S. and Malerba, F. (2005). *Clusters, Networks and Innovation* (Oxford University Press, Oxford, UK).

Brouthers, K. D. and Nakos, G. (2004). SME entry mode choice and performance: a transaction cost perspective. *Entrepreneurship Theory and Practice*, 28(3), pp. 229–247.

Brunswicker, S. and Vanhaverbeke, W. (2015). Open innovation in small and medium sized entreprises (SMEs): external knowledge sourcing strategies and internal organizational facilitators. *Journal of Small Business Management,* 53(4), pp. 1241–1263.

Bugshan, H. (2015). Open innovation using Web 2.0 technologies. *Journal of Enterprise Information Management*, 28(4), pp. 595–607.

Caloghirou, Y., Kastelli, I. and Tsakanikas, A. (2004). Internal capabilities and external knowledge sources: complements or substitutes for innovative performance? *Technovation*, 24(1), pp. 29–39.

Cantner, U., Joel, K. and Schmidt, T. (2009). The use of knowledge management by German innovators. *Journal of Knowledge Management*, 13(4), pp. 187–203.

Ceci, F. and Iubatti, D. (2012). Personal relationships and innovation diffusion in SME networks: a content analysis approach. *Research Policy*, 41(3), pp. 565–579.

Chesbrough, H. (2003). *Open Innovation: The New Imperative for Creating and Profiting from Technology* (Harvard Business School Press, Boston, MA).

Chesbrough, H. and Crowther, A. K. (2006). Beyond high tech: early adopters of open innovation in other industries. *R&D Management*, 36, pp. 229–236.

Chesbrough, H. and Prencipe, A. (2008). Networks of innovation and modularity: a dynamic perspective. *International Journal of Technology Management*, 42(4), pp. 414–425.

Chiaroni, D., Chiesa, V. and Frattini, F. (2009). Investigating the adoption of open innovation in the bio-pharmaceutical industry: a framework and an empirical analysis. *European Journal of Innovation Management*, 12(3), pp. 285–305.

Chiaroni, D., Chiesa, V. and Frattini, F. (2011). The open innovation journey: how firms dynamically implement the emerging innovation management paradigm. *Technovation,* 31(1), pp. 34–43.

Corral De Zubielqui, G., Jones, J. and Lester, L. (2016). Knowledge inflows from market- and science-based actors, absorptive capacity, innovation and performance: a study of SMEs. *International Journal of Innovation Management*, p. 1650055.

Csath, M. (2012). Encouraging innovation in small and medium sized businesses: learning matters. *Development and Learning in Organizations: An International Journal*, 26(5), pp. 9–13.

Dahlander, L. and Frederiksen, L. (2012). The core and cosmopolitans: a relational view of innovation in user communities. *Organization Science*, 23(4), pp. 988–1007.

Dahlander, L. and Gann, D. M. (2010). How open is innovation? *Research Policy*, 39, pp. 699–709.

Danneels, E. and Kleinschmidtb, E. J. (2001). Product innovativeness from the firm's perspective: its dimensions and their relation with project selection and performance. *Journal of Product Innovation Management*, 18(6), pp. 357–373.

De Jong, J. P., Vanhaverbeke, W., Kalvet, T. and Chesbrough, H. (2008). *Policies for Open Innovation: Theory, Framework and Cases* (Research project funded by VISION Era-Net, Helsinki, Finland).

Dodgson, M., Gann, D. and Salter, A. (2006). The role of technology in the shift towards open innovation: the case of Procter & Gamble. *R&D Management,* 36(3), pp. 333–346.

Dufour, J. and Son, P. E. (2015). Open innovation in SMEs — towards formalization of openess. *Journal of Innovation Management,* 3(3), pp. 90–117.

Edwards, T., Delbridge, R. and Munday, M. (2005). Understanding innovation in small and medium-sized enterprises: a process manifest. *Technovation*, 25(10), pp. 1119–1127.

Enkel, E., Kausch, C. and Gassmann, O. (2005). Managing the risk of customer integration. *European Management Journal*, 23(2), pp. 203–213.

Fabrizio, K. R. (2009). Absorptive capacity and the search for innovation. *Research Policy*, 38(2), pp. 255–267.

Filippetti, A. (2011). Innovation modes and design as a source of innovation: a firm-level analysis. *European Journal of Innovation Management*, 14(1), pp. 5–26.

Flowers, S. (2008). Harnessing the hackers: the emergence and exploitation of outlaw innovation. *Research Policy*, 37(2), pp. 177–193.

Franke, N., von Hippel, E. and Schreier, M. (2006). Finding commercially attractive user innovations: a test of lead-user theory. *Journal of Product Innovation Management*, 23(4), pp. 301–315.

Freel, M. and Robson, P. J. (2016). Appropriation strategies and open innovation in SMEs. *International Small Business Journal*, DOI: https://doi.org/10.1177/0266242616654957.

Füller, J., Matzler, K. and Hoppe, M. (2008). Brand community members as a source of innovation. *Journal of Product Innovation Management,* 25(6), pp. 608–619.

Galati, F. and Bigliardi, B. (2016). The unintended effect of the Orphan Drug Act on the adoption of open innovation. *Science and Public Policy*, 43(6), pp. 840–848.

Galati, F., Bigliardi, B. and Petroni, A. (2016). Open innovation in food firms: implementation strategies, drivers and enabling factors. *International Journal of Innovation Management*, 20(3), p. 1650042.

Gans, J. S. and Stern, S. (2003). The product market and the market for "ideas": commercialization strategies for technology entrepreneurs. *Research Policy*, 32(2), pp. 333–350.

Gassmann, O. (2006). Opening up the innovation process: towards an agenda. *R&D Management*, 36(3), pp. 223–228.

Gassmann, O. and Enkel, E. (2004). Towards a theory of open innovation: three core process archetypes. *Proceedings of the R&D Management Conference*, Lisbon, Portugal, July 6–9.

Gassman, O., Enkel, E. and Chesbrough, H. (2010). The future of open innovation. *R&D Management*, 40, pp. 213–221.

Hamel, G., Doz, Y. and Prahalad, C. (1989). Collaborate with your competitors. *Harvard Business Review*, 67(1), pp. 133–139.

Harryson, S. J., Dudkowski, R. and Stern, A. (2008). Transformation networks in innovation alliances — the development of Volvo C70. *Journal of Management Studies*, 45(4), pp. 745–773.

Hart, C. (1998). *Doing a Literature Review: Releasing the Social Science Research Imagination* (Sage Publications, London, UK).

Henkel, J. (2006). Selective revealing in open innovation processes: the case of embedded Linux. *Research Policy,* 35(7), pp. 953–969.

Hernández-Mogollon, R., Cepeda-Carrión, G., Cegarra-Navarro, J. G. and Leal-Millán, A. (2010). The role of cultural barriers in the relationship between open-mindedness and organizational innovation. *Journal of Organizational Change Management*, 23(4), pp. 360–376.

Hoffman, D. L., Novak, T. P. and Schlosser, A. E. (2001). The evolution of the digital divide: examining the relationship of race to Internet access and usage over time. In: Benjamin M. Compaine, ed., *The Digital Divide. Facing a Crisis or Creating a Myth* (MIT Press, Cambridge, MA), pp. 47–97.

Holzmann, T., Sailer, K., Galbraith, B. and Katzy, B. R. (2014). Matchmaking for open innovation-theoretical perspectives based on interaction, rather than transaction. *Technology Analysis & Strategic Management,* 26(6), pp. 595–599.

Hossain, M. (2015). A review of literature on open innovation in small and medium-sized entreprises. *Journal of Global Entrepreneurship Research,* 5–6, pp. 1–12.

Huizingh, E. K. (2011). Open innovation: state of the art and future perspectives. *Technovation*, 31(1), pp. 2–9.

Hurmelinna-Laukkanen, P., Sainio, L. M. and Jauhiainen, T. (2008). Appropriability regime for radical and incremental innovations. *R&D Management*, 38(3), pp. 278–289.

Huston, L. and Sakkab, N. (2006). Connect and develop: inside Procter and Gamble's new model for innovation. *Harvard Business Review*, 84, pp. 58–66.

Kaufmann, A. and Tödtling, F. (2001). Science–industry intercation in the process of innovation: the importance of boundary-crossing between systems. *Research Policy*, 30, pp. 791–804.

Kim, H. and Park, Y. (2010). The effects of open innovation activity on performance of SMEs: the case of Korea. *International Journal of Technology Management,* 52(3/4), pp. 236–256.

Kirschbaum, R. (2005). External technology commercialization-policy guidelines. *International Journal of Technology Management,* 48, pp. 241–254.

Klewitz, J. and Hansen, E. G. (2014). Sustainability-oriented innovation of SMEs: a systematic review. *Journal of Cleaner Production*, 65, pp. 57–75.

Knudsen, M. P. and Mortensen, T. B. (2011). Some immediate — but negative — effects of openness on product development performance. *Technovation*, 31, pp. 54–64.

Kruse, P. (2012). The role of external knowledge in open innovation — A systematic review of literature. *Proceedings of the European Conference on Knowledge Management*. Academic Conferences International Limited, Cartagena, Spain, p. 592.

Lee, S., Park, G., Yoon, B. and Park, J. (2010). Open innovation in SMEs — an intermediated network model. *Research Policy,* 39, pp. 290–300.

Lichtenthaler, U. (2008). Open innovation in practice: an analysis of strategic approaches to technology transactions. *IEEE Transactions on Engineering Management*, 55(1), pp. 148–157.

McAdam, M., McAdam, R., Dunn, A. and McCall, C. (2014). Development of small and medium-sized enterprise horizontal innovation networks: UK agri-food sector study. *International Small Business Journal*, 32(7), pp. 830–853.

Mina, A., Bascavusoglu-Moreau, E. and Hughes, A. (2014). Open service innovation and the firm's search for external knowledge. *Research Policy*, 43(5), pp. 853–866.

Montelisciani, G., Gabelloni, D., Tazzini, G. and Fantoni, G. (2014). Skills and wills: the keys to identify the right team in collaborative innovation platforms. *Technology Analysis & Strategic Management*, 26(6), pp. 687–702.

Nakagaki, P., Aber, J. and Fetterhoff, T. (2012). The challenges in implementing open innovation in a global innovation-driven corporation. *Research-Technology Management*, 55(4), pp. 32–38.

Oakey, R. P. (2013). Open innovation and its relevance to industrial research and development: the case of high-technology small firms. *International Small Business Journal*, 31(3), pp. 319–336.

OECD. (2008). *Open Innovation in Global Networks*. Paris, France: OECD Policy Brief.

Parida, V., Westerberg, M. and Frishammar, J. (2012). Inbound open innovation activities in high-tech SMEs: the impact on innovation performance. *Journal of Small Business Management,* 50(2), pp. 283–309.

Pettigrew, A. M., Arthur, M. B. and Hendry, C. (1990). *Training and Human Resource Management in Small to Medium Sized Enterprises: A Critical Review of The Literature and a Model for Future Research*. (HMSO, London).

Pullen, A., Weerd-Nederhof, P., Groen, A. and Fisscher, O. (2008). Configurations of external SME characteristics to explain differences in innovation performance (PDF retrieved online at http://doc.utwente.nl/73738/1/Pullen.pdf).

Pullen, A. J., Weerd-Nederhof, P. C., Groen, A. J. and Fisscher, O. A. (2012). Open innovation in practice: goal complementarity and closed NPD networks to explain differences in innovation performance for SMEs in the medical devices sector. *Journal of Product Innovation Management*, 29(6), pp. 917–934.

Rahman, H. and Ramos, I. (2010). Open innovation in SMEs: from closed boundaries to networked paradigm. *Issues in Informing Science and Information Technology*, 7(4), pp. 471–487.

Rahman, M. and Ramos, I. (2013). Challenges in adopting open innovation strategies in SMEs: an exploratory study in Portugal. *Issues in Informing Science and Information Technology*, 10, pp. 431–448.

Ramos, I., Cardoso, M., Carvalho, J.V. and Graça, J.I. (2009). An Action Research on Open Knowledge and Technology Transfer. In: Dhillon G., Stahl B.C., Baskerville R. (eds.), *Information Systems — Creativity and Innovation in Small and Medium-Sized Enterprises*. IFIP Advances in Information and Communication Technology (Springer, Berlin, Heidelberg, Germany), vol. 301, pp. 211–223.

Ramsden, M. and Bennett, R. J. (2005). The benefits of external support to SMEs: "hard" versus "soft" outcomes and satisfaction levels. *Journal of Small Business and Enterprise Development*, 12(2), pp. 227–243.

Remneland-Wikhamn, B., Ljungberg, J. A. N., Bergquist, M. and Kuschel, J. (2011). Open innovation, generativity and the supplier as peer: the case of iPhone and Android. *International Journal of Innovation Management*, 15(1), pp. 205–230.

Rocheska, S., Kostoska, O., Angeleski, M. and Mancheski, G. (2014). Determinants for adopting open innovation strategies in SMEs. *SMEs Development and Innovation: Building Competitive Future of South-Eastern Europe*. Ohrid, Macedonia. Retrieved on November 11, 2016 from eprints.uklo.edu.mk.

Santoro, G., Ferraris, A., Giacosa, E. and Giovando, G. (2016). How SMEs engage in open innovation: a survey. *Journal of the Knowledge Economy*, pp. 1–14. doi: https://doi.org/10.1007/s13132-015-0350-8.

Scozzi, B., Garavelli, C. and Crowston, K. (2005). Methods for modeling and supporting innovation processes in SMEs. *European Journal of Innovation Management*, 8(1), pp. 120–137.

Segarra-Oña, M. D. V., Peiró-Signes, A., Albors-Garrigós, J. and Miret-Pastor, P. (2011). Impact of innovative practices in environmentally focused firms: moderating factors. *International Journal of Environmental Research*, 5(2), pp. 425–434.

Shinn, T. and Lamy, E. (2006). Paths of commercial knowledge: forms and consequences of university–enterprise synergy in scientist-sponsored firms. *Research Policy*, 35(10), pp. 1465–1476.

Spithoven, A., Vanhaverbeke, W. and Roijakkers, N. (2013). Open innovation practices in SMEs and large enterprises. *Small Business Economics*, 41(3), pp. 537–562.

Teirlinck, P. and Spithoven, A. (2013). Research collaboration and R&D outsourcing: different R&D personnel requirements in SMEs. *Technovation*, 33(4), pp. 142–153.

Theyel, N. (2013). Extending open innovation throughout the value chain by small and mediumsized manufacturers. *International Small Business Journal*, 31(3), pp. 256–274.

Tödtling, F., Lehner, P. and Kaufmann, A. (2009). Do different types of innovation rely on specific kinds of knowledge interactions? *Technovation*, 29, pp. 59–71.

Torkkeli, M., Kotonen, T. and Ahonen, P. (2007). Regional open innovation system as a platform for SMEs: a survey. *International Journal of Foresight and Innovation Policy*, 3(4), pp. 336–350.

Transfield, D., Denyer, D. and Palminder, S. (2003). Towards a methodology for developing evidence-informed management knowledge by means of systematic review. *British Journal of Management*, 14, pp. 207–222.

Tsai, K. H. (2009). Collaborative networks and product innovation performance: toward a contingency perspective. *Research Policy*, 38(5), pp. 765–778.

Ussman, A., Almeida, A., Ferreira, J., Mendes, L. and Franco, M. (2001). SMEs and innovation: perceived barriers and behavioural patterns. *The International Journal of Entrepreneurship and Innovation*, 2(2), pp. 111–118.

van de Vrande, V., de Jong, J. P. J. and Vanhaverbeke, W. (2009). Open innovation in SMEs: trends, motives and management challenges. *Technovation,* 29, pp. 423–437.

Vanhaverbeke, W. (2017). *Managing Open Innovation in SMEs* (Cambridge University Press, Cambridge, UK).

Vanhaverbeke, W., Van de Vrande, V. and Chesbrough, H. (2008). Understanding the advantages of open innovation practices in corporate venturing in terms of real options. *Creativity and Innovation Management*, 17(4), pp. 251–258.

van Hemert, P., Nijkamp, P. and Masurel, E. (2013). From innovation to commercialization through networks and agglomerations: analysis of sources of innovation, innovation capabilities and performance of Dutch SMEs. *The Annals of Regional Science*, 50(2), pp. 425–452.

Vega-Jurado, J., Gutiérrez-Gracia, A., Fernández-de-Lucio, I. and Manjarrés-Henríquez, L. (2008). The effect of external and internal factors on firms' product innovation. *Research Policy*, 37(4), pp. 616–632.

Verbano, C., Crema, M. and Venturini, K. (2015). The identification and characterization of open innovation profiles in Italian small and medium-sized enterprises. *Journal of Small Business Management*, 53(4), pp. 1052–1075.

von Hippel, E. (1998). Economics of product development by users: the impact of "sticky" local information. *Management Science*, 44(5), pp. 629–644.

West, J. and Gallagher, S. (2006). Challenges of open innovation: the paradox of firm investment in open-source software. *R&D Management,* 36(3), pp. 319–331.

Wynarczyk, P., Piperopoulos, P. and McAdam, M. (2013). Open innovation in small and medium-sized enterprises: an overview. *International Small Business Journal*, 31(3), pp. 240–255.

Xiaobao, P., Wei, S. and Yuzhen, D. (2013). Framework of open innovation in SMEs in an emerging economy: firm characteristics, network openness, and network information. *International Journal of Technology Management*, 62(2/3/4), pp. 223–250.

Yun, J. H. J. and Mohan, A. V. (2012). Exploring open innovation approaches adopted by small and medium firms in emerging/growth industries: case studies from Daegu–Gyeongbuk region of South Korea. *International Journal of Technology, Policy and Management*, 12(1), pp. 1–19.

Zahra, S. A. and Filatotchev, I. (2004). Governance of the entrepreneurial threshold firm: a knowledge-based perspective. *Journal of Management Studies*, 41(5), pp. 885–897.

Chapter 4

Adoption of Open Innovation by Small Firms to Develop Frugal Innovations for Inclusive Development

*Mokter Hossain**

mokter.hossain@aalto.fi

Abstract

People at the grassroots level have been developing a growing number of frugal innovations (FIs). Many of them do not have formal education and access to science and technologies. FIs are playing important roles for inclusive development. Open innovation (OI) has been studied in the context of large firms, small and medium-sized firms, or high-tech industries. However, OI has not been explored in the context of FIs. In this chapter, I explore the role of OI in three FI cases that emerged in rural India. I also explore the role of these cases for inclusive development. The chapter enhances our knowledge about OI and expands the scope of OI

* Corresponding author.

to new application areas. I find that small firms that develop FIs at the grassroots level need more support in the development stage than in the commercialization stage. They need extensive engagement in OI activities, such as networking, collaboration with different partners, and scouting. FIs play a significant role in inclusive development.

4.1. Introduction

Innovation plays a pivotal role in increasing the standard of living, job creation, and inclusive development (Bradley *et al.*, 2012). Open innovation (OI) shows a new way to manage innovation (Chesbrough, 2003). OI is defined in various ways. A latest definition of OI is given by Chesbrough and Bogers (2014, p. 17): "Open innovation is a distributed innovation process based on purposively managed knowledge flows across organizational boundaries, using pecuniary and nonpecuniary mechanisms in line with each organization's business model." Prior research has explored OI initially in the context of large firms (Chesbrough, 2006) and then in the context of small and medium-sized enterprises (SMEs) (Brunswicker and Vanhaverbeke, 2015; de Vrande *et al.*, 2009; Hossain, 2015).

There are studies on OI comparing large firms and SMEs (Spithoven *et al.*, 2013). Yet, small firms that emerge at the grassroots level embrace frugality and offer a unique setting to study OI. However, they have been overlooked in the current OI literature (Hossain, 2016a). One notable exception is a recent book chapter on social OI (Chesbrough and Di Minin, 2014): social OI is closely related with the frugal innovation (FI) phenomenon (Radjou and Prabhu, 2015). Small firms that develop FIs are significantly different from the conventional SMEs along many dimensions (see Levänen *et al.*, 2016). They are innovation-driven with strong outside-the-box thinking, developed by individuals who have no knowledge of or access to formal science and technology. Many FIs emerge from much-unexpected sources (Radjou *et al.*, 2012).

Some large firms are heavily involved in the development of FI as well (Zeschky *et al.*, 2014). FI is the process of reducing the complexity and cost of production. It is defined as "a resource scarce solution (i.e., product, service, process, or business model) that is designed and implemented despite financial, technological, material or other resource constraints, whereby the final outcome is significantly cheaper than competitive offerings (if available) and is good enough to meet the basic needs of customers who would otherwise remain un(der)served" (Hossain *et al.*, 2016, p. 133).

Small firms, such as Embrace, Mitticool, and Jayaashree Industries are showing a sustainable way of serving low-income customers at the grassroots level where innovations need to be developed with a different approach so that they are affordable and appropriate for the underserved customers (Levänen *et al.*, 2016). These small firms need more external knowledge than the traditional firms do because they lack resources, technology expertise, and marketing tools.

Some of the FIs have trickled up, and some others are attempting to trickle up to high-income countries from the low-income countries (Simula *et al.*, 2015). Many FIs, such as Tata Nano — the cheapest car in the world — and GE's Mac 400 ECG machine, are highly cited cases of FIs that show a new way of pursuing innovation and serving customers in both high- and low-income markets (Govindarajan and Ramamurti, 2011).

In this chapter, I intend to contribute to the extant literature by applying OI to the new context of FIs in small, grassroots-level firms. Individuals, SMEs, and large firms are developing FIs (Simula *et al.*, 2015). However, I examine the role of OI in small firms that develop FIs based on the OI principles. The empirical settings are three grassroots-level firms in rural India, which with their introduction of FIs create a positive impact on society and generate profit for the innovating firms. Thus, the chapter points out the role of OI to develop FIs by small, grassroots-level firms for inclusive development.

4.2. Prior Research

In general, the success of innovations has been measured based on their return on investment (Gu, 2016). However, many scholars argue that innovations should increasingly be measured by other outcomes such as societal impact (George *et al.*, 2012; Herrera, 2016). Solving problems at the grassroots level is important for innovations with societal impact (Gupta, 2006). The FIs at the grassroots level in a rural setting has only recently received attention in the academic literature, especially due to their societal impact (Hossain, 2016a). Despite the significance of grassroots-level innovation for sustainable development, its impact is poorly understood (Hossain, 2016b). Embracing OI for FI seems to be an interesting approach for the grassroots-level firms that have very few internal resources and competences and are confronted with very specific innovation requirements from customers given the low-cost setting.

SMEs engage in many OI practices (van de Vrande *et al.*, 2009). In the context of Korean SMEs, Lee *et al.* (2010) found that networking is an important way to facilitate OI practices in SMEs. Among Swedish SMEs, Parida *et al.* (2012) revealed that different OI practices result in different innovation performances: technology sourcing is associated with radical innovation, and technology scouting with incremental innovation performance. Using data of SMEs in Belgium, Spithoven *et al.* (2013) argue that SMEs are more effective (than large firms) in the application of different OI approaches. SMEs' engagement in external knowledge sourcing is beneficial for the successful launching of an innovation and value appropriation from the innovation (Brunswicker and Vanhaverbeke, 2015). OI is a promising option for high-tech SMEs (Su *et al.*, 2016; Vanhaverbeke, 2017). Some SMEs are very successful in outlicensing their technologies (Bianchi *et al.*, 2010). Among Spanish innovative SMEs, Hochleitner *et al.* (2016) found that OI traits are more prevalent in pioneer SMEs than follower SMEs. A study by Huang *et al.* (2015) indicates that external knowledge sources from interfirm networking are more valuable in gaining benefits of OI in Chinese SMEs.

Embracing OI to develop FI at the grassroots level is an uncharted territory (Hossain, 2013). OI was initially studied in the context of developed countries (Chesbrough, 2003) and limitedly explored in Asian countries, such as China (Huang *et al.*, 2015) and South Korea

(Lee *et al.*, 2010). So far, to the best of my knowledge, the extant literature has not considered OI at the grassroots level, especially not in the context of FI.

Anecdotal evidence indicates that OI is also crucial at the grassroots level where characteristics of the innovations and the partners involved with those innovations are different from the other contexts. Hence, exploring OI at the grassroots level is important to enrich our knowledge on OI and to expand its scope.

4.3. Methods

For this study, I selected three cases: Mitticool, Jayaashree Industries, and Ksheera Enterprise. These cases are widely cited in the literature and media. They have reached a mature stage in their life cycle, and therefore they are suitable for an in-depth understanding of FI with the objective to understand the role of OI in their development.

Mitticool has emerged from its flagship product Mitticool — a clay refrigerator that keeps food items fresh for several days without electricity. Jayaashree Industries was established based on the innovation of a low-cost sanitary napkin making machine. The third case — Ksheera Enterprise — developed low-cost milking machines, which are designed for milking one to several animals. Data sources used for this chapter include face-to-face interviews with the three innovators and their key office-bearers, expert interviews, in-field observations in India, secondary interview data, documentaries, press releases, media reports, and other secondary documents. Three cases are considered to be appropriate for an explorative study in a multiple case setting (Eisenhardt, 1989). The interviews with the innovators were long and conducted in several phases. They were partially recorded with due permission, and the interviews were partially transcribed.

Multiple sources of data collection were applied for triangulation, in order to boost the validity of data through cross-verifications (Denzin and Lincoln, 1994). Triangulation — combining different kinds of data — improves the accuracy and reliability of a study

(Jick, 1979). I used an open-ended questionnaire that is appropriate for in-depth exploration of a phenomenon as suggested by the epistemologically constructionist approach (Rubin and Rubin, 2011).

The chapter is based on the exploration of three FI cases. An overview of these three cases has been depicted in the following section.

4.4. Cases

4.4.1. *Mitticool*

Mitticool is a clay-made refrigerator invented by a traditional porter Mansukhbhai Prajapati from a small village of Gujarat state, India. Pottery is his family business by tradition. From childhood, he was used to helping his parents in various household activities, including in pottery works. In 1979, his family lost everything due to the breakdown of Machhu dam; they had to move to a nearby town, and he dropped out from class 10 after a failed attempt to pass. He started working in a tiles manufacturing firm as a teenager to support the family. Sadly, within a month, he got an eye injury while working in the chimney of the tiles manufacturing factory and had to leave the job. When his eyesight improved in 1984, he started a tea lorry near the highway but he had to close down the shop within six months due to continuous sarcastic comments from his acquaintances. However, during his tea business, one of his uncles visited him in the tea lorry and asked for a person who would be interested to work in a rooftop tiles manufacturing unit. He showed his interest, joined the unit as a trainee in 1985, worked for three years, and learned all the jobs of the unit.

He got the idea of his own enterprise while working at the tiles manufacturing unit. In general, a potter works manually and can make about 100 items per day. Seeing that a tiles manufacturing machine with handpress can produce a large quantity of items, he thought that the operating mechanism of a tiles manufacturing machine could be used to make earthen pans. In 1988, he quit the job and took a loan of around US$ 500 from a local money lender to start his own business. With that money, he purchased a small piece of land for his factory and other machinery, such as dyes, presses, soil-mixing machine, and electric pottery wheel. By modifying an existing roof tiles making hand machine he developed a handpress machine with a manufacturing capacity of 700 earthen pans per day. With numerous hurdles and unsuccessful attempts, he gained knowledge over time to produce quality pans with his machine.

He built his factory and started manufacturing 50 pans per day, carried items on his bicycle and sold them in different villages. However, he received complaints from customers that his pots were not properly heated and that they broke easily. He had to lower the price of his pots. Meanwhile, he started experimenting with various proportions of mixtures of raw materials to understand the right combination of materials for making pots. In 1989, he also started using the electric-operated wheel and presses, which enabled him to manufacture a large number of pots in a shorter time.

A turning point of his business was in 1995 when a businessperson was looking for someone who could supply clay water filters, because the businessman needed to deliver them to Nairobi, Kenya. Eventually, he found Mansukhbhai who delivered 500 water filters in eight days to the businessman at a price of US$ 3.30 each, double the initial agreed price as the quality was much better than expected. His inspiration for Mitticool refrigerator came from a caption "the broken fridge of the poor" in a local newspaper in 2001 depicting a water filter manufactured by him. After three years of experimentation, he managed to develop a clay refrigerator. In 2005, he received an order of 100 refrigerators for US$ 3,340 from a civil engineer. This was featured in a local newspaper. All raw materials — mainly sand and sawdust — were locally available. The current price of a Mitticool fridge of 50 liters capacity is about US$ 80.

4.4.2. *Jayaashree industries*

Arunachalam Muruganantham from the southern part of rural India invented a low-cost sanitary napkin manufacturing machine with a grassroots mechanism to solve unhygienic practices of women in rural India during menstruation. He has been recognized as *Time* magazine's 100 most influential people in the world 2014 for his achievement. In 2016, he was awarded Padma Shri — the fourth highest civilian award in the Republic of India — for social work. Presently, he has over 1300 sanitary napkin making machines installed in 27 states across India and seven other countries. Moreover, his plan is to expand his business into 106 countries. Most of the women in India use ashes, newspapers, sand husks, and dried leaves to take care of the menstruation. Consequently, they suffer from various diseases. He aims "creating one million livelihoods for poor women and making India a 100% sanitary napkin using country up from the current level of only 2% in rural areas."

Muruganantham was fascinated by science from his childhood. At an early age, he used to modify and repair farming equipment. He lost his father when he was studying in his ninth grade and needed to start working to support his family. His work life started as a helper in a local workshop which made farming equipment. He also worked as a machine operator, insurance agent, firm laborer, and yarn-selling agent (NIF, 2009). His entrepreneurial journey started soon after his marriage in 1988. He saw his wife carrying a dirty rag, and he inquired about the rag. His wife replied, "it is none of your business." Eventually, he understood that she was using it to take care of her menstruation. Thus, he learned how women treated menstruation in a highly unhygienic way. He realized that it is not only his wife but also most other women who applied the same practices. The next day, he went to buy a packet of sanitary napkin for his wife. He witnessed that the seller wrapped the packet with old newspapers, and he felt as if he is buying something illegal. Also, he checked the package and found that apparently the price of the item was exorbitantly high compared to the cost of materials used, and that triggered him to make low-cost sanitary napkins. However, he needed women to volunteer to experiment his napkins. He approached his wife and then his sisters to volunteer. His wife and other female family members were embarrassed to cooperate, and his wife was very critical about his experiment.

He approached some female medical students to participate in his experiment. Approaching them was very challenging for him as a workshop worker. After insisting for a long time, 20 medical students agreed to participate. When he requested to return the feedback forms, he witnessed that some girls were filling in the forms of the other girls. He then decided to use the napkins himself. He walked and cycled putting animal blood in a football tube and placing the tube in his body to understand menstruation. After one year of trials, he still was not successful in understanding the mechanisms to make sanitary napkins work. Meanwhile, people of his locality thought he was insane. His wife left him, which is rare in Indian culture.

Pretending he was a local municipality employee, he returned to the previously contacted medical students and provided sanitary napkins and asked them to return used sanitary napkins. He collected the used napkins and piled them on the floor of his room to examine them. Seeing this awkward activity, his mother also left him.

With further experiment, he got a basic understanding of the functionalities of sanitary napkins during the next two-and-a-half years. Initially, Muruganantham mistakenly believed that the raw materials used in branded sanitary napkins by multinationals were cotton. He requested a local professor to help him to understand if his belief was true. With the help of the professor, he wrote letters to the multinationals and potential suppliers pretending himself as a businessperson who wanted to establish a business in southern India and who was searching for raw materials. He ordered some samples, subsequently received the samples, and found that the raw materials used for sanitary napkins are wood fibers which are suitable as materials to drain fluids and to maintain the shape of the sanitary napkins. As wood fiber is not that expensive, he felt that the cost of the machines that are used by multinationals to manufacture sanitary napkins is the key reason for the high cost. Muruganantham developed a sanitary napkin manufacturing machine in four-and-a-half years using locally available materials. After numerous trials, finally in 2004, he was successful in developing the first usable version of the machine to make sanitary napkins. There were two models, a manually operated model at a cost of US$ 2000 and a semiautomatic model at US$ 3500 (NIF, 2009).

4.4.3. *Ksheera enterprise*

Ksheera Enterprise is located in a small village called Murulya in southern India. It was established in 2003 to manufacture low-cost milking machines with the brand names MILKMASTER and IMILKER. A farmer and now a retired school teacher Raghava Gowda had a cow that needed to be cleaned regularly and milked manually every day. Often, milking was neglected as there was no enough time for it, and the cow got an udder infection. Gowda also had strained nerve from manually milking. He pondered upon finding an alternative solution of milking the cow. He started to invent a hygienic and ergonomic low-cost milking machine, inspired by the gutter spray pump, which is used to spray insecticides. He also considered the electricity scarcity of different regions. He took a sabbatical year, paying for a substitute teacher, from his school to focus on the development of the machine. With various trial and error methods during four years, he developed a functional milking machine with a capacity of milking 1.5–2.0 liters per minute. He was able to develop several types of machines with different power sources.

He has been recognized with a national award for grassroots innovation from the President of India. So far, over 10,000 machines have been manufactured, and some of them were exported to Sweden, New Zealand, Mexico, and Kenya. In India, various milk unions provide a subsidy to farmers to purchase these low-cost milking machines. Ksheera Enterprise also used dealers to sell machines. At present, it has a capacity to make 250 machines per month and employs 25 full-time technicians.

These three cases show the emergence of different types of innovation and business models. Hence, understanding them from the lens of OI is not only intriguing but also an exploration of new frontiers of OI. OI takes place in three ways — inbound, outbound, and hybrid (West and Bogers, 2014). I examine these cases to see how they apply different OI approaches in their successful journey from ideation to commercialization.

4.5. Analysis of the FI Cases

4.5.1. *Personal traits and source of ideas*

Sourcing of ideas is a key aspect of the OI concept (Dahlander and Gann, 2010). A commonality between all three innovators is that they all are technicians by profession or fascination. Mitticool's innovator has been working in a tiles manufacturing factory where he learned technical skills. The innovator of the sanitary napkin making machine was a workshop assistant. Although the innovator of milking machine was a school teacher and farmer, he was involved in developing various devices for farming and household purposes. As a child, he was well known in his school for his interest in technical things. Hence, deep involvement in technical activities induces individuals to become innovative.

Mansukhbhai got the idea of making Mitticool from a newspaper article where one of his pots was featured as "the fridge of the poor." Muruganantham was inspired by seeing the taboo and hardship females go through in their menstrual period and feeling multinationals are charging high prices for their sanitary napkins despite the low cost of raw materials used. Raghava Gowda developed the milking machine after facing personal problems with milking and seeing others facing similar problems. All of them have identified the problem to be solved firsthand.

4.5.2. *Outside-the-box approach*

Previous research shows that inbound OI is the most widely used OI approach among the three approaches identified by Gassmann and Enkel (2004). This is also the case for grassroots-level innovations, especially for FI. The three innovators, who had no training in formal sciences and technology, developed these FIs. They needed to source knowledge from various external sources. Mitticool, for example, received technical support from research institutes, customers, and individual experts. One noticeable thing is that all three innovators started by getting ideas from other industries. The Mitticool refrigerator was developed by taking an idea from tiles-making mechanism, whereas Ksheera Enterprise's milking machine was developed by observing the mechanism of gutter-sucking pump. A firsthand witness of the problems women face during their menstruation period triggered Muruganantham to develop affordable sanitary napkin making system. Jayaashree Industries is also active in outbound OI: it provides technology and support to other local people to establish new sanitary businesses which are typically owned by women and nongovernment organizations (NGOs). It thus transfers technology along with necessary supporting services to external parties in order to develop new businesses.

4.5.3. *Open business model*

Firms are increasingly adopting open business model — using external sources for strengthening their internal innovation or external paths for commercialization (Chesbrough, 2010). All three firms used external knowledge from industries and research institutes for their technological advancement. They got most of their knowledge without paying any significant amount of money. Several organizations such as National Innovation Fund (NIF) and Indian Institute of Management, Ahmedabad, have provided holistic support — financial, technical, and other support — for bottom-up grassroots innovations with frugality. These firms use dealers and partner with interested groups such as NGOs, community, and state organizations to commercialize their products. The open business model has brought both opportunities and threats for small firms. A unique business model (e.g., Jayaashree Industries) is more challenging to copy than product innovations — Mitticool and Ksheera — as argued by Chesbrough (2010).

4.5.4. *Networking for innovation*

It is important to emphasize that the networking in grassroots organizations is local. All necessary raw materials, technicians, and other experts were locally available. Collaboration for innovation was limited to local partners and institutes. They used a significant amount of time for experimenting with different options for innovation. A large part of the collaborative support from external organizations and individuals took place informally. The innovator of the Mitticool needed to understand the different ingredients that should be used to gain an optimum mixture of raw materials. Furthermore, the heating temperature is also crucial. There were countless attempts with different combinations of materials and heating temperature. Even though local institutes provided some support about material combination and temperature, experimentation in the actual setting was the key to become successful. In the case of Ksheera Enterprise, collaboration with a local workshop and individual experts who were knowledgeable about the properties of different materials was crucial for developing the milking machine. Similarly, Jayaashree Industries also closely collaborated with local partners, such as small workshops and experts. All three firms got support from external entities to develop and test raw materials and products, fine-tune and package the product, and get the intellectual property (IP) right.

4.5.5. *Joint venture and acquisition*

A joint venture is an option for firms that develop new products through OI. However, finding a right partner with strong complementary resources is a challenge for them. Some multinationals approached Mitticool and Jayaashree Industries to acquire, but the offers were not that lucrative. Furthermore, Mitticool, for example, does not want anyone to change its brand name. The Mitticool is a consumer item that competes with the traditional refrigerators. Hence, it is important to understand which type of partner it should find to set up a joint venture. Recently, Mitticool started negotiation with a sanitary manufacturing firm to use the distribution channel of that firm. Ksheera Enterprise wants to expand its business through dealership and joint venture. However, it stumbles with unscrupulous people who produce inferior quality products and offer more commission to dealers. Some large firms showed their willingness to acquire Jayaashree Industries, but it has no interest in such a deal.

4.5.6. *Collaboration for funding*

OI entails collaboration with external partners to source ideas, technologies, and knowledge. Firms also collaborate with external partners for funding purpose. Hence, whether collaboration for funding can be considered as an element of OI phenomenon is interesting to bring into the discussion. Financial support from NIF had given all three firms a great opportunity to pursue their dreams. In all three cases, individual innovators struggled to finance the development and commercialization of their ideas. In the development stage, Mansukhbhai took a bank loan of US$ 11670, which increased to US$ 31670 accruing interest and capital. He had to sell his ancestral family home to pay off the loan. At his critical juncture, he was approached by NIF and received US$ 3340 from the NIF without any collateral. Subsequently, he received bank loan at a very low interest rate with a special loan scheme. Similarly, Muruganantham received financial support from NIF without any collateral or guarantor, and this fund enabled him to install 82 units in 13 states across India (NIF, 2009). Jayaashree Industries collaborated with local financial institutes and NGOs to raise funding. Gowda received funding initially from his family members — his brother and son-in-law — and then from banks for the commercialization phase. Moreover, Ksheera Enterprise closely collaborated with various associations and got financial support from them. Special loans from some associations to farmers to buy milking machine helped Ksheera Enterprise to expand its business. Funding by external parties thus plays a key role in FI. Hence, it is worthwhile to consider funding as an important element of OI in the process of FI.

4.5.7. *Collaboration for commercialization*

Previous studies indicate that SMEs collaborate more for commercialization than product innovation (Hossain, 2015). However, all three firms needed more support in the development phase rather than the commercialization phase. Nevertheless, external support for commercialization is significant. GIAN and NIF played a key role in business development, trademark design, marketing, media coverage, and networking for these three innovations. NIF supported Mitticool in the off-seasons (e.g., in rainy seasons) when the selling record and the cash inflow are both low. It helped Mitticool to sell its products on an online platform called Big Bazaar.

Muruganantham had problems in both financing his business and selling his products. Initially, Muruganantham's wife started selling sanitary napkins among the relatives and neighbors. Inspired by ATMs, he has also developed a vending machine in 2008 with a capacity of 25 sanitary pads with a coin slot that could be set up in colleges, hospitals, and public places to supply sanitary napkins on demand (NIF, 2009).

In these three cases, collaboration in both development and commercialization was significant. These three innovations have received extensive media coverage and recognition at the regional and national levels. It helped the firms to be known by potential customers. Moreover, they have featured in international media, such as *BBC* and *The Economist*. Even though many customers know the innovations, a large section of the target customers is still beyond reach. Moreover, customers from distant locations have no idea how and where to buy a product. An important way to expand the market is through the dealership in different regions. Both Mitticool and Ksheera Enterprise actively collaborate with dealers, whereas Jayaashree collaborates with NGOs, community associations, schools, hospitals, and government organizations to expand their ventures. Muruganantham deals with customers directly. Word of mouth is the main marketing tool used by the three firms. The main sales channel is direct to customers. For example, Mitticool uses its own and other online marketplaces where customers can directly place their product orders. Telephone order is a frequently used means for the three firms to receive orders.

4.5.8. *Patent infringement*

Low-income countries have weak patent protection systems. Hence, innovations are very vulnerable and others can easily copy and sell at a lower price there. Moreover, when the innovation is simple in nature and the market is not well established, patent infringement and copying are very common. Mitticool is a very simple innovation. Because of various constraints, its expansion in the large Indian market is challenging. One option is to sell through dealers, but in order to attract dealers, it is essential to offer a significant commission. Mitticool offered around 25% commission on the selling price. However, other products — inferior in quality and lower in price than the Mitticool — are available in the market. Moreover, dealers get more commission on selling copied items. In low-income markets, price is the key deciding factor. Both customers and dealers are more interested in lower priced, inferior quality products. Ksheera Enterprise faces a similar challenge with its dealers.

Therefore, it was difficult for the firms to expand their businesses at an expected level. Jayaashree Industries face limited challenges in terms of patent infringement. The uniqueness of its sanitary napkin manufacturing machine is that not just the machine but also the raw materials are unique and difficult to copy. For example, raw materials are imported from countries such as Canada and the US. People who want to copy its business model would find it difficult to acquire all necessary materials to build a complete setting.

4.5.9. *Inclusive development*

Unlike many traditional innovations, FIs developed by grassroots-level firms play a pivotal role in inclusive development. For example, Mitticool has generated full-time employment for about 30 local people. The refrigerator works through a natural process and is made mainly of clay: it does not need electricity. Therefore, it contributes to sustainability. The Mitticool has provided communities to work in supporting activities as suppliers, distributors, and salespersons — there are several hundred people indirectly employed. Similarly, Ksheera Enterprise also generated 25 full-time jobs for technicians along with several full-time positions at the management level. The technicians are mostly students of the innovator or students of the local school in which the innovator used to teach. It provides a way to milk in a hygienic manner, as the traditional milking process is unhygienic.

Jayaashree Industries contributes to inclusive innovation in a different manner. Even though it has several direct employees, each machine generates employment for several women in activities such as sales and manufacturing. Therefore, it plays a significant role in women empowerment. Furthermore, it provides hygienic low-cost sanitary napkins to low-income women who traditionally use filthy rags for their menstruation. It creates awareness among people of different backgrounds on the importance of using hygienic sanitary napkins. Thus, these three firms play a pivotal role in supporting sustainability and inclusive development.

4.6. Discussion

This book chapter has explored a new avenue of OI research. It demonstrates the potential of OI to develop FIs at the grassroots level for inclusive development. Small firms at the grassroots level mainly use inbound OI even though there is some evidence from the three cases that they engage in outbound OI as well. All three small firms found their ideas of FIs outside the industries with purely outside-the-box thinking. Due to weak IP regimes in developing countries, appropriation of FI is a pressing challenge. Moreover, the challenge of appropriation is even more important when innovations are just simple products without any high-end sophistications (e.g., Mitticool and Ksheera Enterprise). It results in misappropriation of IP through patent infringement and wide imitation of products. However, a FI with a strong business model is likely to last longer in the market. This is illustrated by the Jayaashree Industries' case.

At the grassroots level, technical skills along with basic education are important to enable individuals to innovate. People at the grassroots level have limited access to science and technologies.

Even though previous studies argue that the small firms embrace OI rather at the commercialization than at the development stage, this is not the case for small firms at the grassroots level developing FIs. These small firms need OI more in the developing stage than in the commercial stage. Furthermore, they need support for basic commercialization activities, such as trademark design, networking, and finding possible selling outlets. Small firms are complacent once they achieve profits. Eventually, they put less emphasis on business expansion. Scaling up is a pressing challenge for these small firms.

Small firms at the grassroots level do not have an established environment, such as science parks, incubators, living labs, and other hubs to get day-to-day support. Government initiatives to nurture these small firms are very limited not only in the development phase but also in the commercial phase. Even though the market share of these small firms is very low, they show the potential of a new way to serve unserved customers and contribute to inclusive development. Moreover, they induce local innovation potential and entrepreneurial spirit. Thus, FIs at the grassroots level contribute to sustainable development in a different way.

4.7. Conclusions

This chapter shows the potential of adopting OI by small firms to develop FIs. It expands the scope of OI literature into a new context. It also connects two research streams — OI and FI — to understand the overarching spirit of innovation at the grassroots level. Small firms at the grassroots level need to engage extensively in OI activities, such as networking, collaborating, and scouting. I am confident that this chapter will inspire other scholars to explore FI at the grassroots level as a new application field of OI.

References

Bianchi, M., Campodall'Orto, S., Frattini, F. and Vercesi, P. (2010). Enabling open innovation in small-and medium-sized enterprises: how to find alternative applications for your technologies. *R&D Management*, 40(4), pp. 414–431.

Bradley, S. W., McMullen, J. S., Artz, K. and Simiyu, E. M. (2012). Capital is not enough: innovation in developing economies. *Journal of Management Studies*, 49(4), pp. 684–717.

Brunswicker, S. and Vanhaverbeke, W. (2015). Open innovation in small and medium-sized enterprises (SMEs): external knowledge sourcing strategies and internal organizational facilitators. *Journal of Small Business Management*, 53(4), pp. 1241–1263.

Chesbrough, H. (2003). *Open Innovation* (Harvard Business School Press, Boston, MA).

Chesbrough, H. W. (2006). *Open Innovation: The New Imperative for Creating and Profiting From Technology* (Harvard Business Press, Harvard, MA).

Chesbrough, H. (2010). Business model innovation: opportunities and barriers. *Long Range Planning*, 43(2), pp. 354–363.

Chesbrough, H. and Bogers, M. (2014). Explicating open innovation: clarifying an emerging paradigm for understanding innovation. *New Frontiers in Open Innovation*. eds. Henry Chesbrough, Wim Vanhaverbeke, and Joel West (Oxford University Press, Oxford).

Chesbrough, H. and Di Minin, A. (2014). Open social innovation. *New Frontiers in Open Innovation*, eds. Henry Chesbrough, Wim Vanhaverbeke, and Joel West (Oxford University Press, Oxford).

Dahlander, L. and Gann, D. M. (2010). How open is innovation? *Research Policy*, 39(6), pp. 699–709.

Denzin, N. K. and Lincoln, Y. S. (1994). *Handbook of Qualitative Research* (Sage Publications, Inc.).

Eisenhardt, K. M. (1989). Building theories from case study research. *Academy of Management Review*, 14(4), pp. 532–550.

Gassmann, O. and Enkel, E. (2004). Towards a theory of open innovation: Three core process archetypes. In: *R&D Management Conference*, Lisabon, Portugal.

George, G., McGahan, A. M. and Prabhu, J. (2012). Innovation for inclusive growth: towards a theoretical framework and a research agenda. *Journal of Management Studies*, 49(4), pp. 661–683.

Govindarajan, V. and Ramamurti, R. (2011). Reverse innovation, emerging markets, and global strategy. *Global Strategy Journal*, 1(3–4), pp. 191–205.

Gu, L. (2016). Product market competition, R&D investment, and stock returns. *Journal of Financial Economics*, 119(2), pp. 441–455.

Gupta, A. K. (2006). From sink to source: The Honey Bee Network documents indigenous knowledge and innovations in India. *Innovations*, 1(3), pp. 49–66.

Herrera, M. E. B. (2016). Innovation for impact: business innovation for inclusive growth. *Journal of Business Research*, 69(5), pp. 1725–1730.

Hochleitner, F. P., Arbussà, A. and Coenders, G. (2016). Inbound open innovation in SMEs: Indicators, non-financial outcomes and entry-timing. *Technology Analysis & Strategic Management*, 29(2), pp. 204–218.

Hossain, M. (2013). *Adopting Open Innovation to Stimulate Frugal Innovation and Reverse Innovation*, 29th EGOS Colloquium Montréal, Canada, July 4–6, 2013.

Hossain, M. (2015). A review of literature on open innovation in small and medium-sized enterprises. *Journal of Global Entrepreneurship Research*, *5*(1), pp. 1–12.

Hossain, M. (2016a). Frugal innovation: a systematic literature review. Available at https://papers.ssrn.com/sol3/papers.cfm?abstract_id=2768254.

Hossain, M. (2016b). Grassroots innovation: a systematic review of two decades of research. *Journal of Cleaner Production*, 137, pp. 973–981

Hossain, M., Simula, H. and Halme, M. (2016). Can frugal go global? Diffusion patterns of frugal innovations. *Technology in Society*. 46, pp. 132–139.

Huang, F., Rice, J. and Martin, N. (2015). Does open innovation apply to China? Exploring the contingent role of external knowledge sources and internal absorptive capacity in Chinese large firms and SMEs. *Journal of Management & Organization*, 21(5), pp. 594–613.

Jick, T. D. (1979). Mixing qualitative and quantitative methods: triangulation in action. *Administrative Science Quarterly*, 24(4), pp. 602–611.

Lee, S., Park, G., Yoon, B. and Park, J. (2010). Open innovation in SMEs — an intermediated network model. *Research Policy*, 39(2), pp. 290–300.

Levänen, J., Hossain, M., Lyytinen, T., Hyvärinen, A., Numminen, S. and Halme, M. (2016). Implications of frugal innovations on sustainable development: evaluating water and energy innovations. *Sustainability*, 8(1), pp. 1–17.

NIF (2009). Mini sanitary napkin making machine http://nif.org.in/upload/innovation/5th/10-mini-sanitary-napkin-making.pdf.

Parida, V., Westerberg, M. and Frishammar, J. (2012). Inbound open innovation activities in high-tech SMEs: the impact on innovation performance. *Journal of Small Business Management*, 50(2), pp. 283–309.

Radjou, N. and Prabhu, J. (2015). *Frugal Innovation: How to do More with Less* (PublicAffairs), (Profiles Book Ltd, London).

Radjou, N., Prabhu, J. and Ahuja, S. (2012). *Jugaad Innovation: Think Frugal, Be Flexible, Generate Breakthrough Growth* (John Wiley & Sons), Jossey-Bass, USA.

Rubin, H. J. and Rubin, I. S. (2011). *Qualitative Interviewing: The Art of Hearing Data* (Sage), (Sage Publishers, London).

Simula, H., Hossain, M. and Halme, M. (2015). Frugal and reverse innovations — Quo Vadis? *Current Science*, 109(9), pp. 1567–1572.

Spithoven, A., Vanhaverbeke, W. and Roijakkers, N. (2013). Open innovation practices in SMEs and large enterprises. *Small Business Economics*, 41(3), pp. 537–562.

Su, Y. S., Hu, H. Y. and Wu, F. S. (2016). How can small firms benefit from open innovation? The case of new drug development in Taiwan. *International Journal of Technology Management*, 72(1–3), pp. 61–82.

West, J. and Bogers, M. (2014). Leveraging external sources of innovation: a review of research on open innovation. *Journal of Product Innovation Management*, 31(4), pp. 814–831.

van de Vrande, V., De Jong, J. P., Vanhaverbeke, W. and De Rochemont, M. (2009). Open innovation in SMEs: trends, motives and management challenges. *Technovation*, 29(6), pp. 423–437.

Vanhaverbeke, W. (2017). *Managing Open Innovation in SMEs* (Cambridge University Press, Cambridge, UK).

Zeschky, M. B., Winterhalter, S. and Gassmann, O. (2014). From cost to frugal and reverse innovation: mapping the field and implications for global competitiveness. *Research-Technology Management*, 57(4), pp. 20–27.

Chapter 5

How Do Entrepreneurial Leaders Promote Open Innovation Adoption in Small Firms?

Joon Mo Ahn, Tim Minshall and Letizia Mortara*

* *jmahn@sognag.ac.kr*

Abstract

This chapter aims to shed light on the role of the top executives in open innovation in small firms, providing a microfoundation framework for open innovation adoption, which future research will be able to empirically test. Recognizing that open innovation adoption implies an organizational routine change, we suggest that individuals' characteristics, such as risk-taking propensity and endurance as well as their knowledge and network, play important roles in moderating open innovation adoption. We argue that the personal characteristics of top executives, such as innovativeness, and the views managers hold about openness can contribute to the development of an open innovation-friendly atmosphere, which, in turn, increases the momentum of organizational change. Also,

* Corresponding author.

top executives' education and career experience can help firms to develop the necessary capabilities and personal relationships for expanding the company's boundary. We validate this interpretation through explorative content analysis on data from 30 small firms.

5.1. Introduction

Since open innovation first became popular, various topics have been covered by the literature (Dahlander and Gann, 2010; West *et al.*, 2014). However, little attention has been placed on its adoption dynamics (Vanhaverbeke and Cloodt, 2014; Vanhaverbeke and Roijakkers, 2014). Open innovation distinguishes itself from traditional approaches in the sense that it advocates for the purposeful management of innovation processes across firm boundaries (Chesbrough, 2003). This suggests that the virtue of openness lies in the dynamic proactive exploitation of diverse knowledge, in the possibility for firms to reallocate their strategic resources and dynamically configure their response to changing environments (Teece, 2007). Hence, open innovation offers new innovation development and commercialization opportunities, which can help firms to establish new sustainable strategy. However, as open innovation brings in new types of knowledge and interaction, its adoption will inevitably involve a certain level of changes in technology, market, business model, and even organizational structure (Mortara *et al.*, 2011, Ahn *et al.*, 2015, 2017b). Therefore, it is highly likely that successful open innovation adoption depends on how a firm effectively manages change and overcomes the challenges involved (Vanhaverbeke, 2017). A firm must deviate from the current innovation routine to introduce newness by opening its boundary, but such deviation will increase tension and anxiety in organizations (Teece *et al.*, 1997). To take advantage of external knowledge, high levels of personal capabilities are required which are built on the individuals' personal cognition, motivations, and actions (Volberda *et al.*, 2010). For instance, coping with cognitive resistance is a critical issue for setting up a new innovation routine, and, in this process, the leadership is expected to play a strong role, particularly in small firms. Along these lines, this chapter attempts to explore the role of entrepreneurial leaders for open innovation adoption in small firms, by viewing this implementation as an organizational change. This is a gap in current literature as, to date, the impact of firm or environmental characteristics on open innovation has been mostly scrutinized, while the strategic leadership and the microfoundation of open innovation adoption have received less scrutiny.

We first discuss why leadership is important in open innovation adoption by comparing different theoretical perspectives. Second, in the following section, we will explore the dynamic aspect of open innovation adoption and then will move onto the discussion of the role of entrepreneurial leaders, which is examined through explorative content analysis. Conclusions and implications will be presented in the last section of the chapter.

5.2. Open Innovation as an Organizational Change

To date, the adoption of open innovation in small firms has remained largely unexplored. At the time of writing, just over a decade has passed since the open innovation paradigm has gathered traction and companies have been experimenting with different open innovation implementation settings and approaches. As noted by others (e.g., Mortara *et al.*, 2011), various changes in technology, market, business model, and organizational structure/procedures are introduced when a firm becomes more systematic in the absorption of external knowledge or expose internal knowledge. These new approaches will become the key foundations of a dynamic capability for "creating, extending or modifying a firm's resource base" (Helfat *et al.*, 2009). The key assets for the development of these "managerial dynamic capabilities" are rooted in cognition, social, and human capital (Helfat and Martin, 2015). Therefore, well-managed coordination and strong internal support are vital in dealing with changes which open innovation will bring in. To cope with an organization's prejudice against changes, a high level of risk-taking propensity and a pragmatic attitude are essential (Chesbrough, 2003).

To explore this issue, it is necessary for us to understand what an organizational routine is and what triggers its change. Various theories can be employed to improve our understanding, but this chapter relies upon the following two seminal works by Cyert and March (1963), and Nelson and Winter (1982). To argue an analogy between open innovation adoption and a firm's routine change, we borrow some core concepts from their research on how organizations build a behavioral routine and change it, such as "organizational routines" and "standard operating procedures."

5.2.1. *Organizational routine*

An organizational routine can be defined in many ways: "a smooth sequence of coordinated behavior which embodies a firm's capabilities" (Nelson and Winter, 1982), "grammars of action" (Pentland and Rueter, 1994), or a "patterned sequence of learned behaviour" (Cohen and Bacdayan, 1994). The key common denominators of these definitions are order, repetition, interdependence, and patterns (Felin and Foss, 2009). As a routine comprises "programmatic steps," where each step triggers the next, it reduces uncertainty by suppressing organizations' choices (Nelson and Winter, 1982). Since an organizational routine is a set of well-established standards and procedures, firms can avoid risks by following these verified procedures (Argote and Greve, 2007; Cyert and March, 1963). Admittedly, numerous choices are embedded in a routine. Yet, as these choices are selected automatically in programmatic steps, firms can reduce the number of further deliberate choices by constraining their behavior into well-tested approaches (Nelson and Winter, 1982). For example, let us assume that there is a firm which recruits new employees annually. This firm may have to make detailed choices, such as those related to the number of interviewers and the structure of questions to ask, but the right architecture will be most likely established by referring to what was done in previous years. By regularly (and repeatedly) conducting recruitment every year, the firm will be capable of building an organizational memory as to how to conduct interviews and select the best candidates. This implies that the firm can obtain the necessary tacit knowledge for recruiting by setting up a well-defined routine and reduce uncertainty by following this established pattern.

A closed system may enable firms to harness the existing routine intensively, but they will lose their capability to engage in new routines (Teece, 1996), and, in a dynamic environment, firms cannot adhere to a single routine. Firms occasionally have to set up a new routine to adapt themselves to a fast changing environment. As noted by Zollo and Winter (2002), the modification of operating routines enables firms to achieve improved effectiveness, but routine change involves the consideration of the following four aspects.

First, the setting up of a new routine can increase risk (Gavetti *et al.*, 2012). Shifting routines can give firms flexibility, diversity, and adaptability, but firms also have to cope with the new uncertainties involved (Nelson and Winter, 1982). In the above example, trust can be one of the risks involved in this routine change. Before working together at least once, the firm cannot exactly evaluate which headhunting company is trustworthy. So, it has to take a certain degree of risk when outsourcing its recruitment process.

Second, routine change can involve internal conflicts (Dosi and Marengo, 2007; Nelson and Winter, 1982). Since routines represent well-setup procedures, which organizational members have already agreed to follow, a strong driving force is necessary in order to win over individuals and divert them from their (former) path-dependent behavior (Nelson and Winter, 1982). For example, if the consensus for outsourcing recruitment is not sufficiently established, some of the internal members will resist the change. For instance, the human resource management (HRM) division might be against outsourcing because of concerns of delegitimizing its role and losing influence.

Third, a new routine establishment demands the development of organizational knowledge. A routine shift is an effective way of dealing with internal exceptions and external changes (Felin and Foss, 2009; Garicano, 2000). However, to set up a well-defined new operating procedure, firms first have to build the necessary capabilities, which are based on individuals' knowledge and experience (Argote and Greve, 2007; Felin and Foss, 2009). Let us assume the firm in the above example finally decided to outsource its recruitment process. In that case, the necessary knowledge will be information about the past performance and costs of headhunting companies rather than the internal interview skills and selection criteria.

Last, a new routine can expand a firm's network. A new routine establishment requires a firm to deviate from the current innovation pattern; so, in this process, firms have to reconfigure their resources and establish new relationships with new stakeholders. Once a routine is changed, a firm arrives at a new equilibrium.

5.2.2. *Analogy with open innovation adoption*

When following the closed innovation process, firms' knowledge generation and selection occur within the innovation funnel. However, this linear process does not occur in the same way in the open innovation paradigm, since for the latter internal and external knowledge will be merged across firm boundaries. The merge of knowledge can happen in many ways (see Ahn *et al.*, 2015; Mortara *et al.*, 2011). Thus, open innovation adoption can be understood as the origin of the establishment of new routines which permit these pathways for knowledge exchange. As shown in Figure 5.1, for a routine to change from status "A" (i.e., current routine) to status "B" (i.e., new routine), firms have to address the knowledge gap and overcome the "risk and resistance hurdles" (Antons and Piller, 2015).

First, although, overall, the adoption of open innovation aims to reduce the risks faced by the firm, as shown in Figure 5.1, an open innovation adoption process transitorily increases the level of risk (Chesbrough *et al.*, 2006; Gans and Stern, 2003; Mortara *et al.*, 2011), which can be reduced after the new routine has been established. At the end, a newly adopted open innovation routine can lower risks/resistance by decreasing uncertainty or by bringing in new accessory benefits capable to offset its possible negative effects

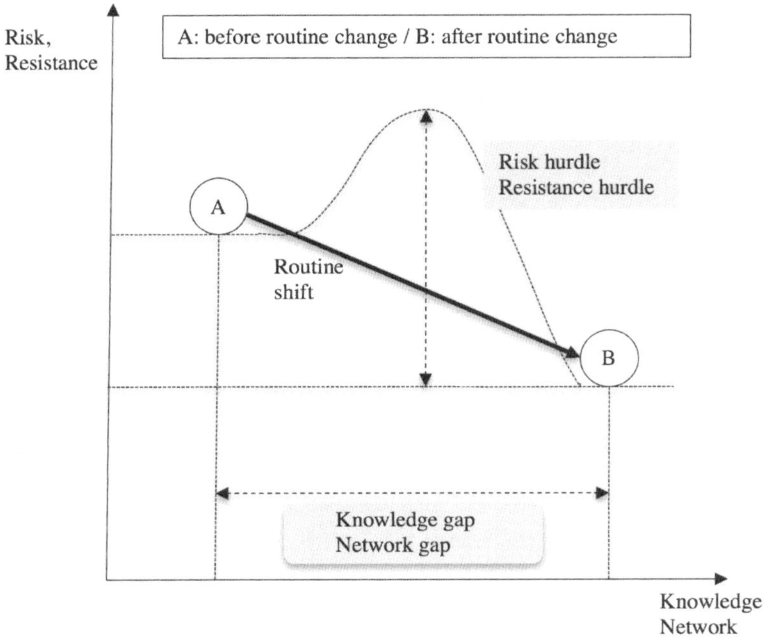

Figure 5.1. Open innovation adoption as a routine change.
Source: The authors.

(e.g., new income, such as royalty from patent licensing). Once firms have successfully adopted open innovation, the risks taken will decrease, as either risks or uncertainties will be distributed among external partners or as innovation strategies will be more diverse. This risk reduction is an important driver and incentive for open innovation adoption. However, it is also worth noticing that open innovation temporarily increases the level of risk during the early stages of adoption. Open innovation adoption presents challenges, such as offering an increased number of choices to managers (Nelson and Winter, 1982) and difficulties in finding trustworthy partners (Dyer and Singh, 1998; Narula, 2004). These will increase uncertainty levels. Further, in inbound open innovation, firms may pay higher transaction costs because of their limited/incomplete information about external knowledge and partners (Arrow, 1974; Coase, 1937; Narula, 2004). They may also have to spend a long time and a large amount of resources on understanding external knowledge, and this effort can be greater if much of the knowledge is tacit (Narula, 2004; Nelson and Winter, 1982; Polanyi, 1966) or cognitively distant (Nooteboom *et al.*, 2007). In outbound open innovation, information leakage is a serious risk, making firms hesitant to open firm boundaries (Laursen and Salter, 2014). Certainly, sharing information with other innovation actors (e.g., in new product development with suppliers, or in exploration phase meetings for alliances/licensing-out) has to be undertaken in order to access external information. Firms have to share at least some aspects of key information to achieve partners' trust and commitment, but this information sharing can threaten firms' technology confidentiality (Bianchi *et al.*, 2011; Laursen and Salter, 2014; Oakey, 2013). In this respect, it can be said that open innovation initially increases the level of risk during the adoption process, even though it has the power to decrease risks and uncertainties in the long term. "The risk hurdle," which is similar to an "activation energy" in a chemical reaction, will be likely to transitorily increase the overall perceived challenges and risks. In chemistry, the "activation energy" is defined as the minimum threshold energy required to start a chemical reaction, and, in open innovation, "the risk hurdle" plays the same role. We argue that open innovation adoption can occur when firms deal adequately with the initial increase in the risk involved in open innovation.

Second, the final level of internal resistance can be lower than the initial level. In fact, open innovation can increase the level of internal resistance while it is being adopted. However, as in the case of risks, once open innovation is adopted, internal resistance may decrease while the adopted open innovation process is being embedded as a new organizational routine. For instance, when a firm tries to collaborate with a new partner or absorbs external information, it can face a negative reaction from the internal staff (Burcharth *et al.*, 2014). This means that overcoming the cognitive and cultural differences between different information/organizations is critical in open innovation adoption (Harryson *et al.*, 2008; Nooteboom *et al.*, 2007). This could be understood considering different circumstances: When acquiring external technology, the

Not-Invented-Here (NIH) syndrome, which is rooted in the prejudice against innovation coming from outside, can hinder the integration of external knowledge (Chesbrough *et al.*, 2006; Katz and Allen, 1982). Inbound open innovation involves multiple steps (e.g., identification–assessment–negotiation–technology transfer; Mortara *et al.*, 2012), and so leveraging external partners requires time-consuming integration and trust-building process (Narula, 2004). Therefore, strong internal support and voluntary engagement are necessary for smooth integration, but the NIH syndrome, by devaluing external knowledge, will impede such integration process (Salter *et al.*, 2014). Similarly, emotional bias, the so-called "Not-Shared-Here (NSH)" syndrome, can inhibit the external disclosure of information, even when it cannot be used internally (Burcharth *et al.*, 2014). Finding a potential receiver of the knowledge is an important issue, but overcoming NSH by persuading internal members who are hesitant to be open will require strong internal consensus. As the exposed knowledge can be used by competitors, it is necessary to convince internal members to overcome their reticence.

Third, open innovation adoption necessitates knowledge building. A lack of appropriate knowledge prevents firms from being open (Drechsler and Natter, 2012); hence, they have to develop the relevant capabilities to initiate changes (Lichtenthaler and Lichtenthaler, 2009). However, the type of necessary knowledge can be different depending on the characteristics of open innovation involved Lichtenthaler and Lichtenthaler, 2009; Robertson *et al.*, 2012; Ahn *et al.*, 2017b). For example, technological knowledge is necessary to identify the appropriate external technology and absorb it (Spithoven *et al.*, 2011), while intellectual property (IP) management skills will be more vital when revealing unused IP throughout-licensing (Bianchi *et al.*, 2011).

Finally, open innovation adoption implies the involvement of new innovation partners. By its definition, open innovation emphasizes the interaction with external partners; so, the expansion of the company network is required, and, at the same time, is a consequence of open innovation adoption. Typical situations include circumstances where the prior network becomes the starting point in innovation collaboration, while in others new networks can be established in the process, by means of external searches or through strategic alliances.

5.3. Theoretical Perspectives on Open Innovation

When recognizing open innovation adoption as an organizational routine change, what determines or drives this change is one of the most important issues. To date, a variety of different perspectives and theories have been employed to explain firms' strategic decisions and the choices involved in innovation adoption, as shown in Table 5.1.

Table 5.1. Key features of the three perspectives on open innovation determinants.

Features	Industrial economic perspectives	New institution theory (NIT)	Strategic leadership (UET)
Core questions	What is the role of firm and industry characteristics in the adoption of open innovation?	What institutional conditions give rise to the adoption of open innovation?	What is the role of top managers in the adoption of open innovation?
Key factors	Firm characteristics and interactions with other organizations	Institutional conditions	Key individual's knowledge, perception, and behavior
Research focus	The effect of organizational characteristics	The effect of external factors	The effect of key individuals
Role of human	Rarely discussed	Rarely discussed	Key individuals initiate and drive the necessary change
Level of analysis	Firm and industry	Industry and country	Individual and firm

On the one hand, the "industrial economic perspective" recognizes organizational factors as key determinants of strategic decisions. Those following this perspective have mainly investigated the role of firm-level characteristics in innovation adoption. The followers of the "new institutional theory (NIT)" recognized the social norms as critical factors determining firms' innovation strategies (Papadakis and Bourantas, 1998). According to these contingency perspectives, a firm's optimal strategic decision depends upon its internal resources or social norms. For example, NIT has perceived a firm as an organization largely influenced by external factors and socially constructed through rules, norms, routines, and schemes (Dimaggio and Powell, 1983). These perspectives have emphasized institutional and organizational factors as the main determinants of a firm's innovation strategy.

In contrast, although this perspective has so far been neglected in the study of open innovation, the "strategic leadership perspective" has emphasized the role of leaders, such as chief executive officers (CEOs) or top managers in innovation (Hambrick and Mason, 1984; Hambrick, 2007). As shown in Table 5.1, strategic leadership, particularly the Upper Echelon Theory (UET), perceives top executives as key agents of change. According to the UET, CEOs or top management are typically unable to fully appreciate all the contextual aspects within which their firms operate (Ahn *et al.*, 2017a). Thus, their limited awareness makes it more likely that they will perceive a business situation based on their own characteristics and/or past experience, and then interpret it in their own way (Hambrick and Mason, 1984). Consequently, the leaders' personal and cognitive characteristics, based on so-called "bounded rationality" (Simon, 1957), can affect a firm's strategic choices.

Although these various theoretical perspectives can be applied, given the characteristics of small organizations shown in Table 5.2, we argue that our understanding on open innovation adoption in small firms can be enhanced when the role of a key individual is fully understood.

First, the simple organizational hierarchy characterizing small firms is likely to increase the influence on the firm's strategy of a few

Table 5.2. Comparison of organizational characteristics.

	Small firms	Large firms
Structure	Simple hierarchy • A few management layers • Organic	Stratified hierarchy • Multiple management layers • Bureaucratic
Procedure	Informal • Low standardization • Heuristic process	Formal • High standardization • Rigid process
Decision-making	Centralized	Decentralized
Advantage	Behavioral advantages • Flexibility	Material advantages • Abundant resources

Source: Adapted from Ahn *et al.* (2017a).

top leaders (Miller *et al.*, 1982; Nooteboom, 1994; Teece, 1996). If multiple management layers are established in large firms, management systems become bureaucratic. However, as in small firms there are fewer management layers, top managers are more frequently involved in everyday business and have to respond more quickly to developments and stimuli (Miller and Toulouse, 1986b; Papadakis and Barwise, 2002; Vossen, 1998). They are also typically involved in both the strategic and operational functions of the firm and are asked to decide on details (Lubatkin *et al.*, 2006).

Second, the scarce resource regime, typifying small firms, contributes to the increase in human influence on the decisions taken. Since small organizations typically lack resources and a structured standardized system, literature poses a greater emphasis on good leadership for their successful outcome (Humphreys *et al.*, 2005; Lubatkin *et al.*, 2006). A low degree of standardization makes innovation procedures more heuristic and informal rather than structurally systemized and formal, which in turn make the decision-making process centralized in the hands of a few key decision-makers.

Third, in small firms, top executives (e.g., CEOs) have longer tenures than in large firms, where the top executives' tenure can be shorter than the period of an innovation project (Teece, 1996) and decisions are more likely to be influenced by Board of Directors (BOD) (Nooteboom, 1994; Rothwell and Dodgson, 1994). It is typically hard to draw clear lines of demarcation between small firms' founders and CEOs, and the BOD is often not systematically established. Hence, the influence of key decision-makers in firms' strategic decision will be substantial (Ahn *et al.*, 2017a).

Owing to the above organizational characteristics, although some collectivists have viewed individuals as unimportant and homogenous when interpreting organizational behavior, the importance of leadership cannot be neglected in a small firm context (Murmann *et al.*, 2003). Therefore, we argue that the perceptions and experience of leaders will be more reflected in small firms' strategy, making the "strategic leadership perspective" suitable for explaining open innovation in small organizations (Miller, 1983; Papadakis and Bourantas, 1998; Hambrick, 2007). However, even if the influence of top managers can outweigh contingent factors when a high level of managerial discretion and job demand exist, we recognize that each perspective has good grounds and that these theories are not mutually exclusive. In fact, to deal with the differences with other theories and reconcile them, Hambrick (2007) suggested two moderators for UET — managerial discretion and job demand. When considering the organizational characteristics of small firms (see Table 5.2), the impact of Hambrick's two moderators become critical, making the key players' perception and experience more reflected in firm-level strategy. Intention and expectation are important drivers of routine change, and the origin of routines arises from individuals' knowledge and interaction (Felin and Foss, 2009). Because a strategy is, by nature, purposive, in small firms, strategic decision, such as a routine change, cannot be separated from a key decision-maker's intention. In this respect, an important change, such as open innovation adoption, will be influenced by leaders whose knowledge and intention are limited by the bounded rationality.

5.4. The Role of Entrepreneurial Leaders

When considering the strong influence of top managers in setting strategic directions in small organizations (Eisenhardt and Schoonhoven, 1990), their impact on an organizational routine change cannot be overlooked. Owing to lack of resources, small organizations have to rely on key individuals in order to bridge a knowledge gap between different routines (Macpherson and Holt, 2007). In this process, top managers' skills arising from their education and experience will play an important role in providing necessary knowledge stock (Colombo and Grilli, 2005). Also, well managed coordination and strong internal support are vital in dealing with the challenges involved in interacting with external partners with different cultures and innovation speeds (Kitchell, 1997; Mortara and Minshall, 2011; Mortara and Minshall, 2014). To cope with an organization's negative prejudice against the exploitation of external knowledge, a high level of risk-taking propensity and a pragmatic attitude are also essential (Chesbrough, 2003). In this context, Chiaroni *et al.* (2011) emphasized the importance of top management in promoting and championing organizational changes in open innovation adoption.

5.4.1. *Leaders' characteristics*

Acknowledging the better fit of UET to small firm context, strong leadership will play an important role in interpreting organizational routine change, i.e., pushing firms to take the necessary risks and mitigating internal prejudices and bias against openness (see Figure 5.1). It follows that we should consider what specific characteristics leaders should possess. As the open innovation literature has paid little attention to leadership so far, there are no guidelines for this. Many characteristics have been typically considered by the strategic leadership literature, but the following characteristics are selected for this chapter based on the analogy between organizational routine change and open innovation adoption.

5.4.1.1. *Psychological Characteristics*

As psychological characteristics reflect the real intentions of individuals, in the sense that psychological characteristics can mirror humans' true intentions and behavioral propensity (Finkelstein *et al.*,

Table 5.3. Psychological characteristics in the literature.

Psychological characteristics	Research	Key findings
Innovativeness	Thong and Yap (1995)	Positive impact on ICT adoption in small firms
	Kirton (1976), Marcati *et al.* (2008)	Positive impact on innovation adoption in small firms
Risk-taking	Lefebvre and Lefebvre (1992)	Positive impact on firms' innovativeness
	Lefebvre *et al.* (1997)	Positive impact on firms' technology policy
	Papadakis and Bourantas (1998)	Insignificant association with both product and process innovation
	Moreno and Casillas (2008)	Positive impact on firms' growth
	Escriba-Esteve *et al.* (2009)	Positive impact on firm performance via firms' strategic orientation
	Rhee *et al.* (2010)	Positive impact on firms' innovativeness and performance
Positive awareness toward openness	Moore and Benbasat (1991), Thong and Yap (1995)	Positive impact on ICT adoption in small firms
	Herrmann and Nadkarni (2014)	Positive impact on the initiation of strategic change
Perseverance	Kitchell (1997)	Positive impact on firms' technological innovativeness

2009; Hambrick, 2007), researchers have attempted to look into the real mind-sets of top executives. The following have been frequently found as significant (see Table 5.3):

- **Innovativeness:** An individual's innovativeness has been an important factor distinguishing between adaptors and innovators (Kirton, 1976), and it is closely related to an entrepreneur's intention to adopt innovation (Marcati *et al.*, 2008). This characteristic has been an important factor in deciding a firm's strategic direction (Escriba-Esteve *et al.*, 2009; Vanhaverbeke, 2017) or establishing entrepreneurial orientation (Moreno and Casillas, 2008).

- **Risk-taking propensity:** The willingness to take risks is an important factor in shaping firms' strategies (Gupta, 1984; Lefebvre *et al.*, 1997; Escriba-Esteve *et al.*, 2009). As many different types of risks and uncertainties are embedded in innovation, decision-makers with a risk-taking mind-set will be more likely to adopt a new innovation Kitchell, 1997; Herrmann and Nadkarni, 2014).
- **Positive awareness:** Key decision-makers' personal perception is an important factor affecting innovation adoption (Moore and Benbasat, 1991). If they are well aware of the benefits of innovation and show high openness, they will strongly engage in the initiation of strategic change (Herrmann and Nadkarni, 2014) and support innovation adoption (Thong and Yap, 1995).
- **Perseverance (endurance):** Firms may not resist technological development, but it is highly likely that they do resist the changes involved (Schein, 1985). This characteristic has been rarely discussed in literature, but Kitchell (1997) pointed out that top executives must persistently deal with resistance. As agents of change, they can continuously break down institutional barriers, which will lead firms to deviate from the current innovation routine.

5.4.1.2. *Demographic Characteristics*

Demographic characteristics have been regarded as good proxies reflecting top executives' cognition and perception of the environment, and, owing to the ease with which this type of data can be obtained, they have been investigated frequently in the literature (Hambrick, 2007; Gerstner *et al.*, 2013). The following demographic characteristics have been widely investigated in prior studies, and the details are summarized in Table 5.4.

- **Education:** Hambrick and Mason (1984) suggested that formal education is a good indicator of top executives' knowledge base

Table 5.4. Demographic characteristics in the literature.

Demographic characteristics	Research	Key findings
Education	Kitchell (1997)	Positive impact of engineering education on firms' technological innovativeness
	Papadakis and Bourantas (1998)	CEOs' formal education affected new product development
	Bantel and Jackson (1989)	Positive impact of educational level on technological innovation
	Wally and Becerra (2001)	Levels of education are significantly associated with openness to change and innovation
	Colombo and Grilli (2005)	Positive impact of managerial and engineering education on firms' growth
	Sundaramurthy *et al.* (2014)	CEOs' science education is positively associated with initial public offering (IPO) performance
Work experience	Bantel and Jackson (1989)	Positive impact of engineering/R&D career on R&D intensity
	Lefebvre and Lefebvre (1992)	Positive impact of engineering/ production experience on firms' innovativeness
	Barker and Mueller (2002)	Positive impact of engineering, R&D, and marketing career on R&D spending
	Colombo and Grilli (2005)	Positive impact of technical work experience on firms' growth
	Escriba-Esteve *et al.* (2009)	Previous experience affects firm performance via firms' strategic orientation
	Sundaramurthy *et al.* (2014)	Positive effect of broad experience on IPO performance

and cognitive preferences, and his/her education has been an important descriptor/predictor for the human/social capital they accrued and which can be affecting a firm's strategy and performance. Birley and Norburn (1987) investigated the differences in a CEO's educational levels between venture companies and companies listed in the journal *Fortune*, and found that educational levels in venture companies are higher than those in large established firms and they influence financial performance in a positive way. Also, Colombo and Grilli (2005) found that there is a positive association between managerial/engineering education of top executives and the firms' growth.

- **Work experience:** Hambrick and Mason (1984) suggested that top executives' former career experience influences a firm's performance as well as the organizational behavior, and top executives will favor a specific business strategy according to their earlier experience. A top executive member who has worked in an R&D department may prefer dealing with product innovation, while those who worked in a customer satisfaction division may concentrate on user innovation.

5.4.2. *How does it work?*

Based upon the analogy between open innovation adoption and a routine change, this chapter suggests the possible mechanism of how the entrepreneurial leaders promote open innovation adoption. In a small firm, in which top executives are strongly embedded in organizations with day-to-day contacts, routine changes (i.e., open innovation adoption) will be affected by the most influential individual in the firm, whose knowledge and intention are limited by bounded rationality (Simon, 1957, 1972). This aspect has been neglected in the literature (Schroll and Mild, 2012; Wynarczyk *et al.*, 2013), but, as noted by Chesbrough and Garman (2009) and Di Minin *et al.* (2010), top executives' leadership and awareness can promote open innovation adoption. As open innovation is "innovating innovation" (Chesbrough, 2003), its adoption will bring in many new management practices and substantial changes. It not only introduces new tasks and job function transitions but also requires different mind-sets (Salter *et al.*, 2014). Therefore, to increase openness, firms have to deviate from existing operating routines and break with conventions (Salter *et al.*, 2014; Colombo *et al.*, 2011). For this transition, establishing an open innovation-friendly culture that highly appreciates interaction and collaboration and building the necessary capability is vital, and entrepreneurial leaders can contribute to this process as in Figure 5.2.

First, leaders' entrepreneurial characteristics establishing strong intention for change can play an important role in coping with risks. Open innovation is a journey seeking a new way of innovation, but a high level of uncertainty exists in its adoption process. Thus, entrepreneurial propensity characteristics, such as "innovativeness" and "risk-taking propensity," are highly likely to push firms to take

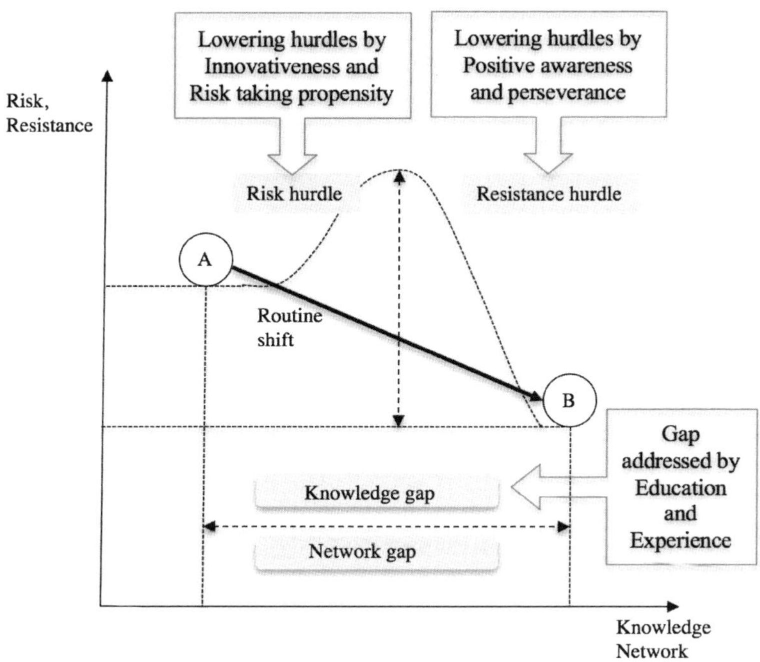

Figure 5.2. The role of leaders' characteristics in open innovation adoption.
Source: The authors.

necessary risks and lower the risk hurdles. Top executives with a high level of these entrepreneurial propensity characteristics are not risk-averse and adopt innovative strategies boldly despite challenges (Howell and Higgins, 1990; Khandwalla, 1976, 1977; Miller and Friesen, 1982; Miller, 1983), thus increasing organizational readiness for change (Hambrick *et al.*, 1993; Jansen, 2000b; Jansen *et al.*, 2009). They play a critical role in establishing a more active entrepreneurial posture in their firms (Miller, 1983; Miller and Toulouse, 1986a), and this can significantly affect innovation methods and tool selection at the firm level (Cassia *et al.*, 2011).

Second, attitude-related characteristics, such as "positive awareness" and "perseverance," are very likely to help firms to reduce resistance in organizations by persuading internal members and mitigating conflicts, as shown in Figure 5.2. As open innovation adoption is an organizational change, it will require a strong proponent to push forward changes and to disseminate the benefits inside the organizations (Smith, 2007). A well-balanced routine implies a form of "truce," in that it suppresses changes by requiring members to comply with a preset routine (Nelson and Winter, 1982). Therefore, in order to break this truce and overcome the friction of change, the role of leaders who can strongly support and facilitate it within their firms becomes critical. Internal resistance (e.g., the NIH and NSH syndromes) springs from sociopsychological bias (Burcharth *et al.*, 2014); so, top executives' positive attitudes toward openness will play a critical role in dealing with such negative prejudice and increasing the momentum for change. Forming a consensus is necessary for coping with internal resistance; so, the positive awareness of key decision-makers will contribute to the establishment of change in a friendly atmosphere. According to the theory of reasoned action, "attitude towards an object is viewed as related to the person's intentions to perform a variety of behaviors with respect to that object" (Fishbein and Ajzen, 1975, p. 14). In this respect, top executives' attitudes can be a good proxy, reflecting the extent of their intention to adopt open innovation. Also, as openness inevitably makes innovation routes heterogeneous (e.g., external knowledge utilization, a new business model or new suborganization), the way in which firms can cope with the heterogeneity-induced uncertainties is critical for a smooth transition. Owing to potentially substantial system differences between organizations, innovation coordination can be very difficult (Dyer and Singh, 1998). When two (or more) organizations with different cultural backgrounds, goals, and ways of doing research collaborate, they have to cope with the challenges presented by progressive adaptation in order to approach each other (Perkmann and Walsh, 2007; Narula, 2004). In this respect, the patience and endurance of top executives can be critical (Kitchell, 1997).

Third, as shown in Figure 5.2, a knowledge gap has to be addressed for a successful routine transition. Thus, "capability building characteristics," such as education and work experience, are very likely to enable firms to develop the open innovation necessary capabilities. This is because knowledge is built up throughout education or past experience. Human capital is closely related to learned abilities, expertise, and knowledge acquired through education, training, and job experience (Becker, 1975), and it can contribute to the development of the necessary organizational capabilities. Because of the scarcity of resources in small firms, the top executives' human capital represents the initial resources invested in the organization (Cooper *et al.*, 1994).

Finally, using their education or their career network, top executives can establish close relationships with external innovation actors. Small firms usually depend more upon entrepreneurial leaders and social/informal networks (Macpherson and Holt, 2007; Vanhaverbeke, 2017); so, when selecting university partners, education-related networks, such as networks of university friends and acquaintances, play an important role in establishing partnerships (Saxenian *et al.*, 2002). Top executives' personal relationships, which are established through their study or employee recruitment, can be effective channels consolidating university–industry links (Cohen *et al.*, 2002; Faulkner and Senker, 1994). If innovation collaboration is initiated using a top executive's personal network (e.g., an alumni relationship), this type of relationship can strengthen network ties because an alumni relationship can be maintained even after the innovation collaboration is officially completed (Perkmann and Walsh, 2007). A similar logic can be expanded to career-based networks. Top managers' work experience can bring numerous benefits to their firms. While working in other firms, not only can they accumulate the necessary managerial knowledge, e.g., how to run an organization and deal with innovation projects, but they can also establish networks with their former organizations. This is because personal relationships are affected by trust, shared values, and mutual objectives (Ceci and Iubatti, 2012), and their strength depends on the duration of the collaboration, the emotional intensity, and reciprocal devotion (Granovetter, 1973). Strong personal relationships enable innovation partners to trust each other and foster knowledge exchange by facilitating tacit knowledge acquisition and informal skills (Asheim, 2000; Mellewigt *et al.*, 2007).

5.5. Content Analysis

5.5.1. *Data and method*

To investigate the validity of the suggested interpretation, explorative content analysis was conducted. Korean government documents covering 30 small firm cases (SMBA, 2011) were analyzed. These documents were originally written to evaluate the efficiency of the government's innovation funding program, thus containing sufficient information, for example, firm names, locations, employee numbers, revenue, firm histories, collaboration types, and innovation partners and record the innovation stories of these firms in a narrative way. Each firm case also includes scripts of interviews with top management and provides their background information. This information covers CEOs' employment as well as their educational background. All the chosen firms were manufacturing firms. Six out of the thirty were ICT, twelve were biotechnology (BT) firms, eleven were traditional manufacturing firms, and one was a chemistry firm. The number of employees ranged from 6 to 250, and the average was 39.3. The firms' revenue ranged between 0.85 and 68 million USD, and the average was 9.6 million USD.

Content analysis is a "research technique for the objective, systematic, and quantitative description of the manifest content of a communication" (Bereleson, 1952, p. 18), and it has been used in social studies to investigate the content of communications, such as political speeches, books, interviews, and newspapers. This method has rarely been adopted in innovation literature (Ceci and Iubatti, 2012). However, the advantage of this method lies in its high level of objectivity and external validity (Ceci and Iubatti, 2012), due to the transformation of qualitative in quantitative data.

According to the guidelines by Krippendorff (2003), a sampling and a context unit of analysis have to be decided first. The sampling units were the firms, in the sense that they can be regarded as individual and independent observation units, and a factual incident was chosen as a context unit. Admittedly, words or sentences are usually selected as a context unit, but due to the limitation of our secondary data which did not provide full and rich interview scripts, an attempt was made to extract factual incidents closely related to our chapter from thematic analysis using theoretical and inductive coding (Fereday and Muir-Cochrane, 2008). As shown in Figure 5.3, the qualitative data of 30 individual small firms (similar key (analysis) units (i.e., codes)) were connected to identify common denominators and to build theories. Some codes were generated from a theoretical understanding of open innovation and UET (e.g., open innovation modes and CEO education) and were used in analyzing and transforming data into two key themes: open innovation in practice and CEO characteristics. Other codes were inductively generated, and they are used to categorize findings arising from the data, for example, for understanding what are the important factors affecting open innovation. These inductively driven codes were used in building the third key theme: firms' success factor in open innovation. Content analysis was conducted in two phases: open coding, which extracts information related to key themes (i.e., open innovation practice, CEO background, and firm characteristics), and axial coding, which attempts to understand how these themes are connected.

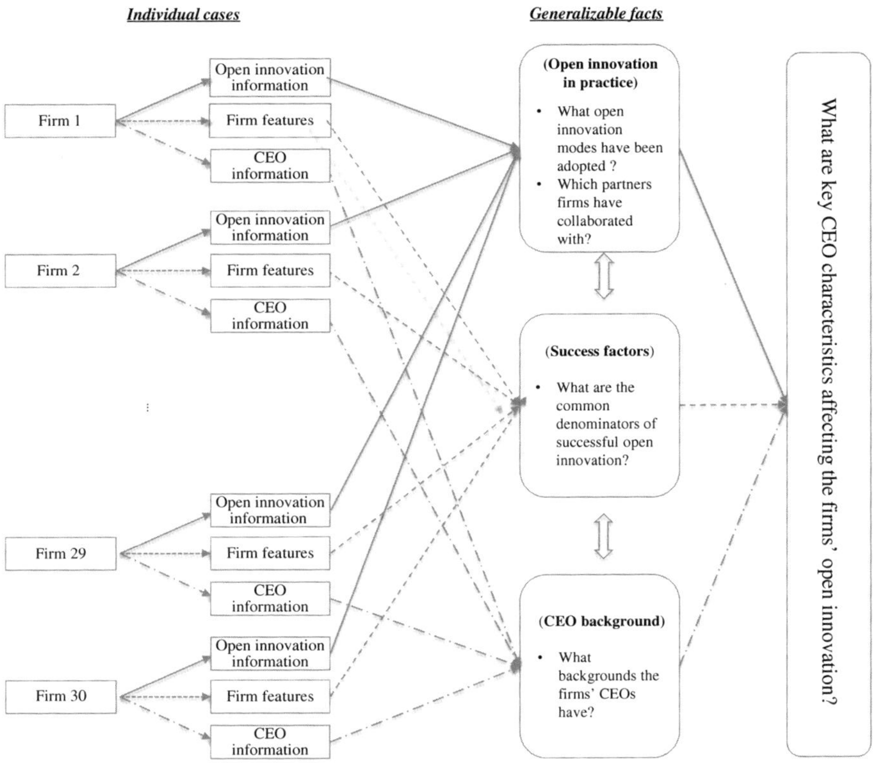

Figure 5.3. Content analysis structure.
Source: The authors.

The data for each firm was categorized into the following three themes: general information about open innovation, specific information about firm features, and CEO information. For open innovation information, the adopted open innovation modes and the types of innovation collaboration partners were investigated. Second, for firm features, an attempt was made to identify the factors affecting successful open innovation adoption by analyzing basic firm information and firm stories. Basic firm information, such as revenue, number of employees, main products, location, and firm history, enabled us to understand the general characteristics of the firms. In addition to this, firm stories, written based on interviews with CEOs and core R&D managers,[1] provided us with a deeper understanding of each firm. With this work, we tried, for example, to get a sense of why the firms adopted open innovation, why they collaborated with the specific external partners and what factors played a vital role in their open innovation implementation. In this way, we could identify important firm-level factors affecting successful open innovation. Third, for CEO information, two different data, CEO backgrounds, and firm stories, were analyzed. CEO backgrounds provide information about the CEOs' educational discipline, the types of academic degree, and prior career experience (if any). Although firm stories were based on secondary interviews, to some extent we were able to identify the linkage between CEO characteristics and open innovation in that the CEOs were also important sources of interview data.

The data analyzed during the first phase were then grouped according to the themes. The open innovation information on individual firms was merged in order to establish the generalizable facts of open innovation practice, and firm features and CEO information were also merged in order to draw generalizable facts of success factors in open innovation and CEO backgrounds. These generalizable facts were cross-analyzed to identify the role of CEOs in open innovation.

5.5.2. *Analysis results*

5.5.2.1. *Open Innovation Adoption*

In terms of open innovation mode, the majority of the sample firms (27 out of 30) employed "R&D collaboration" (see Figure 5.4). "In-sourcing" was adopted by 10 firms (ID 2, 4, 10, 11, 13, 15, 17, 27, 28, and 29) and "M&A" was used in four cases (ID 7, 22, 27, and 28). "Strategic alliance" was used in three cases (ID 13, 16, and 25),

[1] In the original data, professional writers interviewed CEOs and R&D managers, and they reported narrative stories of the firms. In this chapter, we analyzed original interview manuscripts.

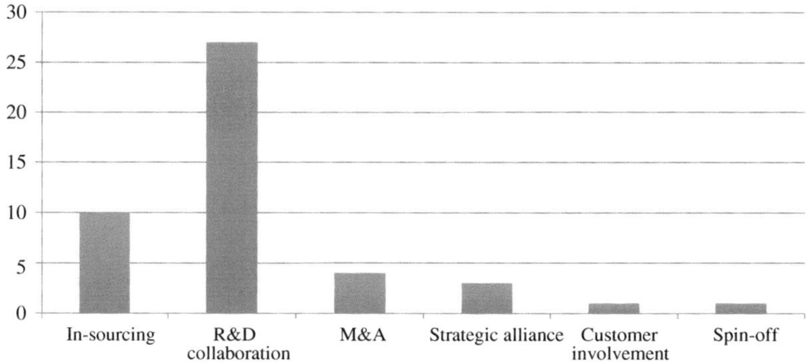

Figure 5.4. Open innovation adoption in sample firms.

while "customer involvement" (ID 12) and "spin-off" (ID 22) were adopted only once. Sixteen firms adopted one open innovation mode, but fourteen firms (i.e., ID 2, 4, 7, 10, 12, 13, 15, 16, 17, 22, 25, 27, 28, and 29) used more than one mode during the year 2010. Except for one instance (ID 22), all firms adopted inbound open innovation.

5.5.2.2. *The Influence of Leaders*

(1) *Reducing risk and resistance hurdles*
From the stories of the firms' open innovation, we were able to see that the CEOs' personal propensities are important factors affecting it. The following three aspects emerge in terms of their importance in innovative small firms. First, the top management were active champions of open innovation activities. Since open innovation is an organizational innovation (Christensen, 2006), the most important driver of this organizational level transformation will be strongly affected by a committed and passionate champion (Di Minin *et al.*, 2010). In our sample, not only did the CEOs engage in internal R&D by being one of the core R&D staff or by applying for their own patents, but they also kept promoting open innovation adoptions. They emphasized the importance of external information and partners and regarded

leadership as a key factor in successful change. For instance, one CEO (ID 7) who did international M&A with a Norwegian vaccine firm stated the importance of the firm's proactive attitude as follows:

> My firm aims to be a "Small Giant." However, for this, to some extent, risk-taking is inevitable. "Not to put too many eggs in one basket" is a fundamental foundation of investment, but small organisations like my firm occasionally have to accept risk by putting all their eggs in one basket. My firm did M&A to broaden its business area, and personally I think that there was no other option at that time. It was a risky move but a necessary one to achieve a leap forward.

Another CEO (ID 14) emphasized the importance of leadership as follows:

> Small firms face endless challenges. To cope with such difficulties, CEOs must show strong leadership. I am personally trying to be a multi-function CEO who can engage in both R&D and management and also make tough decisions quickly. Only when a leader shows full support for innovation, can employees implement it without hesitation.

As the success of open innovation adoption highly relies upon the coordination and harmonization of the various perspectives and interests of different suborganizations, whether senior executives have strong leadership in promoting organizational changes plays an important role (Chesbrough and Garman, 2009).

(2) *Broadening innovation network*
In our sample, top management individuals were at the center of the firms' innovation/business network establishment. We found that in many cases collaboration was initiated by the CEOs' personal networks. For instance, one CEO (ID 15) participated in

many academic conferences to find appropriate collaboration partners holding necessary technologies. Also, one firm (ID 29) used its CEO's personal network (i.e., his Naval Academy alumni network) to diversify its sales channels, and another firm (ID 16) collaborated with a large firm where its CEO had worked before. In addition to network initiation, CEOs played important roles in building trustworthy partnerships. For instance, some of the firms (ID 2, 8, 10, and 20) collaborated with the universities where their CEOs studied (or were studying, in ID 2 and 10). These firms were able to establish close relationships with the collaboration universities using the CEOs' alumni or academic networks. Particularly, by being PhD students at the universities, CEOs of firms with ID 2 and ID 10 were not only able to accumulate the necessary knowledge, but also involved their university supervisors and research teams in the firms' R&D projects. This dual role of the CEO, as a student and company leader, made it easy for the small firms to identify trustworthy partners and form strong bonds with them.

(3) *Addressing knowledge gap*

From the information gleaned from the data, we can extract important CEO characteristics and their roles in innovation. Figure 5.5 shows the clustering results of CEO education and experience. There are four quadrants in the diagram. The horizontal axis shows educational characteristics (i.e., whether CEOs have an engineering background or not), while the vertical axis indicates work experience (i.e., whether they have previous work experience or not).

In terms of education, 73% of the CEOs (i.e., 22 in groups 1 and 4 in Figure 5.5) had an engineering background, while only 27% (in groups 2 and 3) had non-engineering backgrounds. As noted by Kitchell (1997), CEOs with an engineering background are able to understand detailed technologies. Since most of the CEOs had an engineering background, this might have easily helped them make decisions related to innovation, such as identifying the necessary technologies to be absorbed. The fact that

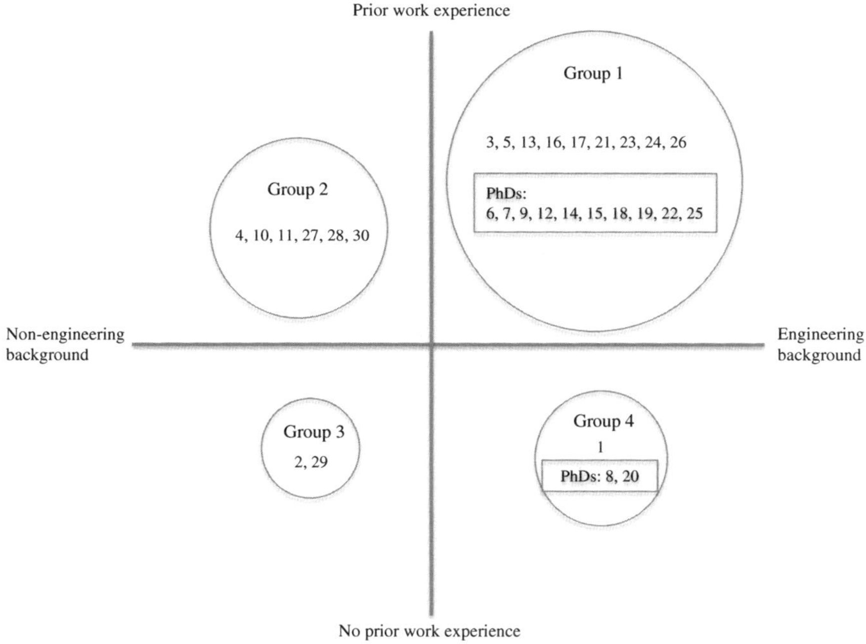

Figure 5.5. Education and work experience.

Note: Numbers indicate firm ID.

most of the firms in groups 1 and 4 were R&D-intensive, firms with high R&D investment and R&D staff ratio may corroborate this. Further, 12 CEOs (10 CEOs in group 1, and 2 in group 2) had a PhD degree (or were studying for a PhD in 2010). As they had expert-level knowledge in the technology field of their firms, it seems consequent that they played critical roles in the firms' innovation. For example, some of the CEOs with a PhD degree (ID 7, 9, 12, 18, and 25) were their firms' patent applicants, and one of the CEOs (ID 12) was a renowned author of academic books and papers in the firm's technology field. Some of the CEOs with a non-engineering background also tried to enhance their firms' innovation competitiveness. Through high investment in R&D (ID 30) or employment of R&D staff (ID 4, 11, and 27), these firms were able to hold many patents or international certificates.

In order to increase their knowledge, two CEOs (ID 2 and 10) even studied for postgraduate degrees at the collaboration partner university. In terms of open innovation modes, most of the firms in groups 2 and 3 adopted multiple open innovation modes, and, interestingly, two firms (ID 27 and 28) where their CEOs had a management background adopted "M&A" as their open innovation mode.

More than half of the CEOs (i.e., 16 of 30 CEOs) had worked in large firms, while only one CEO (ID 7) had worked in a public research institution. Five CEOs (ID 1, 2, 8, 20, and 29) had no prior career experience, and eight CEOs (ID 4, 9, 10, 15, 16, 23, 27, and 30) had worked in other small firms before they established their firms. Consequently, most of the CEOs (83%) had prior work experience. In fact, CEOs' work experience can provide numerous benefits to their firms. While working in other firms, not only can they accumulate the necessary managerial knowledge, e.g., how to run an organization and deal with innovation projects, but they can also secure stable supply contracts with their former organizations. For example, based on the CEOs' personal relationships, one firm (ID 3) used the large firm's distribution chain to develop sales channels, and two other firms (ID 19 and 20) signed exclusive contracts (original development manufacturing, ODM) with their CEOs' former organizations. Considering that marketing is one of the most challenging factors in small firm's growth (Narula, 2004), this close relationship with large firms may have helped them to avoid the "death valley" in the early stages of business.

5.6. Conclusions

This chapter investigated the influence of the human characteristics on open innovation adoption in a small firm context. This chapter was motivated by the desire to complement current work on the organizational (and environmental) factors (Salter *et al.*, 2014; Schroll and Mild, 2012; Wynarczyk *et al.*, 2013; Chesbrough and

Bogers, 2014), which are shaping open innovation adoption. In order to address these research gaps, the current study paid attention to the similarities between open innovation adoption and changes in organizational routines. To increase openness, firms have to deviate from existing operating routines and break with conventions Salter *et al.*, 2014; Colombo *et al.*, 2011), and we found similarities between a routine change and open innovation adoption. As illustrated in Figure 5.1, in an analogy with chemical reactions, the risk/resistance hurdle plays the role of the "activation energy" which has to be overcome to reach the next routine status. The hurdle can be explained by the challenges, the energy, and effort involved in open innovation adoption. Also, the analogy suggests that a routine change is achieved when the necessary knowledge and network expansion are addressed, and we argue that the characteristics of key decision-makers, such as CEOs, can play important roles in mitigating hurdles and addressing the necessary capabilities. As shown in Figure 5.2, top executives' psychological characteristics, such as innovativeness and risk-taking propensity, are highly likely to encourage and support firms to take the necessary risks and help them to reduce resistance by persuading internal members and mitigating conflicts, thus enabling them to deal with risk/resistance hurdles. Also, a positive attitude toward openness and perseverance can play an important role in establishing a decision-maker's perception and awareness of open innovation. These in turn will contribute to a decrease in organizational resistance and an increase in the momentum of change while helping firms to wait for tangible results achieved through openness (Jansen, 2000b; Hambrick *et al.*, 1993). Capability-related characteristics, such as education and work experience, are very likely to enable firms to develop the open innovation capabilities and are expected to bridge the knowledge gap, and these characteristics can also help firms to expand their network.

This interpretation suggests the importance of human elements in open innovation adoption. Top executives' knowledge is an important initial human capital in small firms (Colombo and Grilli,

2005), and their perception and propensity can shape the direction of strategic change (Datta *et al.*, 2003; Herrmann and Nadkarni, 2014). As key agents for change, they will substantially influence the resistance, readiness, and momentum of organizational change (Jansen, 2000b; Hambrick *et al.*, 1993), and this is also valid in the context of open innovation. Although open innovation is a collective-level construct, its microfoundation is rooted in an individual's intentional actions, experience, and preferences. Thus, by paying more attention to key individuals, we will be able to obtain a better and clearer understanding of open innovation.

This chapter provides some theoretical and practical implications. First, the chapter has attempted to extend the boundary of open innovation studies by linking open innovation with other bodies of relevant literature. The heterogeneity and evolving concept of open innovation have made it difficult for researchers to investigate. As noted by Vanhaverbeke and Cloodt (2014), the complex nature of open innovation has made us employ many different theories. Acknowledging this necessity, we broadened the boundary of open innovation. The main focus of the research was on open innovation adoption, but this chapter also embraced various ideas and concepts: the field of management innovation and organizational routine changes led us to including another body of literature, organizational behavior, and, more broadly, organization science. The chapter also focused on key individuals; so, the literature in the field of entrepreneurship, UET, and, more broadly, strategic leadership was investigated. This wider literature coverage may contribute to broadening scholars' horizon in these fields.

Second, top management in small firms have to recognize that they are at the epicenter of open innovation adoption. They have to understand that it is influenced not only by firm and environmental factors but also by their own characteristics. Whether they are well aware of the advantages of open innovation, are willing to take risks, and have established the necessary skills which will reflect considerably on the adoption of open innovation. In fact, even in large firms, a top-down approach had better results in open innovation adoption and implementation (Mortara and Minshall, 2011). For instance, since the CEO of LG Household and Health Care had previously worked in P&G, a firm actively adopting open innovation (Huston and Sakkab, 2006), he understood exactly the benefits and challenges of open innovation and could push it forward overcoming internal resistance (Kim *et al.*, 2008). This transfer of experience could be even more significant in small firms where top management influence is stronger. Thus, acknowledging their roles as initiators and facilitators for organizational change (Jansen, 2000a; Hambrick *et al.*, 1993), top managers have to try to recognize the various benefits open innovation will bring. Then, their positive attitudes based on sufficient understanding will facilitate open innovation adoption in their firms.

References

Ahn, J. M., Minshall, T. and Mortara, L. (2017a). CEOs in innovative SMEs: open innovation initiators and facilitators. In: Salampasis, D. & Mention, A.-L. (eds.). *Open Innovation: Unveiling the Power of the Human Element.* London: World Scientific Publishing.

Ahn, J. M., Minshall, T. and Mortara, L. (2017b). Understanding the human side of openness: the fit between open innovation modes and CEO characteristics. *R&D Management*, 47, pp. 727–740.

Ahn, J. M., Mortara, L. and Minshall, T. (2015). Open innovation: a new classification and its impact on firm performance in innovative SMEs. *Journal of Innovation Management,* 3, pp. 33–54.

Antons, D. and Piller, F. T. (2015). Opening the black box of "Not Invented Here": attitudes, decision biases, and behavioral consequences. *The Academy of Management Perspectives,* 29, pp. 193–217.

Argote, L. and Greve, H. R. (2007). A behavioral theory of the firm — 40 years and counting: introduction and impact. *Organization Science,* 18, pp. 337–349.

Arrow, K. J. (1974). *The Limits of Organization.* New York, NY: W. W. Norton & Company, Inc.

Asheim, B. T. (2000). Industrial district: the contributions of Marshall and beyond. In: Clark, G., Feldman, M. and Gertler, M. (eds.). *The Oxford Handbook of Economics Geography.* Oxford: Oxford University Press.

Bantel, K. A. and Jackson, S. E. (1989). Top management and innovations in banking — does the composition of the top team make a difference. *Strategic Management Journal,* 10, pp. 107–124.

Barker, V. L. and Mueller, G. C. (2002). CEO characteristics and firm R&D spending. *Management Science,* 48, pp. 782–801.

Becker, G. (1975). *Human Capital.* New York, NY: Columbia University Press.

Bereleson, B. (1952). *Content Analysis in Communication Research.* Glencoe, IL: Free Press.

Bianchi, M., Chiaroni, D., Chiesa, V. and Frattini, F. (2011). Exploring the role of human resources in technology out-licensing: an empirical analysis of biotech new technology-based firms. *Technology Analysis & Strategic Management,* 23, pp. 825–849.

Birley, S. and Norburn, D. (1987). Owners and Managers — the Venture-100 Vs the Fortune-500. *Journal of Business Venturing,* 2, pp. 351–363.

Burcharth, A. L. D. A., Knudsen, M. P. and Søndergaard, H. A. (2014). Neither invented nor shared here: the impact and management of attitudes for the adoption of open innovation practices. *Technovation,* 34, pp. 149–161.

Cassia, L., Minola, T. and Paleari, S. (2011). Entrepreneurship, technology and change: a review and proposal for an interpretative framework. In: Cassia, L., Minola, T. and Paleari, S. (eds.). *Entrepreneurship and Technological Change.* Cheltenham: Edward Elgar Publishing Limited.

Ceci, F. and Iubatti, D. (2012). Personal relationships and innovation diffusion in SME networks: a content analysis approach. *Research Policy,* 41, pp. 565–579.

Chesbrough, H. (2003). *Open Innovation: The New Imperative for Creating and Profiting from Technology.* Boston, MA: Harvard Business School Press.

Chesbrough, H. and Bogers, M. (2014). Explicating open innovation: clarifying an emerging paradigm for understanding innovation. In: Chesbrough, H., Vanhaverbeke, W. and West, J. (eds.). *New Frontiers in Open Innovation.* Oxford, UK: Oxford University Press.

Chesbrough, H. W. and Garman, A. R. (2009). Use open innovation to cope in a downturn. *Harvard Business Review,* 87, pp. 1–10.

Chesbrough, H., Vanhaverbeke, W. and West, J. (2006). *Open innovation: Researching a New Paradigm.* Oxford: Oxford University Press.

Chiaroni, D., Chiesa, V. and Frattini, F. (2011). The open innovation journey: how firms dynamically implement the emerging innovation management paradigm. *Technovation,* 31, pp. 34–43.

Christensen, J. F. (2006). Wither core competency for the large corporation in an open innovation World? In: Chesbrough, H., Vanhaverbeke, W. and West, J. (eds.). *Open Innovation: Researching a New Paradigm.* Oxford: Oxford University Press.

Coase, R. H. (1937). The nature of the firm. *Economica,* 4, pp. 386–405.

Cohen, M. D. and Bacdayan, P. (1994). Organizational routines are stored as procedural memory: evidence from a laboratory study. *Organization Science,* 5, pp. 554–568.

Cohen, W. M., Nelson, R. R. and Walsh, J. P. (2002). Links and impacts: the influence of public research on industrial R&D. *Management Science,* 48, pp. 1–23.

Colombo, M. G. and Grilli, L. (2005). Founders' human capital and the growth of new technology-based firms: a competence-based view. *Research Policy,* 34, pp. 795–816.

Colombo, M. G., Rabbiosi, L. and Reichstein, T. (2011). Organizing for external knowledge sourcing. *European Management Review,* 8, pp. 111–116.

Cooper, A. C., Gimenogascon, F. J. and Woo, C. Y. (1994). Initial human and financial capital as predictors of new venture performance. *Journal of Business Venturing,* 9, pp. 371–395.

Cyert, R. M. and March, J. (1963). *The Behavioral Theory of the Firm.* Englewood Cliff, NJ: Prentice Hall.

Dahlander, L. and Gann, D. M. (2010). How open is innovation? *Research Policy,* 39, pp. 699–709.

Datta, D. K., Rajagopalan, N. and Zhang, Y. (2003). New CEO openness to change and strategic persistence: the moderating role of industry characteristics. *British Journal of Management,* 14, pp. 101–114.

Dimaggio, P. J. and Powell, W. W. (1983). The iron cage revisited — institutional isomorphism and collective rationality in organizational fields. *American Sociological Review,* 48, pp. 147–160.

Di Minin, A., Frattini, F. and Piccaluga, A. (2010). Fiat: open innovation in a downturn (1993–2003). *California Management Review,* 52, pp. 132–159.

Dosi, G. and Marengo, L. (2007). On the evolutionary and behavioral theories of organizations: a tentative roadmap. *Organization Science,* 18, pp. 491–502.

Drechsler, W. and Natter, M. (2012). Understanding a firm's openness decisions in innovation. *Journal of Business Research,* 65, pp. 438–445.

Dyer, J. H. and Singh, H. (1998). Strategic supplier segmentation: the next "best practice" in supply chain management. *California Management Review,* 40, pp. 660–679.

Eisenhardt, K. M. and Schoonhoven, C. B. (1990). Organizational growth — linking founding team, strategy, environment, and growth among United-States semiconductor ventures, 1978–1988. *Administrative Science Quarterly,* 35, pp. 504–529.

Escriba-Esteve, A., Sanchez-Peinado, L. and Sanchez-Peinado, E. (2009). The influence of top management teams in the strategic orientation and performance of small and medium-sized enterprises. *British Journal of Management,* 20, pp. 581–597.

Faulkner, W. and Senker, J. (1994). Making sense of diversity — public–private sector research linkage in 3 technologies. *Research Policy,* 23, pp. 673–695.

Felin, T. and Foss, N. J. (2009). Organizational routines and capabilities: historical drift and a course-correction toward microfoundations. *Scandinavian Journal of Management,* 25, pp. 157–167.

Fereday, J. and Muir-Cochrane, E. (2008). Demonstrating rigor using thematic analysis: a hybrid approach of inductive and deductive coding and theme development. *International Journal of Qualitative Methods,* 5, pp. 80–92.

Finkelstein, S., Hambrick, D. C. and Cannella, A. A. (2009). *Strategic Leadership: Theory and Research on Executives, Top Management Teams, and Boards.* Oxford, UK: Oxford University Press.

Fishbein, M. and Ajzen, I. (1975). *Belief, Attitude, Intention and Behavior: An Introduction to Theory and Research.* Reading, MA: Addison-Wesley.

Gans, J. S. and Stern, S. (2003). The product market and the market for "ideas": commercialization strategies for technology entrepreneurs. *Research Policy,* 32, pp. 333–350.

Garicano, L. (2000). Hierarchies and the organization of knowledge in production. *Journal of Political Economy,* 108, pp. 874–904.

Gavetti, G., Greve, H. R., Levinthal, D. A. and Ocasio, W. (2012). The behavioral theory of the firm: assessment and prospects. *Academy of Management Annals,* 6, pp. 1–40.

Gerstner, W.-C., König, A., Enders, A. and Hambrick, D. C. (2013). CEO narcissism, audience engagement, and organizational adoption of technological discontinuities. *Administrative Science Quarterly,* 58, pp. 257–291.

Granovetter, M. S. (1973). The strength of weak ties. *American Journal of Sociology*, 78, pp. 1360–1380.

Gupta, A. K. (1984). Contingency linkage between strategy and general manager characteristics: a conceptual examination. *The Academy of Management Review,* 9, pp. 399–412.

Hambrick, D. C. (2007). Upper echelons theory: an update. *Academy of Management Review,* 32, pp. 334–343.

Hambrick, D. C. and Mason, P. A. (1984). Upper Echelons: the organization as a reflection of its top managers. *Academy of Management Review,* 9, pp. 193–206.

Hambrick, D. C., Geletkanycz, M. A. and Fredrickson, J. W. (1993). Top executive commitment to the status-quo — some tests of its determinants. *Strategic Management Journal,* 14, pp. 401–418.

Harryson, S., Kliknaite, S. and Dudkowski, R. (2008). Flexibility in innovation through external learning: exploring two models for enhanced industry-university collaboration. *International Journal of Technology Management,* 41, pp. 109–137.

Helfat, C. E. and Martin, J. A. (2015). Dynamic managerial capabilities: review and assessment of managerial impact on strategic change. *Journal of Management,* 41, pp. 1281–1312.

Helfat, C. E., Finkelstein, S., Mitchell, W., Peteraf, M., Singh, H., Teece, D. and Winter, S. G. (2009). *Dynamic Capabilities: Understanding Strategic Change in Organizations.* New York: John Wiley & Sons.

Herrmann, P. and Nadkarni, S. (2014). Managing strategic change: the duality of CEO personality. *Strategic Management Journal,* 35, pp. 1318–1342.

Howell, J. M. and Higgins, C. A. (1990). Champions of technological innovation. *Administrative Science Quarterly,* 35, pp. 317–341.

Humphreys, P., McAdam, R. and Leckey, J. (2005). Longitudinal evaluation of innovation implementation in SMEs. *European Journal of Innovation Management,* 8, pp. 283–304.

Huston, L. and Sakkab, N. (2006). Connect and develop: inside Procter & Gamble's new model for innovation. *Harvard Business Review,* 84, pp. 58–66.

Jansen, J. J. P., Vera, D. and Crossan, M. (2009). Strategic leadership for exploration and exploitation: the moderating role of environmental dynamism. *Leadership Quarterly,* 20, pp. 5–18.

Jansen, K. J. (2000a). The emerging dynamics of change: resistance, readiness, and momentum. *Human Resource Planning,* 23, pp. 53–55.

Jansen, K. J. (2000b). The emerging dynamics of change: resistance, readiness, and momentum. *People and Strategy,* 23, p. 53.

Katz, R. and Allen, T. J. (1982). Investigating the Not Invented Here (NIH) syndrome — a look at the performance, tenure, and communication patterns of 50 R-and-D project groups. *R&D Management,* 12, pp. 7–19.

Khandwalla, P. N. (1976/1977). Some top management styles, their context and performance. *Organization and Administrative Science,* 7, pp. 21–51.

Kim, S. K., Ahn, D. H., Song, J. G., Lee, J. W., Bae, Y. H. and Jun, J. Y. (2008). *Building Open Innovation System: Theory, Practice and Policy Implication.* Seoul: STEPI.

Kirton, M. (1976). Adaptors and innovators — description and measure. *Journal of Applied Psychology,* 61, pp. 622–629.

Kitchell, S. (1997). CEO characteristics and technological innovativeness: a Canadian perspective. *Canadian Journal of Administrative Sciences — Revue Canadienne Des Sciences De L Administration,* 14, pp. 111–125.

Krippendorff, K. (2003). *Content Analysis: An Introduction to its Methodology.* London: Sage.

Laursen, K. and Salter, A. J. (2014). The paradox of openness: appropriability, external search and collaboration. *Research Policy,* 43, pp. 867–878.

Lefebvre, E. and Lefebvre, L. A. (1992). Firm innovativeness and CEO characteristics in small manufacturing firms. *Journal of Engineering and Technology Management,* 9, pp. 243–277.

Lefebvre, L. A., Mason, R. and Lefebvre, E. (1997). The influence prism in SMEs: the power of CEOs' perceptions on technology policy and its organizational impacts. *Management Science,* 43, pp. 856–878.

Lichtenthaler, U. and Lichtenthaler, E. (2009). A capability-based framework for open innovation: Complementing absorptive capacity. *Journal of Management Studies,* 46, pp. 1315–1338.

Lubatkin, M. H., Simsek, Z., Ling, Y. and Veiga, J. F. (2006). Ambidexterity and performance in small- to medium-sized firms: the pivotal role of top management team behavioral integration. *Journal of Management,* 32, pp. 646–672.

Macpherson, A. and Holt, R. (2007). Knowledge, learning and small firm growth: a systematic review of the evidence. *Research Policy,* 36, pp. 172–192.

Marcati, A., Guido, G. and Peluso, A. M. (2008). The role of SME entrepreneurs' innovativeness and personality in the adoption of innovations. *Research Policy,* 37, pp. 1579–1590.

Mellewigt, T., Madhok, A. and Weibel, A. (2007). Trust and formal contracts in interorganizational relationships — substitutes and complements. *Managerial and Decision Economics,* 28, pp. 833–847.

Miller, D. (1983). The correlates of entrepreneurship in 3 types of firms. *Management Science,* 29, pp. 770–791.

Miller, D. and Friesen, P. H. (1982). Innovation in conservative and entrepreneurial firms — 2 models of strategic momentum. *Strategic Management Journal,* 3, pp. 1–25.

Miller, D. and Toulouse, J. (1986a). Chief executive personality and corporate strategy and structure in small firms. *Management Science,* 32, pp. 1389–1409.

Miller, D. and Toulouse, J. (1986b). Strategy, structure, CEO personality and performance in small firms. *American Journal of Small Business,* Winter.

Miller, D., Devries, M. F. R. K. and Toulouse, J. M. (1982). Top executive locus of control and its relationship to strategy-making, structure, and environment. *Academy of Management Journal,* 25, pp. 237–253.

Moore, G. C. and Benbasat, I. (1991). Development of an instrument to measure the perceptions of adopting an information technology innovation. *Information Systems Research,* 2, pp. 192–221.

Moreno, A. M. and Casillas, J. C. (2008). Entrepreneurial orientation and growth of SMEs: a causal model. *Entrepreneurship Theory and Practice,* 32, pp. 507–528.

Mortara, L. and Minshall, T. (2011). How do large multinational companies implement open innovation? *Technovation,* 31, pp. 586–597.

Mortara, L. and Minshall, T. (2014). Patterns of implementation of OI in MNCs. In: Chesbrough, H., Vanhaverbeke, W., and West, J. (eds.). *New Frontiers in Open Innovation.* Oxford: Oxford University Press.

Mortara, L., Ford, S., Probert, D. and Minshall, T. (2012). Inbound open innovation: a technology acquisition process model. *Open Innovation, New Insights and Evidence.* London: Imperial College.

Mortara, L., Napp, J., Ford, S. and Minshall, T. (2011). Open innovation activities to foster corporate entrepreneurship. In: Cassia, L., Minola, T. and Paleari, S. (eds.). *Entrepreneurship and Technological Change.* Cheltenham, UK: Edward Elgar.

Murmann, J. P., Aldrich, H. E., Levinthal, D. and Winter, S. G. (2003). Evolutionary thought in management and organization theory at the beginning of the new millennium a symposium on the state of the art and opportunities for future research. *Journal of Management Inquiry,* 12, pp. 22–40.

Narula, R. (2004). R&D collaboration by SMEs: new opportunities and limitations in the face of globalisation. *Technovation,* 24, pp. 153–161.

Nelson, R. and Winter, S. (1982). *An Evolutionary Theory of Economic Change,* Cambridge, MA: Belknap Harvard Press.

Nooteboom, B. (1994). Innovation and diffusion in small firms — theory and evidence. *Small Business Economics,* 6, pp. 327–347.

Nooteboom, B., Vanhaverbeke, W., Duysters, G., Gilsing, V. and Van den Oord, A. (2007). Optimal cognitive distance and absorptive capacity. *Research Policy,* 36, pp. 1016–1034.

Oakey, R. P. (2013). Open innovation and its relevance to industrial research and development: the case of high-technology small firms. *International Small Business Journal,* 31, pp. 319–336.

Papadakis, V. and Bourantas, D. (1998). The chief executive officer as corporate champion of technological innovation: an empirical investigation. *Technology Analysis & Strategic Management,* 10, pp. 89–109.

Papadakis, V. M. and Barwise, P. (2002). How much do CEOs and top managers matter in strategic decision-making? *British Journal of Management,* 13, pp. 83–95.

Pentland, B. T. and Rueter, H. H. (1994). Organizational routines as grammars of action. *Administrative Science Quarterly,* 39, pp. 484–510.

Perkmann, M. and Walsh, K. (2007). University-industry relationships and open innovation: towards a research agenda. *International Journal of Management Reviews,* 9, pp. 259–280.

Polanyi, M. (1966). *The Tacit Dimension.* New York, NY: Doubleday.

Rhee, J., Park, T. and Lee, D. H. (2010). Drivers of innovativeness and performance for innovative SMEs in South Korea: mediation of learning orientation. *Technovation,* 30, pp. 65–75.

Robertson, P. L., Casali, G. L. and Jacobson, D. (2012). Managing open incremental process innovation: absorptive capacity and distributed learning. *Research policy,* 41, pp. 822–832.

Rothwell, R. and Dodgson, M. (1994). Innovation and size of firm. In: Dodgson, M. (ed.). *Handbook of Industrial Innovation.* Aldershot: Edward Elgar.

Salter, A., Criscuolo, P. and Ter Wal, A. L. (2014). Coping with open innovation. *California Management Review,* 56, pp. 77–94.

Saxenian, A. L., Motoyama, Y. and Quan, X. (2002). *Local and Global Networks of Immigrant Professionals in Silicon Valley.* San Francisco: Public Policy Institute of California.

Schein, E. (1985). *Organizational Culture and Leadership.* San Francisco: Jossey Nass.

Schroll, A. and Mild, A. (2012). A critical review of empirical research on open innovation adoption. *Journal fur Betriebswirtschaft,* 62, pp. 85–118.

Simon, H. A. (1957). *Models of Man.* New York, NY: Wiley.

Simon, H. A. (1972). Theories of bounded rationality. In: McGuire, C. B. & Radncr, R. (eds.). *Decision and Organization.* New York, NY: North-Holland Publishing Company.

SMBA (2011). No Innovation, No Future, Daejon: Small and Medium Business Adminstration.

Smith, D. (2007). The politics of innovation: why innovations need a godfather. *Technovation,* 27, pp. 95–104.

Spithoven, A., Clarysse, B. and Knockaert, M. (2011). Building absorptive capacity to organise inbound open innovation in traditional industries. *Technovation,* 31, pp. 10–21.

Sundaramurthy, C., Pukthuanthong, K. and Kor, Y. (2014). Positive and negative synergies between the CEO's and the corporate board's human and social capital: a study of biotechnology firms. *Strategic Management Journal,* 35, pp. 845–868.

Teece, D. J. (1996). Firm organization, industrial structure, and technological innovation. *Journal of Economic Behavior & Organization,* 31, pp. 193–224.

Teece, D. J. (2007). Explicating dynamic capabilities: the nature and microfoundations of (sustainable) enterprise performance. *Strategic Management Journal,* 28, pp. 1319–1350.

Teece, D. J., Pisano, G. and Shuen, A. (1997). Dynamic capabilities and strategic management. *Strategic Management Journal,* 18, pp. 509–533.

Thong, J. Y. L. and Yap, C. S. (1995). CEO characteristics, organizational characteristics and information technology adoption in small businesses. *Omega-International Journal of Management Science,* 23, pp. 429–442.

Vanhaverbeke, W. (2017). *Managing Open Innovation in SMEs.* Cambridge, UK: Cambridge University Press.

Vanhaverbeke, W. and Cloodt, M. (2014). Theories of the firm and open innovation. In: Chesbrough, H., Vanhaverbeke, W. & West, J. (eds.). *New Frontiers in Open Innovation.* Oxford: Oxford University Press.

Vanhaverbeke, W. and Roijakkers, N. (2014). Enriching open innovation theory and practice by strengthening the relationship with strategic thinking. In: Pfeffermann, N., Minshall, T. and Mortara, L. (eds.). *Strategy and Communication for Innovation.* 2nd edition. London: Springer.

Volberda, H. W., Foss, N. J. and Lyles, M. A. (2010). Perspective-absorbing the concept of absorptive capacity: how to realize its potential in the organization field. *Organization Science,* 21, pp. 931–951.

Vossen, R. W. (1998). Research note — relative strengths and weaknesses of small firms in innovation. *International Small Business Journal,* 16, pp. 88–94.

Wally, S. and Becerra, M. (2001). Top management team characteristics and strategic changes in international diversification the case of US multinationals in the European community. *Group & Organization Management,* 26, pp. 165–188.

West, J., Salter, A., Vanhaverbeke, W. and Chesbrough, H. (2014). Open innovation: the next decade. *Research Policy,* 43, pp. 805–811.

Wynarczyk, P., Piperopoulos, P. and McAdam, M. (2013). Open innovation in small and medium-sized enterprises: an overview. *International Small Business Journal,* 31, pp. 240–255.

Zollo, M. and Winter, S. G. (2002). Deliberate learning and the evolution of dynamic capabilities. *Organization Science,* 13, pp. 339–351.

Chapter 6

How do SMEs Use Open Innovation When Developing New Business Models?

Robert William Anderson[*,‡], *Nuran Acur*[†,§] *and Jonathan Corney*[*,¶]

[*] *Design, Manufacture and Engineering Management University of Strathclyde, Glasgow, G1 1XJ, UK*
[†]*Adam Smith Business School University of Glasgow, UK*
[‡]*r.w.anderson@strath.ac.uk*
[§]*nuran.acur@glasgow.ac.uk*
[¶]*jonathan.corney@strath.ac.uk*

Abstract

Access to external knowledge to accelerate innovation is becoming a key business process as firms continue to recognize the importance of going beyond their boundaries when advancing products and services. This chapter considers the increasingly important case of technology-driven service innovation in the context of

[‡] Corresponding author.

crowdsourcing for innovation. Drawing on an in-depth crowdsourcing for innovation project, this chapter analyzes the development of service innovation with the crowd. The data is anchored and grounded in observations, interviews, workshops, and an online idea platform. We contribute to an acknowledged aspect, neglected in the literature, understanding how small and medium-sized enterprises (SMEs) actually use crowds in service innovation, when moving from a product-based to service-based business.

6.1. Introduction

Although the adoption of open innovation methodologies is frequently motivated by the desire to disrupt "corporate group think," the approach does not have to be limited to large companies. Indeed, the effectively "closed innovation" approaches found in many small and medium-sized enterprises (SMEs) can create more profound challenges than those seen in larger enterprises (who can dedicate resources to open innovation projects) (Van de Vrande *et al.*, 2009; Parida *et al.*, 2012).

Innovation in SMEs and their impact on industries and countries' economies is undisputable. In the UK, small businesses account for 99.3% of private sector firms; 15.7 million people are employed in SMEs in the UK covering 60% of all private sector employment in the UK (SBS, 2016). These figures do not fluctuate much across Europe and other developed nations. In an attempt to stimulate innovation, more SMEs are leveraging external sources of knowledge and technology (Chesbrough, 2006; Vanhaverbeke, 2017). This research aims to support SMEs in the search, collection, and assimilation of ideas from external sources to commercialize their technology in new markets and industries, within or outside the SMEs' current product and services and/or markets. SMEs are naturally short of the resources and competences required to take their technology outside current industries and markets (Bianchi *et al.*, 2010). Open innovation is now widely considered by managers as a powerful paradigm for commercialization in which firms gain benefits from the "purposive inflows and outflows of knowledge to accelerate innovation, and to expand the markets for external use of innovation, respectively" (Chesbrough *et al.*, 2006, p. 1). During open innovation, firm boundaries become highly permeable and dynamic, as ideas, resources, and information flow in and out of the firm. As such, it is a challenging task for SMEs to protect their technology while exploiting their current technology due to firm smallness, closed boundary decision-making processes, distinguishing internal from external ideas, limited resources, and personnel. This research helps SMEs who would like to commercialize their technologies and overcome these obstacles by developing a model to facilitate the open business model process via crowdsourcing which is a fast, cost-efficient, and flexible process for external idea assimilation inside the SMEs.

Firm smallness is both a strength and weakness of an SME. It is a strength because change can be rapid and ideas can be easily tested, but it is also a weakness because the close working relationships of SME staff almost inevitably lead to a shared consensus rather than a diversity of radical ideas. One area where the tension between "speed to market" and "radical product concept" is evident is in the creation of business models for the SME that can be employed to unlock the value embedded in new technologies. This chapter describes how SMEs are using open innovation to overcome their size and generate radical new business models.

Scholars generally agree that firms commercialize innovative ideas through their business models; more recently, business model innovation has been gaining acceptance as a new subject of innovation involving new ways of collaboration (Zott *et al.*, 2011). However, while a number of studies point to business model innovation as a method for firm transformation (e.g., Chesbrough and Rosenbloom, 2002; Demil *et al.*, 2015), a comprehensive and empirical account of using outsiders in business model innovation development in technology commercialization and new market development is still in its infancy. We fill this void by examining the following question: "How does open business model innovation create value for the SME's technology through turning outsiders (i.e., the crowd) into collaborators?"

Crowdsourcing is an effective way of getting others involved in the organization and can be used as an effective tool to facilitate the open innovation model as well as open innovation ecosystem development. We define crowdsourcing innovation in a crowdsourcing context as "the public generation of innovative solutions to a complex problem posed by the company sponsoring the challenge call" (Majchrzak and Malhotra, 2013, p. 258). In the last few years, however, a number of studies have highlighted how firms are relying on the ideas and discussions of external individuals (e.g., Antorini *et al.*, 2012; Mount and Martinez, 2014) among other open innovation processes (Chesbrough, 2003), to support the exchange and understanding of knowledge flows into and out of the firm. This is because they jointly engage in the less investigated form of crowdsourcing for business model innovation, establishing cooperative innovation processes, and subsequently the commercialization of ideas. However, while the notion of crowdsourcing and business model innovation is acknowledged by scholars, we still know little about how the crowd can help firms "do" business model innovation and commercialization. Past studies have focused on specific areas within crowdsourcing e.g., crowd behavior, motivation or managing projects, but an integrated framework detailing the interplay between the firm, the crowd, and the process of business model innovation is still missing.

Technology-driven services are becoming the driving force in developed countries through enhancing employment growth and innovation as products and services become increasingly commoditized (Poulsson and Kale, 2004; Gupta and Vajic, 2000; Pine and Gilmore, 1999). Technology-driven service innovations often create new markets or shift existing market structures and require user learning as they often induce significant behavioral changes on the side of the users (Lettl *et al.*, 2006). Accordingly, services-led growth and experience provision emerged as alternatives for firms that wanted to continuously add value through technology differentiation (Pine and Gilmore, 1998). End-users' increasing sophistication has compelled a growing number of innovative companies to build portfolios of experiences to support their technologies via offering new business models (Berry *et al.*, 2002; Voss *et al.*, 2008).

Therefore, the question of how SMEs should be designed to generate and exploit new forms of experience for their technology is a very important one. Unlike technology innovation in manufacturing SME organizations (Bianchi *et al.*, 2010), the drivers of business model innovation and new service development in experience-intensive technology-based businesses are often not restricted to physical entities, but rather to a combination of processes, experience, end-user involvement and needs, organizational learning, and different contextual elements (e.g., physical setting, social actors, service facilitators, etc.) that should be appropriately integrated to offer memorable, meaningful experience services for the SMEs' current technology (Gupta and Vajic, 2000; Hoch and Deighton, 1989; Vanhaverbeke, 2017). As such, service development life cycles should consider integrating into the current value offerings for technology with newly obtained and/or created knowledge and experience in technology. The theoretical motivation of this chapter therefore is to provide a more complete account of how SMEs can use current technology to design new business models with the crowd which emerge and become embedded into organizations over their life cycle. In this chapter, we analyze how the experience-creation process is organized to unlock technology value and how different social actors participate in the creation.

To date, research has focused on the application of crowdsourcing at the front end of the innovation process (Bayus, 2013; Huang *et al.*, 2014; Lauto *et al.*, 2013) for service/product or technological innovation development. While there is a general understanding of the link between crowd involvement in value creation activities, no research to our knowledge exists that focuses on how crowds can affect the experience design of patented e-technology. For example, Soloviev *et al.* (2010) and Chanal and Caron-Fasan (2010) studied how firms use crowds (i.e., a large community of developers) to develop firms' value creation, performance, and competitive advantage. They conceptualize the business model based on activities or the activity system. The available research mainly associates business models with strategy by following Teece's (2007) business model conceptualization which exposes a "hypothesis about what customers want, and how an enterprise can best meet those needs, and get paid for doing so" (p. 1329). Our research considers technology as an enabler of the business model and embraces how crowds can be used as a means of facilitating the design of experience of innovation/ technology.

Furthermore, there are barriers to managers engaging in such open service business model innovation. Chesbrough (2010) identifies two such barriers as the "barrier of confusion," where it is "far from clear to them even what the right business model ought to be" and the "barrier of obstruction," whereby "managers readily recognize the right business model but its development is resisted due to its conflicts with the prevailing business model, managers resist experiments that might threaten the ongoing value to their company." A process of business model innovation needs to facilitate in overcoming these barriers and move toward what Chesbrough (2010) terms "a commitment to experimentation." To tackle these barriers, our research develops a framework based on a crowd-led collaborative business model framework. Our framework encompasses designing a process for engagement with various stakeholders through the crowd — including customers or lead users of the SME — as well as various tools for analyzing the existing business model, understanding how it has evolved, understanding models emerging in the industry of the SME and monitoring the consequences of changes being made to the business model in simulated exercises. Our framework has five phases of business model innovation as seen in Figure 6.1. The phase, (1) *Initiate*, is concerned with facilitating the definition of SMEs' problems, obstacles, and opportunities; (2) *Discover, Understand,* and *Define* is a social process of dialogue and investigation, out of which new understandings of stakeholder needs emerge; (3) *Ideate* is an online social process in which the crowd create solutions in response to the crowd challenge established in the *Initiate* phase and explained in the *Discover, Understand,* and *Define* phase; (4) *Design and Form* is the process of establishing the value of ideas and knowledge from the *Ideate* phase to design new ways of offering technology commercialization and/or new market development; and (5) *Value proposition*

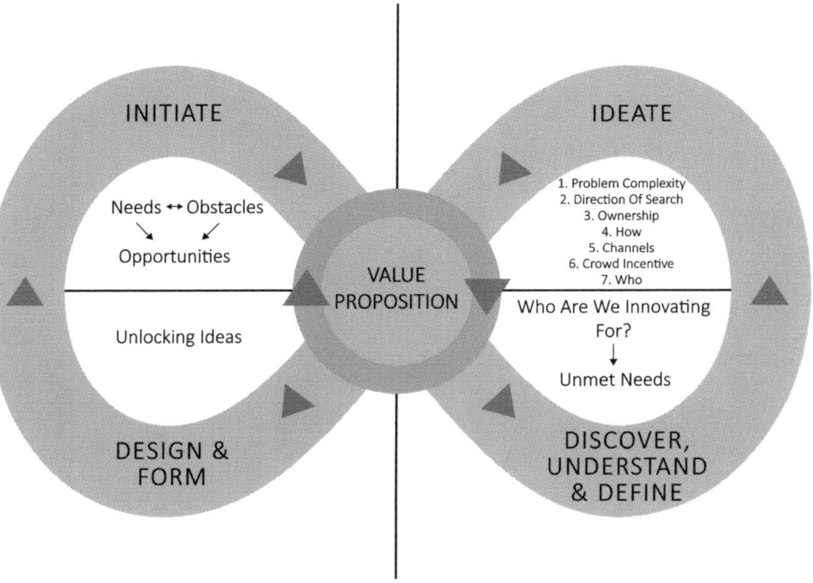

Figure 6.1. Crowdsourcing for business model innovation framework.

involves the firm and the crowd iterating between using and building on the knowledge generated in each phase creating investible propositions. Informed by Anderson *et al.*'s (2016) research, we consider technology "as an enabler of the business model rather than as a part of the concept *per se*" (Zott *et al.*, 2011, p. 1034).

6.2. Research Approach and Methods

Our study combined participant observation with theory building to explore how the involvement of crowds can lead to a new service business model innovation and new market creation via experience design. This research adopted a three-phase process described by Eisenhardt (1989) and Yin (1984) to examine crowd-based business model development in SMEs. The case study method is well suited for building understanding inductively from data on contextual, relational, and cognitive factors underlying business model components (Eisenhardt, 1989; Strauss, 1987).

To improve our understanding of the role of crowdsourcing in business model innovation development, we carried out a case study of business model development with the crowd in a small biomedical firm. We considered the setting appropriate because previously reported crowdsourcing and business model innovation research is limited to large, research-intensive firms. At the time of the research, the firm held patented technology with a need to design experience process for their technologies via business models.

Interviews were conducted with the SME managers to gain a better understanding of the firm's current value propositions, its market, technology characteristics, and its competitive advantages. The aim of the second exploratory phase was to describe the business model innovation design elements (i.e., who, how, what, where) and how they are related. The third phase involves designing and using the crowdsourcing platform to obtain external idea development, discussions, and voting across a number of Internet-enabled devices. All recorded audiovisual data were transcribed verbatim. The study includes data from a SME research site PhysicalHealing UK Ltd. The name of SME is a pseudonym to comply with confidentiality requirements.

PhysicalHealing UK Ltd is a rehabilitation firm using a combination of engineering and therapy to treat people suffering from catastrophic injuries or illnesses. At the time of the study, PhysicalHealing was headquartered in Glasgow, UK, employing four directors, one company secretary, and a number of mobile field service technicians. The directors were responsible for the overall strategy and organization of the firm, with the aim of providing rehabilitation engineering technology and related services within the UK. This involved working with a number of partners throughout the UK and abroad, including the National Health Service (NHS) and local governmental organizations. In order to capitalize on the firm's expert knowledge and know-how, the directors decided that including services associated with the products they provide should play a central role in the business. Providing services would give the firm new growth opportunities and an opportunity to have profit growth independent of their larger partners.

6.2.1. *Data collection*

Data collection followed the recommendations for case study work (Eisenhardt, 1989; Yin, 1984), and combined informational searches, participant observation, semistructured interviews, and casual interactions. Over a period of one-and-a-half years, the first author was responsible for the overall project management of the firm and crowd engagement, taking part in project meetings and performing interviews, in addition to organizing and attending crowd workshops, helping set up and moderating the crowd platform, and having informal conversations. The other researchers acted as "outsiders" in questioning the information gathering, data analysis, and emerging themes in the analysis. All authors were involved in the workshops and the crowdsourcing platform.

The study is supported by 14 interviews, 2 meetings, 2 workshops, 45 workshop participants, 1 online platform, and 14 hours of audiovisual recordings. Our interviews intended to develop a comprehensive picture of how crowdsourcing can lead to business model innovation from those involved. Two research workshops were conducted. The Initial workshop was to understand the open innovation requirements where participants, alongside firm managers, worked on understanding the end-user and stakeholders' problem definition to take to the crowd. This was followed by a developmental workshop analyzing crowd ideas and their associated business models.

6.2.2. *Data analysis*

The research draws upon a case study approach via a combination of secondary and primary data. This research is grounded in the application of data from crowdsourcing platform interactions in the ecosystem development as well as face-to-face interviews with key actors in the ecosystem and observations over the project. A netnography approach, or online ethnography, refers to a number of related online case research methods that adapt to the study of communities and cultures that are created via computer-mediated social interactions (Kozinets, 2002, 2010). Netnography follows the four main stages of entrée, data collection, analysis, and interpretation, and supplying ethical standards (Kozinets, 2002, 2010). We collected data in netnography, which boils down to the downloading of online content analyzed for theorizing purposes. This online crowdsourcing content contained innovation idea text, videos, photos, chats, collaboration ideas, etc.

Our analysis involved a descriptive analysis of raw interview transcript data based on simple coding, and a conceptual model testing phase to relate emergent empirical and key theoretical concepts across all data sources. In accordance with qualitative research methods, coding of interview data is considered part of data analysis (Miles and Huberman, 1994). Our interview data were analyzed initially using open coding and subsequently using selective coding based on key concepts from our area of study (i.e., business model development, crowdsourcing for innovation, and technological innovation).

6.3. Results

The study includes data from one SME research site, *PhysicalHealing UK Ltd*. Our research focused on crowdsourcing projects to develop business models, the development of a market via service experience design of a newly developed technology. The main aim of the project was to provide rehabilitation engineering technology and related services within the UK. Working with industrial partners in the UK and abroad, the National Health Service (NHS) and a wide-ranging professional network, PhysicalHealing UK Ltd., builds on its vast knowledge and experience to deliver cutting-edge innovative products and clinical knowledge to those in need of orthotic apparatus, diabetic foot care, as well as stroke and spinal cord rehabilitation in the UK market. The SME has a new high-value technology range, which is used by individuals who have a health condition, e.g., recovering from a stroke, or MS, or spinal cord injury. There is a high market potential due to the number of people in the community who suffer from these conditions.

6.3.1. *Component 1 (Initiate): How are we currently creating value with our technologies? What is changing?*

In the initiate phase, directors collaborated with the "university team" to build a common understanding of the objectives of the project and, in particular, on how to make sense of the business problem to take to the crowd. We had two meetings with managers and analyzed secondary data and company reports. In this phase, we defined the needs, obstacles, and opportunities of and for *PhysicalHealing*.

6.3.1.1. *Needs*

PhysicalHealing wanted to find a business model to allow for their expansion into new markets with their current and new technology range. The directors were responsible for the overall strategy and organization of the firm, with the aim of providing rehabilitation engineering technology and related services within the UK. This involved working with a number of partners throughout the UK and abroad, including the NHS and local governmental organizations. In order to capitalize on the firm's expert knowledge and know-how, the directors decided that including services associated with the technology they provide should play a central role in the business. Providing services would give the firm new growth opportunities and an opportunity for profit growth independent of their larger partners. The firm traditionally has diverse ways of delivering services and products. For example, some products were sold directly to the NHS, while other products were sold to businesses who resell products as an expert customer. The problem the firm faced was in understanding how to develop these diverse ways of delivering services and experiences into a scalable business with different experiences and market offerings to customers. Previously, PhysicalHealing's 100% market for its product was orthotics and NHS. Although the firm started with rehabilitation engineering, they saw vulnerability in what they were doing. Hence, they added more specialist rehabilitation technologies that go into the private healthcare market in addition to the NHS. The real challenge faced by the SMEs is that they offer specialist products needing specialist knowledge along with them to be as effective as they can be; for that reason, they faced difficulties in scaling the business. The difficulties related to expanding their reach to cover the whole of the UK, providing high-valued products to people at the far end of the country, while providing the same standard of service. So, scaling their business had been a problem.

As such, the market curve has gradually changed within six years to 70% of new technology for orthotics and NHS, and the remaining 30% of new technology going to people using the technology at home. In the past six years, PhysicalHealing managed to enter a new market by aiming to reach end-users of their products. The following needs were identified in the first two meetings with the management team:

(1) *Scaling the business*: How does PhysicalHealing scale the niche products to reach more mainstream markets, partnering through existing suppliers or forming new partnerships or selling directly to the NHS or others?
(2) *Developing a new service experience for new technology*: Although PhysicalHealing has really interesting products, how can services be developed that complement those products or make those products more complete or attractive by adding services on top of the products? This is what needs to be defined.
(3) *Establishing new markets or conditions*: To comply with the British Government policy and new health conditions, PhysicalHealing needs to develop a new market for the business in a number of different ways.

6.3.1.2. *Obstacles*

PhysicalHealing works daily with patients who experience chronic and debilitating illnesses and injuries. To ease their daily living, the firm applies specialist knowledge with an extensive product line. The business challenge has been to connect externally and develop limited working partnerships with healthcare professionals and carers, families, and end-users of theirs and similar products to gain insight, knowledge, and new ideas of how to develop new services to enhance their existing products. While it is envisaged that the relationship with the NHS will be enhanced, the company will also understand their market threats and weaknesses. The managing director stated, "there's a lot of repeat business, its unfortunately quite concentrated in relatively few businesses to business transactions."

In addition, six years ago, PhysicalHealing started investigating niche technologies in the rehabilitation sector to offer the UK market and developed new, value-adding niche technologies. They now have two high-value technologies, which could serve both the business-to-business market and the business-to-consumer market. As such, these technologies require different ways of trading than what the firm is used to. For these newly introduced technologies, managers stated their challenge by saying "I'm nervous about increasing the volume of prospects we have to deal with unless we can handle what happens at the ground you know the interface with the client. One of the limits to that is how to deliver that personal service effectively at scale at a distance."

The managers also recognize other challenges in how to grow and manage the growth with their high value-adding technologies. This is highlighted by the Managing Director who indicated that "It's very easy to say well, you just hire more people, but it's very difficult to borrow money at reasonable rates, for growth, and we've always, through the whole time of business, we've grown steadily, without external finance, it's just like literally nothing there but the boot straps, so probably we could have grown faster with external finance, but we also would have been in a greater risk inherent with that." So, the main threat for the future for PhysicalHealing is not just managing growth but also managing to scale the business despite competition and changes in the marketplace.

6.3.1.3. *Opportunities*

Due to the new UK Government policy, where healthcare is increasingly being delivered in the community, PhysicalHealing has an opportunity to reach end-users. End-users include patients, NHS managers, or third-party suppliers. This new policy enables PhysicalHealing to offer a new service experience for the patient who is treated at home with their new technology. This is highlighted by the Managing Director who indicated that "if you look at it 5 years ago the coin was flipped because there were people starting to go out into the community to deliver care, you're hardly in hospital now before you're out of hospital. What keeps you at home…is healthcare delivered in the community."

PhysicalHealing has an opportunity to meet patients' expectations, filling the gap of what the NHS is unable to cover, which is reflected by the MD saying "the reality is at the moment there's a gap between the expectations that people have when they are rehabilitated. People have always and are led to believe that the NHS will do everything for you and so if you've had a stroke you'll be treated, but unfortunately the NHS doesn't have the resources to allow people to achieve the potential for recovery. So, there's a gap between people's expectation and the reality of what the NHS can deliver."

6.3.2. Component 2: *Discover and understand and define — Who are we aiming to create value for?*

In this phase, we discover, understand, and define PhysicalHealing's problems and how it creates value for their customers. We held the first workshop, where the managers engaged in learning activity with people from different backgrounds, around the available choices, to inform the best way the firm would work with the crowd (workshop activities, crowdsourcing platform, crowd motivation, crowd activities, and employee/crowd interactions). We engaged in sourcing new ideas for PhysicalHealing's new high value-adding technologies with the participants in the first workshop in order to begin internalizing the ideas through an open dialogue in understanding new directions in value creation.

The workshop aimed to understand the nature of the customers and their problems and try to develop solutions for their problems. A total of 28 participants with diverse backgrounds attended the workshop, including healthcare professionals, biotechnology professionals, designers, inventors, mechanical engineers, academics, marketing professionals, and so on. The workshop started a dialogue with the participants to define key stakeholders and understand the stakeholders' unmet needs widely. We divided the participants into different groups. We had three different activities. In the first activity, each group defined stakeholders and by using the Persona empathy tool (see Figure 6.2) aim to understand how stakeholders are thinking and feeling about the health and work problems they face. Representatives of the stakeholders were involved in the workshop. We asked workshop participants to temporarily place themselves into a sensitive health experience and imagine how somebody (i.e., patient, doctor, NHS employee, patient carer, and so on) else might think and feel in a particular context. We asked workshop participants to discuss and complete a worksheet that reflects upon how a patient (our persona) might feel, think, and do in certain situations.

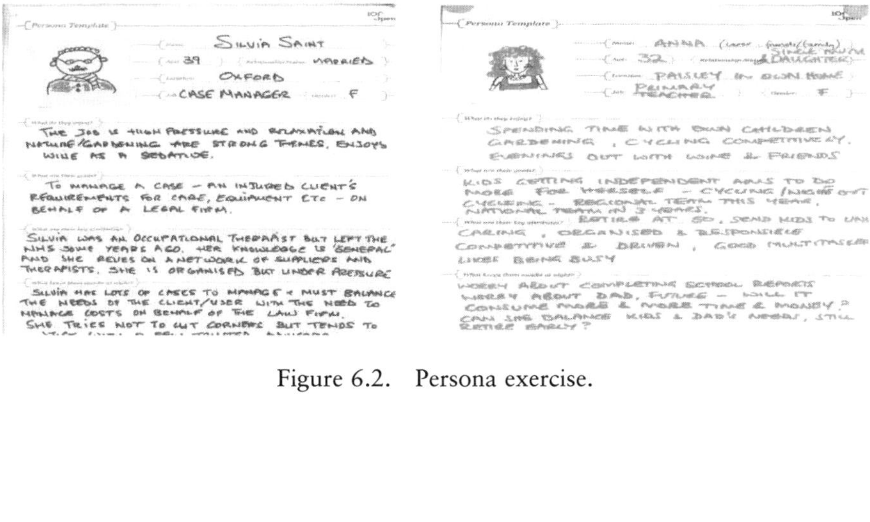

Figure 6.2. Persona exercise.

Personas identify the key stakeholders' problems, expectations, and frustrations, and bring stakeholders to existence by providing them names, personalities, and often drawings.

Participants needed to consider stakeholders' motivations and challenges: What problems do our stakeholders face? Can we find new ways of cooperating and collaborating with the new technology? Why are our stakeholder problems important? Why have their problems not been solved yet? How can you connect your technology to stakeholder problems?

The participants defined the four key stakeholders and their unmet needs as follows:

- **Service Users:** This category consists of the actual end-users. The participants define the following unmet needs for this category:
  - Understanding recovery progress through measurement to motivate service users and family
  - Understanding the trend of users' performance to reduce anxiety while increasing motivation
  - Involving family in the recovery process
  - NHS treatment is often not measured, making it difficult to see progress
- **Carers:** This category consists of people who are friends and families that are providing care to end-users. Their unmet needs are as follows:
  - Availability of appropriate rehabilitation technology within the home
  - Training on the rehabilitation technology
  - Cost of rehabilitation technology
- **Healthcare professional category:** This category consists of doctors, nurses, and others who administer professional healthcare. Their unmet needs are as follows:
  - Limited business knowledge of the wider NHS operation
  - Difficulties in managing stakeholder expectations
  - Wider communication and connection

- **Healthcare managers from medical supplies companies:** Their unmet needs are as follows:
  - Lack of training, education, and experience
  - Access to the right people and equipment at the right time
  - Up-to-date information
  - Managing cases efficiently and effectively
  - Appropriate management systems
  - Personal skills update
  - Increased awareness of new ideas

6.3.3. *Ideate: How do you organize a collaborative, online community?*

In this stage, we used the online platform, "PhysicalHealing.crowdicity.com," to obtain suggestions from the crowd. Crowdicity is a UK-based SME that provides online crowdsourcing tools for facilitating innovation and cocreation among internal and external crowds that vary in size.

The platform ran for six weeks in total. There was one challenge displayed on the platform, the "service experience innovation challenge," and the platform was organized in three phases. First, the crowd defined unsolved problem areas, similar to that described in the second component. We used three problem definitions from the workshop to start conversations between the participants. In the following weeks, users were asked to define solutions to those problems, involving brainstorming and building on each others' ideas; in the final week, the online participants started thinking about implementing the ideas.

PhysicalHealing defined the governance of the platform based on seven criteria (see Table 6.1). The managers decided that their problem was complex, requiring a directional search, with PhysicalHealing holding the ownership of ideas and information on the online platform. Users/crowd were motivated with small gift incentives. PhysicalHealing decided to only invite trusted participants onto the platform; this included trustworthy customers, healthcare professionals, suppliers,

Table 6.1. Online project elements.

Problem complexity — Complex	
Direction of search	Theoretical university team facilitated the conversation
Ownership	Firm — Crowd sign a contract when you enter the platform
How	Collaboration — Crowd could build each others' idea
Channels	Lateral, IT backed, widely advertised
Crowd incentive	Predominantly intrinsic
Who	Broad coverage of those holding relevant knowledge, e.g., Facebook language translation

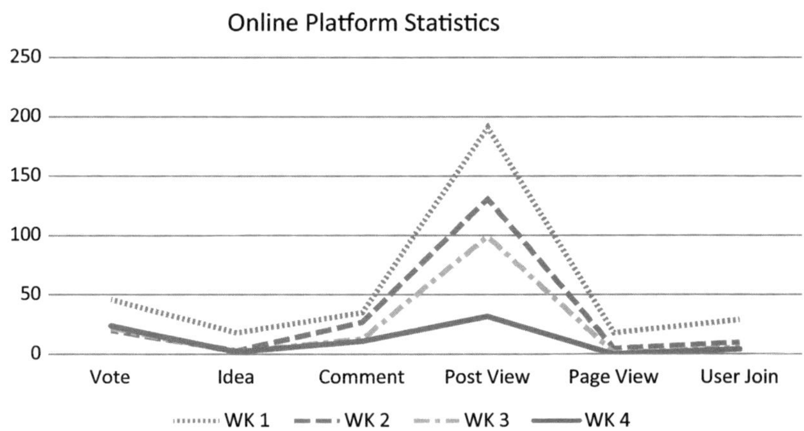

Figure 6.3. Online platform statistics.

university students, and academics from different fields, government partners, and external industry experts.

The platform's architecture design features provided participants two complementary options: (1) Users were able to submit entirely new problems/solutions and ideas and/or (2) build on existing problem statements/solutions, and ideas. All participants were able to comment on existing ideas, which drove cooperation and cocreation. Figure 6.3 shows the platform statistics; when the platform closed, there were 25 ideas generated, with 112 votes for ideas.

The four top voted ideas were as follows:

Rehashare: No one uses a rehabilitation machine all the time; so, it would make sense to share the cost of it with others. But, of course, the problem is finding people with the same therapeutic need in the same region. So, perhaps the company website could help facilitate the organization of some form of "group purchase" of machines or therapy. Patients (or perhaps therapists) could register an interest in being part of a group purchase in the same way sites such as http://www.buyinggroups.co.uk/ do for various consumer items. In the US, there appears to be a site attempting to facilitate the group purchase of "drug rehabilitation" services http://www.rehabs.com/deals/ and others orientated more toward healthcare professionals rather than patients http://www.healthcarebuyinggroup.com/.

Second life of used equipment — remanufacturing: The high value products can be remanufactured. Remanufacturing is one of the product recovery options in which a used product can be brought back to a condition at least equal to that of a new product in terms of quality, performance, and warranty. Used products can be disassembled into components, which then will be cleaned, inspected, and tested to verify that they meet the required quality standards to be reused without further work. Those that do not meet the requirements can be reprocessed via remanufacturing. If this is not possible due to technological issues or economic reasons, they will be replaced with new parts. The remanufactured parts will then be reassembled together with new parts into the product.

Ananet and Anaphone: In order to "delight our clients with service and support at distance" and to scale up, the business needs to create Ananet — a space where product information, therapist training and support materials, and information for carers can be held and disseminated. Creating a portal with a mix of documentation and video content accessible to "business customers," "therapists," "end-users' carers," and "end-users" would create a space for both business-to-business and user community interaction, company mediation, and more. This platform will offer many services. Anaphone is a service that would allow those whose issues were more urgent or less well met by community support to access a specialist or peer by telephone for support and advice. This could be managed by the company, a contractor or via a network of distributed therapists on a paid for, call-back basis.

PhysicalHealing@hospital: Could PhysicalHealing have bases in the hospitals that deal with, for example, spinal cord injury to be closer to the patients and medical staff? A more joined up and fluid service could be formed around the patient who will be able to access the expertise and knowledge within the care facility before going back to the community.

6.3.4. *Design and form: How do the ideas unlock the value embedded in the technology and translate into new experience design for new market offerings?*

In this phase, we organized a workshop to review and assess the crowd ideas value for the new service experience design for SMEs' new high-value technologies. We invited SMEs' managers, three governmental department employees (Innovation Division managers), academics, as well as people from the sectors where the top ideas originated. There were 15 participants. We divided the participants into the three groups.

We asked each group to map the service touchpoints (see Figure 6.4) in order to understand the end-user's journey as well as his/her needs at a certain time and his/her interaction with technology-driven services (see Figure 6.4). Risdon (2013) defined touchpoint as "all of the contact points between the customer and the service provider, which involve an interaction with a human need in specific time and place." Touchpoints are things that people interact with all the time. Web, print, phone, people, anything they can touch, see, hear, taste can be described as a touchpoint. By following the framework of

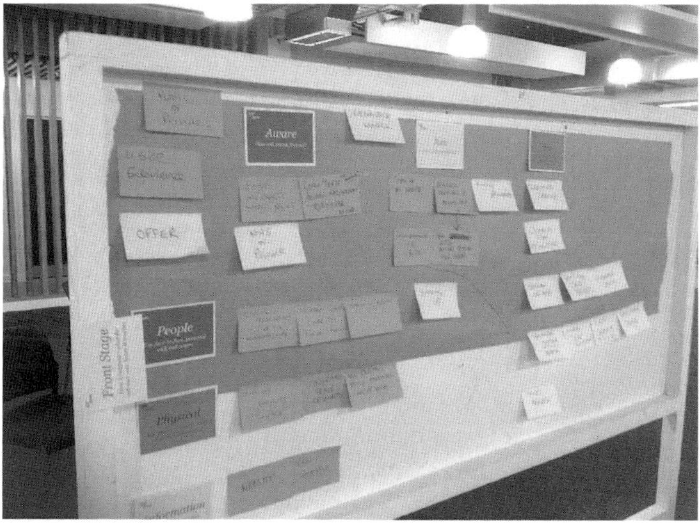

Figure 6.4. Service touchpoints.

Bitner *et al.* (2008), we used the service blueprinting approach to understand where value could be created in different touchpoints by defining five key stages; (1) aware, (2) join, (3) use, (4) continue, and (5) leave. Participants defined each key stage of a service. Amazon's touch points were described as: signing up through the web platform, engaging with the service through browsing, purchasing and delivery. All of these operations are key to making the service work effectively. So, there are five generic stages to service life. Starting with *aware,* how does someone initially engage with our technology-driven service?

Let us consider an idea of how participants develop service blueprints for the SME's technology. For "Ananet," mentioned previously, the intention was to provide an online facility that would deal with many of the pre-issues of service delivery and support people who would seek to interact with PhysicalHealing's technologies. This service could offer different experiences to the end-users, such as Ananetwork, Anaphone, etc. The portal allowed a range of media, text media, images, audiovisuals, and podcasts that would support the communities of practice, with Ananetwork providing a cluster point where information could be gathered, shared, and disseminated in an overlapping set to different communities of interest and practice, and could have supporting infrastructure that would allow discussion, conversation, question, and answer. Anaphone is a guide; so it could be anaSkype or anaVisualCommunityHangout, so that people can have a conversation. So, the idea is very much about creating infrastructure around which the technology can be disseminated, and that portal could support communities; it could support the network, it could support product sale, and it could support remanufacture. It is just a cluster point for that information, but with support of real people doing real stuff with the product and on a 25% billing budget.

Ananet was very much about creating communities of practice, communities of purpose, and discourse within the user base. Ananet could be a facilitated community of users, carers, therapists, and clinicians, who either support the users of the equipment or who would influence the purchase of equipment, but in effect, are a community of interest who have an involvement and an engagement, and that might be facilitated in the other idea. One of the participant stated that "I noticed somebody had communities and community use, which is a kind of enactment of that in the real world, this was more a virtual community of online computer aided supported at a distance, but it ties in, I think there is also an idea coming in at therapists." The Director of PhysicalHealing is conscious about "how to make it vibrant, and with all our communities it's a challenge to get that momentum to see them grow."

6.3.5. *Value proposition*

How do we commercialize our technology in new ways? What are the markets for technology? Who should we serve? How should we meet end-user needs?

Let us consider what, who, and how questions for three ideas (Rehashare, remanufacturing, and Ananet) to develop a value proposition for the SME (see Table 6.2).

Table 6.2. Value proposition questions.

Who should we serve?

- Current customers, who no longer need the equipment or would like to upgrade or have been advised to start work on another part of their body.
- New users who have been advised to use rehabilitation equipment but are price-sensitive.
- Healthcare specialists (doctor, nurse, organizations that provide services, care, etc.).
- Online information platform for equipment users, offering community support, access to healthcare professionals, informational videos, and documents.

What should we offer?

- Providing dynamic, fluid service.
- Easy access to expensive rehabilitation equipment via providing product access at the right time and place.
- Competitive pricing.
- Convenient locations.
- Professional support.
- Risk reduction.
- Opportunity to sell back used equipment.
- Accessible to a wider range of people.
- Convenient and usable service.
- Easily accessible specialized information via Ananet.
- Ease of access to expensive rehabilitation equipment.
- Range of payment options.
- No health board waiting list — flexible booking system.

How should we offer?

- Define different customer touchpoints and develop services for each touchpoint.
- Collaboration with stakeholders, for example, partnering with healthcare specialists, pharmacy chains, current hospital partnerships, support groups for space, and access to customers, local authorities for space, and so on.
- Using key resources (e.g., physical space and human resources) and developing key competitiveness around each touchpoint service.
- Defining different places to use services.
- New marketing strategy for the "old" and "new" products.
- Develop human resource competences to monitor, update, and communicate across the firm, customers and partners, linking to social media.

In explaining how *PhysicalHealing* developed value propositions for the top ideas following the crowd-based business model, we find that the firm is able to propose *new offerings* by combining the ideas (reshare, Ananet, remanufacture) to offer new services for its technology. The idea behind Ananet could be linked to the remanufacture idea via the production idea or sharing idea. As such, offering a community of practice web portal would allow the SME to apply the reuse of equipment idea based on the "sharing economy" principle. The sharing economy principle is about building socioeconomic ecosystems in order to share ideas, experiences, physical resources (i.e., nonuse equipment), as well as intellectual resources (i.e., knowledge about the particular health conditions and available technologies and their usage). The equipment may no longer have utility to its owner due to the changing condition (i.e., sadly people die or their conditions worsen or people recover). As such, *PhysicalHealing* can offer a web portal to develop more economies of scale/more leverage to negotiate/preserve price.

Through ideas from the crowd, *PhysicalHealing* can reach new customers and enter new market segments. This enables the firm to apply a diversification strategy to offer new services in addition to new technology. This also includes *new channels*; *PhysicalHealing* finds new channels to sell its technology via offering a new experience design through community of practices as well as sharing technology concept. To enable the Anatrade community of practice deployment to happen, *PhysicalHealing* considers including several organizations, which have the ability to act as channels to system designers, producers, integrators, and potential end-users (patients, carers, health specialists). These channel partners will be intimately involved in the test phase of the project and will also act as the primary stage-gate manager in order to ensure that the evolving outputs are relevant to the wider healthcare sector.

- **New offerings:** *PhysicalHealing* combined three ideas (reshare, Ananet, remanufacture) to offer new services for its technology. As such, offering a community of practice web portal would allow SME to apply the reuse of equipment idea based on the "sharing economy" principle. The sharing economy principle is about building socioeconomic ecosystems in order to share ideas, experiences, physical resources (i.e., nonuse equipment) as well as intellectual resources (i.e., knowledge about the particular health conditions and available technologies and their usage). The equipment may no longer have utility to its owner due to the changing conditions (i.e., sadly people die or their conditions worsen or people recover). As such, *PhysicalHealing* offers a web portal to develop more economies of scale/more leverage to negotiate/preserve price.

- **New customers and segments:** *PhysicalHealing* is able to apply a diversification strategy and begin offering new services for users in addition to technology.
- **New channels:** *PhysicalHealing* finds a new channel to sell its technology via offering a new experience design through community of practices as well as sharing technology concept. To enable the Anatrade community of practice deployment to happen, *PhysicalHealing* considers including several organizations which have the ability to act as channels to system designers, producers, integrators, and potential end-users (patients, carers, health specialists). These channel partners will be intimately involved in the test phase of the project and will also act as the primary stage-gate manager in order to ensure that the evolving outputs are relevant to the wider healthcare sector.

6.4. Discussion and Conclusions

Our chapter analyzed how the crowd-based service experience creation process is organized and how different social actors (i.e., crowd) participate in the creation. Our resulting insights contribute to the theory of a business model innovation process. Through developing a crowdsourcing approach, our approach facilitated the owners and managers of an SME to seek sources of inspiration and expertise outside of their normal circle when constructing or reconstructing their business models — and, indeed, when implementing them. We believe our framework can enable SMEs in the effective visualization, communication, and tracking of their business model by allowing SMEs to structure the various components of the business model and map the dynamic relations/links between these components.

The focus on crowdsourcing as one of the options for developing new business models shows that the project does not merely seek to stimulate innovation across and within SMEs but also to be innovative in the stimulating of this process of strategic and commercial change via introducing experience service design. Moreover, collective intelligence helped to understand what firms can share and where the technological platform improves the flow of information inspiration, and resources, across innovation ecosystems. Also well-validated information may be more important than greater flows and connections. We could claim that the impact of the crowd-based open innovation is designed to cut across the three main flagship initiatives embedded within new service design in the following ways:

- **Smart growth:** By designing a technology-driven service design, the SME can manage to start offering new experience via different collaborations.
- **Sustainable growth:** To foster a community practice that is both socially.
- **Inclusive growth:** To develop new innovation capabilities and skills via the community of practice idea that can operate as an effective platform by including patients, carers, and healthcare professionals.

Reflecting back to literature, we can conclude that our framework has three generative mechanisms that explain the business model innovation process by which new services are constructed: context, cognition, and practice (Gupta and Vajic, 2000). Typically, our new service development involves more than one mechanism. Context refers to physical and relational settings where social interactions and activities take place (Bitner, 1990; Gupta and Vajic, 2000). This is captured in the first workshop through the Persona's activity. Cognition refers to "social process and locates it in integrated whole agency, action, and setting" (Gupta and Vajic, 2000,

p. 41). Cognition (i.e., the process of constructing knowledge) and context (e.g., physical, social, and cultural background) generate and transform each other, within a dialectical relationship. Our crowdsourcing platform develops the cognition mechanism. Finally, practice captures the interrelationship between context and cognition and enables the configuration of an individual's service experience to be observed. This is captured via the service blueprinting design. It means understanding socially defined tasks organized into specific cultural settings (Johnston *et al.*, 2000). These mechanisms broaden the new service development process from individual activity to social activity, from contextual enhancement to cognitive processing that is inseparable from the ongoing actions.

The framework is particularly powerful in strengthening professional connections between members of the community. These interactions within the community are conducive to nurturing new knowledge, stimulating innovation, and sharing existing tacit knowledge within the group. A community of practice has three structural elements: a domain of knowledge, a community of people who care about this domain, and the shared practice that they are developing to be effective in their domain (Wenger *et al.*, 2002). The domain creates common ground and a sense of common identity and purpose. A well-identified domain legitimizes the community by affirming its purpose and the related value for participants. It inspires members to contribute and guides their learning process. This is reflected in the idea "*Ananet*" to launch a community of practice via linking other two ideas (i.e., reshare, remanufacture) to Ananet. The term refers to formal and informal groups of practitioners into which newcomers enter to acquire the sociocultural practices of the community. More recently, communities of practice have become associated with knowledge management. The community model recognizes that individuals (i.e., doctors, researchers, health specialists) have a strong intrinsic motivation to share ideas. Its communitarian approach builds on coownership and common vision. It values knowledge already present at the regional level.

The directors of PhysicalHealing were very satisfied with the program that was put in place; from the outset they identified that they were ready to "open their minds" and that's "why we're standing here today" to "develop not only within Scotland but into the rest of the United Kingdom into Europe." The research has demonstrated this through constant interaction between the company, Strathclyde University, and various actors like Scottish Enterprise, Edinburgh Council, and even starting new complementary collaborations with other healthcare firms from the process. This is an area which the SME has started to "take much more seriously," and through the conversations with the Strathclyde partners we're able to face the "increasing business challenges with confidence." The learning and future delivery of the program has been so successful that "we are all victims of our own patterns, habits and ways of thinking" so the chance for PhysicalHealing to share the experience, open their boundaries, and use the collective intelligence of many was "immensely valuable in that you have changed our thinking enough to make a positive difference to the company, this is very different to what we've ever done, we never sit down and think in such a radical way and it has given us the opportunity to step away from what we do every day, learn and come up with better ideas...this has been very valuable, something we've learned from."

Acknowledgments

The authors are grateful for the useful comments, insights, and guidance gathered from the co-owner of 100% Open, Roland Harwood. This study benefited from funding from the Scottish Enterprise Innovation Scheme (HE Grant Line: 123104 Strategic Funding — Knowledge Exchange Project Code: H08009 Innovation Vouchers and EPSRC doctoral training program).

References

Anderson, R. W., Acur Bakir, N. and Corney, J. (2016). Working with the crowd in business model innovation: *Crossing Borders and Boundaries: The Changing Role of Innovation/Entrepreneurship*. University of Strathclyde, Glasgow, 12–14 June. European Institute for Advanced Studies in Management, Brussels, Belgium.

Antorini, Y. M., Muñiz Jr, A. M. and Askildsen, T. (2012). Collaborating with customer communities: Lessons from the LEGO Group. *MIT Sloan Management Review*, 53(3), p. 73.

Bayus, B. L. (2013). Crowdsourcing new product ideas over time: an analysis of the Dell IdeaStorm community. *Management Science*, 59(1), pp. 226–244.

Berry, L. L., Carbone, L. P. and Haeckel, S. H. (2002). Managing the total customer experience. *MIT Sloan Management Review*, 43(3), p. 85.

Bianchi, M., Campodall'Orto, S. C., Frattini, F. and Vercesi, P. (2010). Enabling open innovation in small and medium sized enterprises: how to find alternative applications for your technologies. *R&D Management*, 40, pp. 414–431

Bitner, M. J., Ostrom, A. L. and Morgan, F. N. (2008). Service blueprinting: a practical technique for service innovation, *California Management Review*, 50, p. 66.

Chanal, V. and Caron-Fasan, M.L. (2010). The difficulties involved in developing business models open to innovation communities: the case of a crowdsourcing platform. *Management*, 13(4), pp. 318–340.

Chesbrough, H. W. (2006). *Open Innovation: The New Imperative for Creating and Profiting from Technology*. Harvard Business Press. United States of America.

Chesbrough, H. (2010). Business model innovation: opportunities and barriers. *Long Range Planning*, 43(2), pp. 354–363.

Chesbrough, H. and Rosenbloom, R. S. (2002). The role of the business model in capturing value from innovation: evidence from Xerox Corporation's technology spin-off companies. *Industrial and Corporate Change*, 11(3), pp. 529–555.

Chesbrough, H., Vanhaverbeke, W. and West, J. (2006). *Open Innovation: Researching a New Paradigm*. Oxford University Press, on demand.

Demil, B., Lecocq, X., Ricart, J.E. and Zott, C. (2015). Introduction to the SEJ special issue on business models: business models within the domain of strategic entrepreneurship. *Strategic Entrepreneurship Journal*, 9(1), pp. 1–11.

Eisenhardt, K. M. (1989). Building theories from case study research. *Academy of Management Review*, 14(4), pp. 532–550.

Gupta, S. and Vajic, M. (2000). The contextual and dialectical nature of experiences. *New Service Development: Creating Memorable Experiences*, In: Fitzsimmons, J. A. and Fitzsimmons, M. J. (ed.) New Service Development, Creating Memorable Experiences. USA. Sage. pp. 33–51.

Henry, C., 2003. *Open Innovation: The New Imperative for Creating and Profiting from Technology*. Boston, MA, Harvard Business School.

Hoch, S.J. and Deighton, J. (1989). Managing what consumers learn from experience. *The Journal of Marketing*, 53(2), pp. 1–20.

Huang, Y., Vir Singh, P. and Srinivasan, K. (2014). Crowdsourcing new product ideas under consumer learning. *Management Science*, 60(9), pp. 2138–2159.

Johnston C. G., James R. H., Lye J. N. and McDonald I. M. (2000). An evaluation of collaborative problem solving for learning economics. *Journal of Economic Education*, 31(1), pp. 13–29.

Kozinets, R. V. (2002). The field behind the screen: using netnography for marketing research in online communities. *Journal of Marketing Research*, 39(1), pp. 61–72.

Kozinets, R. V. (2010). *Netnography: Doing Ethnographic Research Online*. Sage Publications, London.

Lauto, G., Valentin, F., Hatzack, F. and Carlsen, M. (2013). Managing front-end innovation through idea markets at novozymes. *Research-Technology Management*, 56(4), pp.17–26.

Lettl, C., Herstatt, C. and Gemuenden, H. G. (2006). Users' contributions to radical innovation: evidence from four cases in the field of medical equipment technology. *R&D Management*, 36(3), pp. 251–272.

Majchrzak, A. and Malhotra, A. (2013). Towards an information systems perspective and research agenda on crowdsourcing for innovation. *The Journal of Strategic Information Systems*, 22(4), pp. 257–268.

Miles, M. B. and Huberman, A. M. (1994). *Qualitative Data Analysis: A Sourcebook*. Beverly Hills, CA, Sage Publications.

Mount, M. and Martinez, M. G. (2014). Social media. *California Management Review*, 56(4), pp. 124–143.

Parida, V., Westerberg, M. and Frishammar, J. (2012). Inbound open innovation activities in high-tech SMEs: the impact on innovation performance. *Journal of Small Business Management*, 50(2), pp. 283–309.

Pine, B. J. and Gilmore, J. H. (1998). Welcome to the experience economy. *Harvard Business Review*, 76, pp. 97–105.

Pine, I. I., B. J. and Gilmore, J. H. (1999). *The Experience Economy: Work is Theatre & Every Business a Stage*. Harvard Business School Press. Boston, Massachusetts.

Poulsson, S. H. and Kale, S. H. (2004). The experience economy and commercial experiences. *The Marketing Review*, 4(3), pp. 267–277.

Risdon, C. (2013, December 2). Unsucking the touchpoint. Retrieved June 20, 2015, from http://www.adaptivepath.com/ideas/un-sucking-the-touchpoint/.

Soloviev, V. I., Kurochkin, P. A., Rendiuk, A. V. and Zazuk, A. V. (2010). Innovative business models in the media industry. *Annales Universitatis Apulensis Series Oeconomica*, 12(2), p. 2010.

Strauss, A. L. (1987). *Qualitative Analysis for Social Scientists*. Cambridge University Press, Cambridge.

Teece, D. J. (2010). Business models, business strategy and innovation. *Long Range Planning*, 43(2), pp. 172–194.

Van de Vrande, V., De Jong, J. P., Vanhaverbeke, W. and De Rochemont, M. (2009). Open innovation in SMEs: trends, motives and management challenges. *Technovation*, 29(6), pp. 423–437.

Vanhaverbeke, W. (2017). *Managing Open Innovation in SMEs*. Cambridge, Cambridge University Press, Cambridge.

Voss, C., Roth, A. V. and Chase, R. B. (2008). Experience, service operations strategy, and services as destinations: foundations and exploratory investigation. *Production and Operations Management*, 17(3), pp. 247–266.

Wenger, E., McDermott, R. A. and Snyder, W. (2002). *Cultivating Communities of Practice: A Guide to Managing Knowledge*. Harvard Business Press, Boston, Massachuesetts.

Yin, R. (1984). *Case Study Research*. Beverly Hills, CA, Sage Publications.

Zott, C., Amit, R. and Massa, L. (2011). The business model: recent developments and future research. *Journal of Management*, 37(4), pp. 1019–1042.

Part III

SMEs in Inter-organizational Networks and Innovation Ecosystems

Chapter 7

Managing SMEs' Collaboration Across Organizational Boundaries Within a Regional Business Ecosystem

Agnieszka Radziwon and Marcel Bogers*

**agra@btech.au.dk*

Abstract

Small and medium-sized enterprises (SMEs) often engage in relationships with external partners that complement their activities in innovating their products and services in order to maintain their market share in a competitive landscape. In this chapter, we present a case that focuses on SMEs' perspective on managing and organizing intercompany collaboration within a regional business. We explore how purposefully managed mutual knowledge flows across organizational boundaries applied by SMEs contribute to the development of

* Corresponding author.

the ecosystem they are immersed in. Our key findings include insights into the advantages offered by being embedded within regional ecosystem boundaries, such as low transaction costs and reduced risks of opportunistic behaviors, environmental opportunities for external knowledge sourcing, or increase of the sphere of influence. We also highlight accompanied challenges, such as low autonomy and control over the partners, and indicate shared issues and responsibilities crucial for further ecosystem development. As a conclusion, we offer a set of recommendations both for managers and policymakers concerning general organizational requirements and governing structures.

7.1. Introduction

Scholars have been exploring the notion of "ecosystems" from various perspectives. Studies in this domain propose a conceptual basis (Moore, 1993), discuss the role of large firms as orchestrators (Rohrbeck *et al.*, 2009), or investigate the interaction between different stakeholders more generally (van der Borgh *et al.*, 2012). However, research has so far put only limited attention on the specific role of SMEs (Bogers *et al.*, 2017; Brunswicker and van de Vrande, 2014; Vanhaverbeke, 2017), even though these may often represent the largest group of ecosystem stakeholders and significantly contribute to the economic activities in any region. In 2015 alone, small and medium-sized enterprises (SMEs) employed 90 million people in the European Union (EU), which constituted two-thirds of the EU employment and generated value added of €3.9 trillion that accounted for almost three-fifths of the EU value added in the nonfinancial business sector (Muller *et al.*, 2016). Therefore, increasing understanding of the role of SMEs and ways of how they could contribute to the development of their ecosystems is of significant interest to academics, practitioners, and policymakers.

The innovation management literature identifies four main types of ecosystems: business ecosystems (Moore, 1993), innovation ecosystems (Adner, 2006), entrepreneurial/entrepreneurship ecosystems (Prahalad, 2005), and knowledge-based ecosystems (van der Borgh *et al.*, 2012). Additionally, putting the natural sciences references aside, one can also come across service ecosystems (Lusch and Vargo, 2014) in marketing and technology/digital ecosystems (Feijóo *et al.*, 2009) in computer science. In this chapter, we provide a study of a regional business ecosystem, with emphasis on interfirm relationships between various SMEs. We conceptualize the notion of business ecosystems as loose networks of interdependent stakeholders (Iansiti and Levien, 2004) who jointly coevolve their capabilities and roles (Moore, 1996) around a new innovation (Moore, 1993). More specifically, we delineate a regional business ecosystem (from now on referred to as ecosystem) through four factors: interdependencies (expressed through the presence of strong ties between both the firm's immediate and extended customer–supplier networks), coevolution (expressed by both formal and informal previous collaboration experiences), proximity (which covers the regional aspect through its geographical dimension), and the presence of an orchestrator. Even though a territory where the ecosystem is located does not need to be limited to a geographical region, prior research shows that knowledge flows are indeed geographically localized (Glückler *et al.*, 2017). Thus, the regional perspective allows scrutinizing open innovation practices performed both on the interfirm and ecosystem levels. Our research builds on prior findings of embracing the open innovation paradigm (Rohrbeck *et al.*, 2009), value creation (van der Borgh *et al.*, 2012), and value capturing mechanisms in the ecosystem context (Ritala *et al.*, 2013).

While Chesbrough and Bogers (2014) recently clarified and developed the open innovation definition as "a distributed innovation process based on purposefully managed knowledge flows across organizational boundaries" (p. 17), the notion of boundaries itself has not received much attention in this context. Given the definition, considering the boundary notion puts emphasis on the knowledge as a medium of interaction (Nambisan and Baron, 2013) and potential source of competitive advantage (Zahra and Nambisan, 2012), which may be favored by spatial proximity (Tödtling and Kaufmann, 2001) and thus constituting an inherent part of any ecosystem. Due to the size (small) and overall complexity of operational activities (relatively low in comparison to large firms), focusing on SMEs (SME level of analysis) allows us to distinguish between and analyze organizational boundaries along with the boundaries of an ecosystem (ecosystem level of analysis). For this purpose, in order to analyze how open innovation practices are applied by SMEs in an ecosystem, we build on Santos and Eisenhardt's (2005) concept of organizational boundaries as an innovation frontier of any organization.

Since the boundary issue and its understanding are of particular importance to any system or organization (Gulati *et al.*, 2012; Ritala and Almpanopoulou, 2017), Santos and Eisenhardt (2005) not only develop four boundary conceptions (efficiency, power, competence, and identity) along with their distinctive features, but also indicate that the investigation of these boundaries in an ecosystem is one of the possible future research directions. More specifically, we are interested in understanding where exactly these boundaries between ecosystem members intersect as well as what would be the managerial implications of those interactions. We explore these questions using a "problem-driven boundary phenomena" approach (Santos and Eisenhardt, 2005) by exploring SMEs' perspective of an ecosystem, through inductive exploratory research. The research question we address is: How do purposefully managed mutual knowledge flows across organizational boundaries applied by SMEs contribute to the development of the ecosystem they are immersed in? The research specifically explores the ecosystem and its selected members, namely SMEs, by addressing their role in the development of this ecosystem of complementary partners (Adner and Kapoor, 2010; Moore, 1993; Santos and Eisenhardt, 2005). We analyze an ecosystem with a focus on six embedded SME cases. Our primary data are 23 semistructured interviews with chief executive officers (CEOs), managing directors, or co-owners of these firms. We complement it with other field data, observations, and publicly available secondary data.

Our findings inform the general conceptualization and specific intersections of organizational boundaries where open innovation may successfully take place in a safe ecosystem environment. The primary contribution of this chapter is to improve the understanding of the type and expression of organizational boundaries as innovation frontiers of both the firm and the ecosystem. Our findings support the claims of Gulati *et al.* (2009) about relational capabilities that could be enriched by partners' accumulated experience. In the case of ecosystems, it is expressed even stronger than in alliances or joint ventures, due to interdependencies and coevolution history among ecosystem members. Our findings highlight the importance of dedicated management of interorganizational collaboration across and within regional ecosystem boundaries. Therefore, we offer a set of practical guidelines for managers, which could help them in establishing new collaborative initiatives in a safe, open innovation environment. Our results suggest that properly identified ecosystem boundaries could be an important element of strategic planning for the organizations that are part of it.

This chapter continues with reviewing the conceptual context of the study and the applied research strategy. Subsequently, we report the results of our analyses followed by a discussion of the findings. We conclude with highlights of how these insights extend current thinking and briefly discuss the research implications, limitations, and further research outreach.

7.2. Background

7.2.1. *Collaborative innovation in ecosystems*

Most contemporary ecosystem concepts are conceived as communities of interconnected and interdependent stakeholders (Adner and Kapoor, 2010; Adner, 2006; Iansiti and Levien, 2004; Moore, 1993; van der Borgh *et al.*, 2012) located at a particular territory (Iansiti and Levien, 2004), which is not necessarily limited to a region (Autio *et al.*, 2014), but also could be a virtual platform (Mäkinen *et al.*, 2014).

Many ecosystems are centered on a platform that ecosystem members use to stimulate their innovative performance (Iansiti and Levien, 2004). Competitors and complementors collaborate to create value both for them (as firms) and for the ecosystem. Usually there is an orchestrator (Adner, 2006) or a keystone (Iansiti and Levien, 2004) that supports both the development of the ecosystem at large and the enhancement of its performance through innovation (Zahra and Nambisan, 2012). This role could be played by a large firm (Adner and Kapoor, 2010), a university, a public research organization, or a governmental institution (van der Borgh *et al.*, 2012), which does not fully focus on ecosystem management, but rather establishes a fruitful environment through business-friendly policies. Ecosystems consist of different stakeholders, such as suppliers, customers, competitors, universities, and other complementors, who contribute to the process of creating value (Adner and Kapoor, 2010; Eisenhardt and Galunic, 2000; West and Bogers, 2014). These stakeholders (called loosely interconnected participants (Iansiti and Levien, 2004, p. 5)) usually have relatively strong connections to the extent that they may strongly "depend on one another for their effectiveness and survival" (Iansiti and Levien, 2004, p. 5).

The regional aspect is mostly tackled by (regional) innovation systems or the (regional) clusters literature, which highlights the benefits of localized learning reaped from clustering (Malmberg and Maskell, 2002). This special territorial configuration can enhance knowledge inflows and outflows focused on improving innovation performance (Chesbrough, 2003). As the innovation system literature focuses on knowledge flows developed by research institutes and universities (Asheim and Coenen, 2005) as well as clusters, the ecosystem perspective that we present in this chapter covers all sources of knowledge (including interfirm spillovers). What is more, we take into consideration not only formal but also informal partnerships, which constitute an integral part of ecosystems. In this sense, critical-mass industrial clusters (Oh *et al.*, 2016), which usually are formed by a top-down approach from policymakers, driven or mediated by governmental and nongovernmental actors, and have more or less formal structures, could be part of an organically grown ecosystem.

According to Valkokari (2015), ecosystem boundaries could be indicated via dimensions like geographical (e.g., local, regional, or global), temporal (e.g., static or dynamic), permeable (e.g., open or closed), or flow types (e.g., knowledge, value, or material). We conceptualize the notion of an ecosystem that helps us to identify a proper research setup based on four factors derived from the literature, which sum up this section. These are interdependencies (expressed through the presence of strong ties both between a firm's immediate and extended customer–supplier networks), coevolution (expressed by both formal and informal previous collaboration experiences), proximity (which covers the regional aspect through its geographical dimension), and the presence of an orchestrator. What is more, we focus on an open system with a free knowledge flow. These factors serve as guidelines to identify initial ecosystem boundaries.

Nevertheless, Moore's (1993) key argument behind the introduction of an ecosystem was a lack of guidance for managers seeking to explore and better understand the relationships between innovation, firms, and change in already existing frameworks related to networks, in the context of strategic alliances and virtual organizations. A natural evolution of alliances, portfolios, and networks into ecosystem or the so-called meta-organizations (Gulati *et al.*, 2012; Leroi-Werelds *et al.*, 2017) can be observed in the increasing number and type of involvement of partners (including both competitors and complementors), the degree of joint strategizing, and the formalization of partnerships as well as the duration of collaboration. Ecosystems are comprised of interorganizational networks that ensure resources and knowledge exchange in order to provide competitive advantage (Zahra and Nambisan, 2012). Studies have shown that interorganizational collaboration can be an important driver for innovation performance (Powell *et al.*, 1996). Accordingly, firms embedded in networks can leverage their external environment to achieve better innovative output (Shan *et al.*, 1994).

In the context of open innovation, knowledge becomes an important medium of interaction between the members of an ecosystem. Firms, especially SMEs, located in an ecosystem build their competitive advantage through collaboration with complementary partners (Adner and Kapoor, 2010; Nambisan and Baron, 2013) or thanks to the application of knowledge available in the region (van der Borgh *et al.*, 2012). The most important resource for innovation and its exchange — knowledge — is favored by spatial proximity (Tödtling and Kaufmann, 2001); thus, a regional ecosystem may constitute a fruitful ground to study open innovation in general, especially in a loose network of interdependent SMEs, which jointly coevolve their capabilities in order to develop new innovative products and services. There may be a particular challenge for SMEs to develop such a network given that they may, to a large extent, become dependent on external sources of innovation to complement their internal knowledge base. Existing ecosystem literature does not fully explain the role of SMEs in a wider network of complementary partners. Small firms are usually suppliers to the keystone or orchestrator, which typically constitutes the center of researchers' attention. As such, SMEs are challenged to find new ways to organize their innovation activities in a larger context than their current operational scale (van de Vrande *et al.*, 2009; Vanhaverbeke, 2017). Embeddedness in a regional ecosystem may significantly lower the risk of opportunistic behaviors, thus encouraging more open knowledge flows within the ecosystem.

Successful innovation increasingly relies on a more open approach toward obtaining, integrating, and commercializing external sources of knowledge (Chesbrough, 2006; Laursen and Salter, 2006; West and Bogers, 2014). Since "innovative businesses can't evolve in a vacuum" and "they must attract resources of all sorts, drawing in capital, partners, suppliers, and customers to create cooperative networks" (Moore, 1993, p. 75), open innovation practices may be easier to implement in an ecosystem environment. Close interactions as well as mutual interdependencies between ecosystem actors should assure the creation of innovative outcomes developed in value cocreation processes (Gastaldi *et al.*, 2015; Mäkinen *et al.*, 2014; Vargo and Lusch, 2011). This process should be stimulated and supported by policymakers and entrepreneur support networks (Autio *et al.*, 2014).

7.2.2. Organizational and ecosystem boundaries

Managing organizational boundaries (of both ecosystems and companies that constitute its members) may be critical in creating an overall context in which economic activities are organized (Aldrich, 1999). Santos and Eisenhardt (2005) develop four boundary conceptions (efficiency, power, competence, and identity) (see Table 7.1), which cover the span from complementary to coevolutionary and synergistic relationships.

Efficiency conception with a legal-ownership view of boundary decisions (Santos and Eisenhardt, 2005) is strongly linked to various types of transactional attributes, which both in the short and long run affect the efficiency of governance mechanisms (Felin and Zenger, 2014). Taking into consideration that SMEs are often the largest group of actors in ecosystems, one has to deal with a paradox of high assets specificity (Williamson, 1981) and market governance occurring thanks to reduced environmental uncertainty. Usually, high asset specificity increases the number of opportunistic agents; however, ecosystem embeddedness enforces fair play within the ecosystem boundaries. Interdependence among ecosystem partners (Adner, 2006; Moore, 1993) and transactions could also, in the case of ecosystems, lead to market governance instead of traditionally assumed hierarchical governance (Nickerson and Zenger, 2004) due to interconnection advantages and needless of any special protection mechanisms within the ecosystem boundaries.

Power conception, in contrast to efficiency, emphasizes the sphere of organizational influence, and is thus strongly linked to control of both direct and indirect exchange relations (Santos and Eisenhardt, 2005). Organizations, which naturally aim at uncertainty reduction, should try to set the boundaries "at the point that maximizes strategic control over crucial external forces" (Santos and Eisenhardt, 2005, p. 495). Reducing dependence and increasing power may be very challenging especially for SMEs in an ecosystem context, which provides benefits on the grounds of strong interdependencies of its members. Therefore, instead of ownership mechanisms that may be

Table 7.1. Organizational boundary concept.

Conception	Key definition(s)	Key focus and tools	Expression
Boundaries of efficiency	"The efficiency conception asks whether a transaction should be governed by a market or organization" (Santos and Eisenhardt, 2005, p. 492)	Conducting transaction inside or outside an organization should focus on minimization of governance costs of exchange. Boundary management tools center around hierarchical vs. market mechanisms.	Transaction attributes: • Environmental uncertainty • Behavioral uncertainty • Technological uncertainty • Interdependence • Asset specificity
Boundaries of power	"The power conception asks what is the appropriate sphere of organizational influence" (Santos and Eisenhardt, 2005, p. 495)	Direct or indirect involvement in exchange relations should focus on maximization of control over strategic relationships. Boundary management tools center around ownership vs. control mechanisms.	• Degree of dependence • Network position
Boundaries of competence	"The competence conception asks what resources the organization should possess." (Santos and Eisenhardt, 2005, p. 497)	Resource configuration and coevolution with environmental opportunities should focus on maximization of the value of firms' resource portfolio. Boundary management tools center around externally and internally oriented dynamic capabilities.	• Resource proximity • Dynamic capabilities • Knowledge resources
Boundaries of identity	"The identity conception asks who the organization is." (Santos and Eisenhardt, 2005, p. 500)	Resolving identity inconsistencies between organizations and its activities should focus on coherence. Boundary management tools center around conscious vs. unconscious mechanisms.	• Cognitive frames and coherence • Emotional coherence • Employee rotation process

relevant for large players, SMEs may choose to influence other organizations through a set of nonownership mechanisms including both collaborative and cooperative behaviors (Powell *et al.*, 1996). In this way, the sphere of organizational influence could be extended without extending the legal boundaries (Pfeffer and Salancik, 1978).

The key conceptual considerations relate to the key competencies that an organization should possess, thus focusing on the resource portfolio in general as well as its related configuration (Santos and Eisenhardt, 2005). Since a lack of resources is usually listed as one of the key challenges of SMEs, environmental opportunities that help to coevolve resources may constitute an important boundary-related decision for these ecosystem members (Gulati *et al.*, 2009). These environmental opportunities may emerge along with new ecosystem initiatives. Nevertheless, access to new knowledge and resources may incentivize SMEs to join these activities, but at the same time could be one of the biggest challenges.

The concept of identity centers on the mind-set by which organizational members understand who they are. Santos and Eisenhardt (2005, p. 11) indicate that the organizational boundaries should assure "coherence between the identity of the organization and its activities." The perception of identity and in particular collective sense-making (Weick, 1995) may have a very subjective dimension. It is usually driven both by managerial cognition as well as shared values and norms that serve as information filters (Prahalad and Bettis, 1986). Since social relationships formed both inside firms and inside the ecosystem are important for determining knowledge flow patterns (Agrawal *et al.*, 2006), the presence and location of the boundary of identity can significantly stimulate or hinder open innovation practiced in an entire ecosystem.

There is an ongoing debate around the organization and management of activities inside and outside the boundaries of the firm (Kapoor and Lee, 2013). Scholars have started to turn their attention to innovation networks of actors involved in developing and in commercializing innovations. Moreover, Bogers *et al.* (2017) underline the challenge to innovate and preserve the collective welfare of the ecosystem, while enhancing the individual performance of its members. That is why, we focus on exploring how open innovation practices are applied by SMEs located in an ecosystem. Analyzing the organizational boundaries of the smallest ecosystem members as well as the regional context of this research enables us to zoom into the exact places where these boundaries between ecosystem members intersect and processes of managing and organizing interorganizational collaboration, which could potentially result in further development of an ecosystem.

7.3. Research Design

In order to gain a deeper understanding of both the organizational and the ecosystem boundaries, we considered a case study approach as the most suitable research method that will help to identify patterns among various relationships as well as to help incorporate these into a theoretical model (Eisenhardt, 1989; Eisenhardt and Graebner, 2007). The theoretical sample was chosen based on several criteria. First, the main empirical basis of this study was planned to be face-to-face interviews (ideally combined with a company/factory visit, which could allow for further observations) conducted with an organization that will be part of an ecosystem. Therefore, the organizations invited for participation in this research had to be located within certain proximity of the central city in the region (Freel, 2003; Sternberg, 1999), which in this case is Sønderborg, located in the Region of Southern Denmark. Second, we focused on small and medium-sized manufacturing companies (covering stages from development, production to commercialization and related services). Due to a lack of research focused on the "small end" of the SME range, we have targeted companies with up to 100 employees. Third, in order to assure that these firms are a part of an ecosystem (fulfilling the interdependency and potentially also coevolution criteria described in Section 7.2.1), the selected organizations should actively engage in some of the local networking activities (like clusters, university-related projects, etc.). The company database was created based on the consultation with the local southern Danish clusters and industry contact point at the university. Altogether, 12 organizations decided to accept our invitation and took part in our study.

7.3.1. *Data collection*

The data collection protocols were carefully designed in order to strengthen reliability and internal and external validity by: applying methods, interviewers, and data triangulation; collecting feedback to interview transcripts and general findings; developing formal interview guidelines; and assigning different roles to different researchers (Eisenhardt, 1989; Yin, 2009). As for the data collection methods, we (1) primarily conducted semistructured interviews (Davis and Eisenhardt, 2011; Eisenhardt, 1989); (2) made personal notes during both company and manufacturing facility visits; (3) collected and studied publicly available secondary data related both to the interviewed companies individually and to the regional activities where some of these firms played an active role (these included also region-related reports and presentations), and data received from the companies.

In order to avoid cold calling, we presented our research idea at one of the meetings of the local cluster and followed up with telephone conversations inviting identified firms to participate in the research. From the very beginning, all of the firms were aware of the data collection procedures: two rounds of interviews, which will be recorded and kept confidential. There was no interest in signing a nondisclosure agreement. All the transcripts were sent to an interviewed company, and after the study completion, each company will get an individual innovation-related report based on the interview data. Also, the overall findings will be presented at one of the cluster's meetings, where all members will get a chance to ask questions and comment on the results.

Our primary informants were managers who play an important role in the decision-making process of the firm (Corley and Gioia,

2011), that is, CEOs, managing directors, or co-owners of the firms. In total, we conducted 23 face-to-face on-site interviews that typically lasted between 60 and 120 minutes. We used two different interview protocols for each of the interview rounds, where the second one was based on the initial findings from the first round. Despite the relatively small size of the initial sample, we reached data saturation very quickly. Therefore, we not only decided not to contact any more companies from our list, but also focused our analysis on a theoretical sample of six cases that would exhibit the greatest diversity of opinions and perspectives. The remaining organizations did not demonstrate any patterns that deviate from the presented findings. In our initial list of companies, we also included one medium and one large firm for test purposes. We wanted to make sure that we develop a good understanding of all perspectives (small, medium, and large firms) as well as get a chance to clarify some of the emerging concerns directly with a potentially interested party. Having the six SME cases embedded in the larger case of the regional ecosystem enabled us to predict similar results across the cases (literal replication) and predict contrasting results for predictable reasons (theoretical replication), in line with the case study methodology (Yin, 2009).

Table 7.2. presents an overview of the collected data and case descriptions.

Table 7.2. Overview of interviews.

Company	Strategic profile	Number of interviews	Interviewee position
A	Micro-enterprise	2	CEO
C	SME	2	Managing Director
D	SME	2	Development Manager
E	SME	2	Managing Director
F	Micro-enterprise	2	CEO
G	SME	2	Managing Director

7.3.2. *Data analysis*

The data analysis relates to an inductive qualitative study with grounded theory elements, where researchers closely follow guidelines for comparison techniques (Eisenhardt, 1989; Yin, 2009). In this inductive analysis, we first identify themes that emerged within the embedded cases and then compare them across cases. The grounded theory aspects are reflected in the grouping of these themes into categories and then by interconnecting these categories and building a story upon which the final research results are based. The construction of categories was an iterative process that involved theoretical arguments, which helped to establish a common meaning across multiple observations (Eisenhardt, 1989; Locke, 2001). The assessment was conducted by multiple researchers separately. First, each company case was analyzed separately by a different researcher, and then an independent researcher conducted a cross-case analysis, which was later combined into one framework. Grounded theory principles guided the initial coding and development of the first-order codes as well as theoretical subcategories. The theory that helped us in explaining the findings is based on four theoretical dimensions of the concept of organizational boundaries proposed by Santos and Eisenhardt (2005). These dimensions cover the boundaries of identity, the boundaries of efficiency, the boundaries of competence, and the boundaries of power, as summarized in Table 7.1. The final constructive themes emerged along with familiarizing with the concept proposed by Santos and Eisenhardt (2005). The analysis process had an iterative nature, and the organizational boundaries lens helped us in structured sense-making of our empirical evidence, which led us to the answers to our research questions.

7.4. Findings

Based on the results that emerged from the data, we will start with highlighting the findings related to the boundaries of efficiency and power that relate mostly to the organizational level (of the firms) and conclude with boundaries of competencies and boundaries of identity that concern not only the firms, but also the entire ecosystem. For all of the emerged theoretical categories that constitute theoretical dimensions, we provide key illustrative quotes from the interviews.

7.4.1. *Boundaries of efficiency*

According to Santos and Eisenhardt (2005, p. 492), the organizational boundaries "should be set at a point that minimizes the cost of governing activities." Nevertheless, in the case of SMEs, opportunities for conducting transactions inside the organization maybe rather low, especially due to relatively less amount of time and resources; therefore, they need to choose to conduct them externally through a market exchange. What is more, presence in an ecosystem may significantly lower some transactional costs such as exchange uncertainty or contractual costs. This happens due to familiarity and trust among ecosystem members, which lower the chance of opportunistic behaviors inside the ecosystem.

Coordination costs (Conner and Prahalad, 1996) may still constitute a meaningful part of the market exchange process, especially in consideration with the partners from outside the ecosystem. Nevertheless, ecosystem members may choose to share them through joint internationalization activities that will be locally facilitated. In order to increase its success rate, the interviewed companies indicated several prerequisites that should support both value creation and capturing (e.g., through new joint activities or projects) for the entire ecosystem. Among these, one can find commitment, shared business objectives, and innovation-driven goals (Adner, 2006; Adner and Kapoor, 2010; Ritala *et al.*, 2013) (Figure 7.1: Company A2, F4; Figure 7.3: G2). What is more, the common goal has to be not only known and understood by every member but also be able to incentivize companies to the extent that they stimulate the action. This could apply both to the entire ecosystem as well as to a particular initiative or project that is conducted within it.

Gulati (1995), while exploring factors that explain the choice of governance structures in interfirm alliances, found evidence

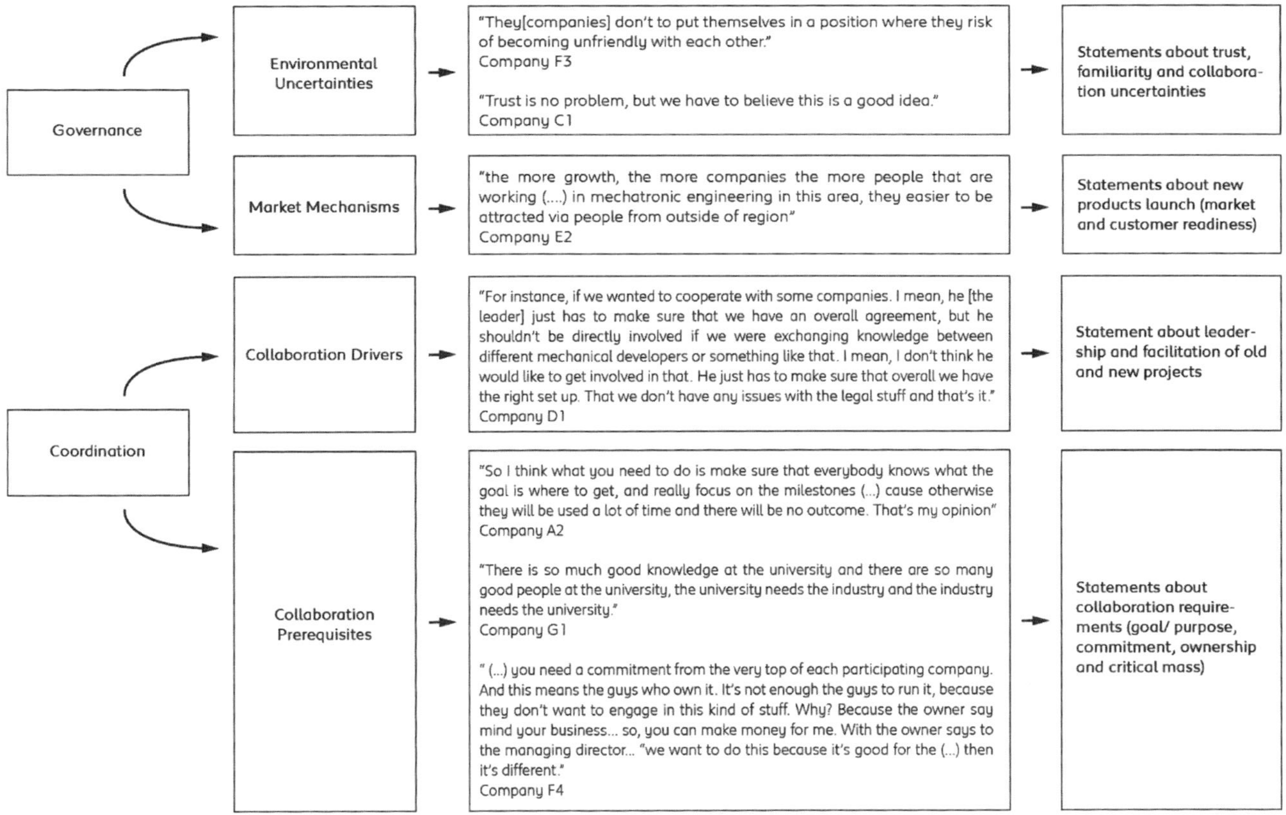

Figure 7.1. Boundaries of efficiency.

supporting the fact that trust (Figure 7.1: Company C1) and familiarity are very important factors in interfirm cooperation. Therefore, a company embedded in an ecosystem could benefit from lowering transactional costs related to the search of potential collaboration partners within the ecosystem. Nevertheless, the same familiarity and already existing good relationships may also keep some of the companies from getting involved in new initiatives. This could be related to potential risks that may emerge together with new activities. This means that some of the companies may hesitate to get into closer collaboration (Figure 7.1: Company F3), because if it is unsuccessful, they may worsen their current, good relationships with other ecosystem members. What is more, the closer the companies are, the more impact it could have for their future development. Conversely, the same relations and interdependencies could stimulate ecosystem members to stay transparent in their actions and make sure that they do not lose their current credibility within the ecosystem. Any forms of unfair attempts to gain competitive advantage could thus result in lowering chances for potential partnerships and alliances (Gulati *et al.*, 2009).

The ecosystem literature indicates the importance of a leader, or the so-called "central ecological contributor," which is the driving force of the whole community (Moore, 1993, p. 81). A natural leadership role in various ecosystem examples is taken by a large company, which not only has the necessary resources, but usually also has technology leadership (Adner and Kapoor, 2010; Iansiti and Levien, 2004; Moore, 1993). Interviewees pointed out that they definitely see the need of having one person in charge of potential ecosystem activities (Figure 7.1: Company D1; Figure 7.2: Company G3). Some of them mentioned a leader, manager, or just a kind of coordinator. What they mentioned as important tasks that this person should fulfill was taking care of the realization of the initiative's goals, assuring the right setup, and taking care of legal matters along with, in general, being in control of what is happening. Even though SMEs tend to undertake an open and collaboration focused approach, they also see potential barriers related to closer collaboration with large enterprises (Figure 7.2: Company G3). What is more,

Figure 7.2. Boundaries of power.

despite the fact that they also recognize a different working style, and other potential challenges, a university is also seen as an important partner (Figure 7.1: Company G1). Both a university and the (local) government through different initiatives could be a binding force of various types of collaborations inside the ecosystem, which could be a way of combining interests of various stakeholders of the ecosystem. A particular way of how SMEs, startups, and research projects could benefit from ecosystem embeddedness is through collaborations initiated and strongly supported by entrepreneurial universities (Etzkowitz, 2003), which try to enhance active technology transfer to the industry as well as stimulate the process of new industry creation. Brem and Radziwon (2017) highlight transparency, trust, and credibility that are stimulated by already established relationships as important ecosystem features that discourage any form of opportunistic behaviors, and thus provide a safe developmental environment for the smallest and potentially the most vulnerable ecosystem members.

In sum, organizations may need to compare the costs between internal and external governance, both of the firm and of an ecosystem. Due to strong interdependencies, which enforce fair play within the ecosystem boundaries, joint initiatives directed toward the market inside of the ecosystem are characterized by low transactional costs and generally reduced environmental uncertainty. Initiatives directed toward the market outside of the ecosystem could also benefit from significantly lower overall transaction and coordination governance costs, if only ecosystem members assure clear purpose, commitment, and good leadership and coordination of these activities between involved ecosystem members. The biggest advantages of such initiatives formed by ecosystem members could be a high level of initial trust and familiarity, which imply partial uncertainty reduction as well as potentially better utilization of local assets.

7.4.2. *Boundaries of power*

A strong network of partners inside an ecosystem can be particularly helpful in influencing expansion outside of the ecosystem boundaries, while increasing the dependence leading to the benefits of ecosystem participation. A significant part of ecosystem interdependencies may already have its roots in past collaboration experience that could increase the propensity of joining new collaborative activities (Gulati *et al.*, 2009). These activities could differ depending on the asset-specificity of the ecosystem members. In general, literature reports that technological collaboration is a popular domain of interaction between large and small firms. Typical modes of collaboration are licensing agreements, joint ventures, and sponsored spinouts (Rothwell and Dodgson, 1991). This is also supported by one of the interviewees who suggested joint technology development as a potential and promising knowledge-sharing area (Figure 7.2: Company D2).

However, there are also different opinions about potential collaboration areas. Another interviewee (Figure 7.2: Company E4) suggests that between SMEs this "collaboration goal" should not be on the technology level, since it may be too difficult to find a suitable technology and turn it into a reasonable project in terms of time span. Instead, the collaboration between SMEs may involve joint product development. Another potential collaboration idea is to share resources, for example, manpower (Figure 7.2: Company D2, D4). Ecosystem members could create joint groups of people using the same hardware/software or mechanics, and maybe also share employees in noncompetitive areas.

7.4.3. *Boundaries of competencies*

Maintaining an interest and keeping partners incentivized to invest their time and money in a particular initiative is crucial in order to achieve any long-term common goals. Lack of motivation could easily result in loss of interest in further involvement, especially in the case of SMEs. This is a function of the general level of their resources (especially money, time, and man power), which may be significantly lower than in the case of larger companies. This also implies higher risks when SMEs get involved in activities, which do not directly contribute to their core activities. The echo of potential financial challenges associated with establishing new initiatives could be also heard during the interviews. Therefore, some of the SMEs would prefer to have some external funding available to support new projects and initiatives, since they may not be able to finance them themselves (Figure 7.3: Company A3, C2). Research shows that due to financial instability, SMEs may be exposed to difficulties in expanding their activities (van de Vrande *et al.*, 2009). Therefore, availability of additional funding could significantly lower the risk related to their involvement in any collaborative activities (Blomqvist *et al.*, 2005; Sjödin *et al.*, 2011).

Some of the interviewees (Figure 7.3: Company B1, F5, F6) brought up the topic of competencies, which are very specific to the southern Danish region and present in this ecosystem.

Their way of thinking seems to align with the "smart specialization" concept, which focuses on macro-level of analysis, thus touching upon activities, which could lead the entire region to specialize both in R&D and innovation (Foray *et al.*, 2009). The concept has a strong link to regional policies, which should prioritize the development of "distinctive and original areas of specialization" (Foray *et al.*, 2009, p. 101). These conscious boundaries of special, regional competencies link strongly to the last theoretical dimension, which are the boundaries of identity.

7.4.4. *Boundaries of identity*

Our data indicate two different levels of identity: the firm identity and the ecosystem identity. Due to a constant flow of knowledge and information in the ecosystem that is partly caused by employees' mobility, the researchers did not observe any strong firm identity, whereas the ecosystem identity seemed to inspire commitment and emotional attachment to the region (Kogut, 2000).

This particular type of ecosystem identity could have evolved over time and have been shaped by various types of interactions (Porac *et al.*, 1995) between ecosystem members, like experience in working or being suppliers for the largest players of the region or shared challenges. They have developed a sense of responsibility to

Figure 7.3. Boundaries of competencies.

the region. Despite the concerns related to collaboration with large firms, due to potential discrepancies in shared business objectives and different styles of work, some of the interviewed SMEs think that large firms should directly or indirectly support regional growth and thus develop the ecosystem (Figure 7.4: Company A1, F2). These thoughts may by supported by a set of common interests, which seems to be shared between the ecosystem members. The most mentioned is the attraction of skilled manpower to the region (Figure 7.4: Company D4, E1). According to one of the ecosystem members, the companies in this region compete only based on human resources (Figure 7.3: Company D2); however, there could also be an observed rotation of people between different ecosystem members (Figure 7.4: Company F1).

Santos and Eisenhardt (2005) refer to the hiring of new employees as an important boundary change mechanism. In the context of an ecosystem, it may have a much bigger importance than just hiring a person that could fit into a single firm, especially while considering the hiring process of senior staff members. For an ecosystem, regardless of who is hiring, there is a strong awareness that there probably exists an informal competition in the area of human resources, but at the same time, also acceptance that an engineer hired in one firm could over time change his employer. That is why during the interviews most of the interviewees focused on hiring manpower for the ecosystem rather than for their organization in particular.

The boundaries of identity may be coevolutionary with the other boundaries, especially with efficiency and competencies boundaries. However, consistent identity may be a strong decision-making driver that could potentially dominate both the boundaries of power and competencies, which means that ecosystem members may give up some power or decide to open up for knowledge exchange, because this would positively influence and be of added value to the entire ecosystem. Nevertheless, due to both unconscious boundaries of the mind as well as emotional attachment and approach of organization members, the organizational identity may sometimes have a quite unpredictable influence on the decision-making process. As a consequence, decision-makers that are locally well established may decide

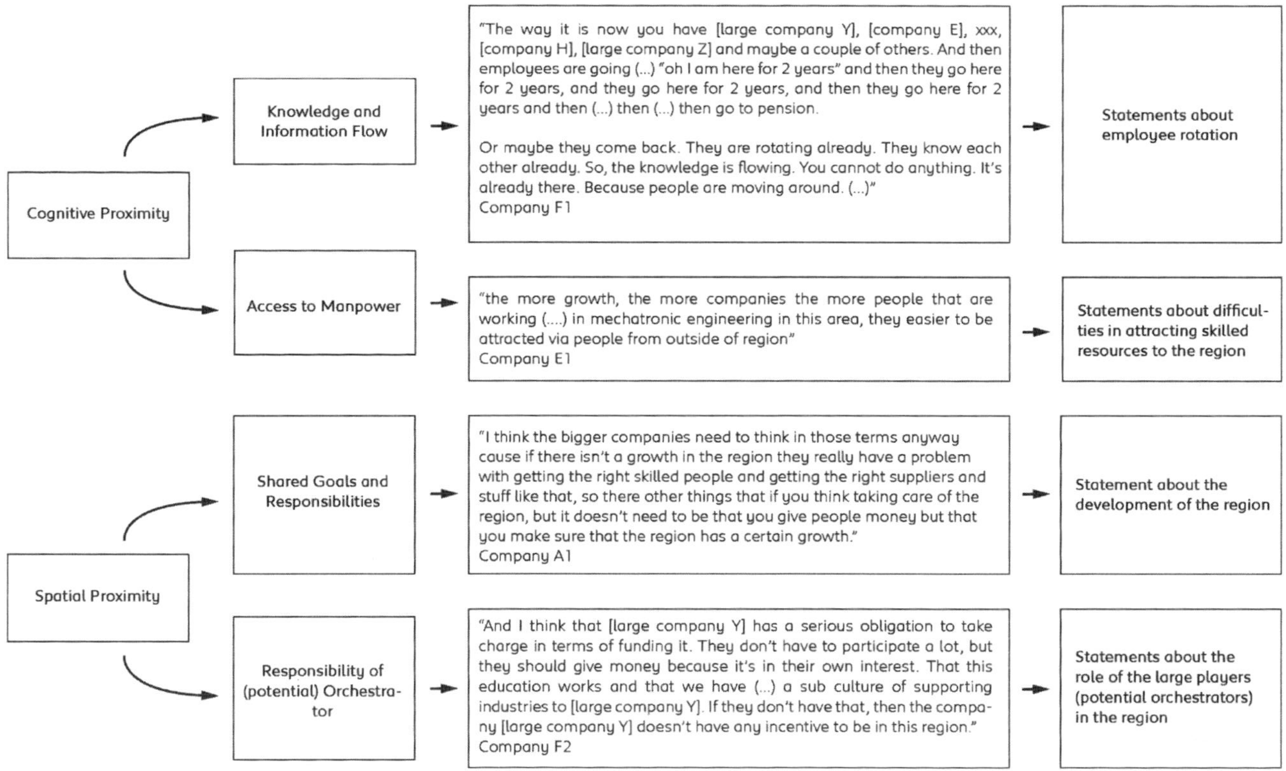

Figure 7.4. Boundaries of identity.

in favor of the organization and the ecosystem, whereas a newer resident or relative outsider's decisions may depend on the stronger emotional attachment.

7.5. Discussion and Conclusions

This study explores purposefully managed mutual knowledge flows across organizational boundaries applied by SMEs that contribute to the development of the ecosystem they are immersed in. The study adapts the Santos and Eisenhardt (2005) concept of organizational boundaries to analyze and discuss where firm level and ecosystem boundaries intersect. The research findings contribute to the understanding of the organizational boundaries of the smallest ecosystem players as well as to the ecosystem as a whole.

Setting boundaries of efficiency in the right place could significantly increase open innovation activities, thus positively influencing ecosystem members' overall performance. Ecosystem embeddedness could provide a special environment where, due to strong interdependencies, which reduce risks of opportunistic behaviors, transaction costs may significantly decrease. Such environments could serve as a good ground for forming new collaborative initiatives with current or potential partners, where coordination costs (Conner and Prahalad, 1996) may shift from internal to external (to the ecosystem). In this way, SMEs could benefit in two ways. First, low transactional costs and risks of opportunistic behaviors could stimulate purposeful inflows and outflows of knowledge and thus enable pursuing new projects with a focus not only on incremental but also radical innovations. Second, as the coordination costs will be shifted to the ecosystem as a whole, in the case of need for knowledge sourcing from outside of the ecosystem, SMEs could either share the coordination costs with their local partners or try to leverage the network of contacts of other larger and more established ecosystem members. The latter in particular could help in establishing successful initiatives directed toward further market expansion of the ecosystem members. Nevertheless, in order to set appropriate ecosystem governance structures (either formally or informally), ecosystem members expect good communication across the ecosystem as well as a driving force of joint activities. What is more, shared goals and business objectives are important prerequisites for value creation and capturing in the ecosystem (Adner, 2006; Adner and Kapoor, 2010; Ritala *et al.*, 2013). These findings also indirectly imply that leadership of the ecosystem is an important part of managing intercompany initiatives (Adner and Kapoor, 2010; Moore, 1993).

Boundaries of power focus on exchange relationships between ecosystem members. On the one hand, strong embeddedness may decrease the autonomy and control over the partners (internal to the ecosystem), but on the other hand, due to interdependencies (among others, resource dependence) through strong network and coevolution history, organizations may develop and maintain stronger relationships and expand their sphere of influence. Nonownership mechanisms may then play a much more important role in the day-to-day ecosystem management than ownership relationships, namely developing cooperative and coopetitive relationships as well as forming friendly ties. In the ecosystem context, some of these relationships may be already established thanks to previous collaboration history, which serve as a good ground for establishing new initiatives through which ecosystem members could coevolve together. Some of these initiatives could concern new or further technology, product or processes development, as well as resource sharing.

The organizational boundaries of competencies could be expressed both on a firm and ecosystem level, thus serving as a source of competitive strength. In order to maximize the value of existing resources as well as coevolve resource configurations (Gulati *et al.*, 2009), SMEs could leverage low transactional cost advantages by opening up to external sources of knowledge. These could be related to partners from their personal and professional network as well as through indirect network ties (Wynarczyk *et al.*, 2013). One of the resource-related challenges that SMEs experience is related to funding both for their operational activities and for future collaboration initiatives. These issues could be either solved or highly supported by policymakers as well as national and regional governments. Since open innovation-based ecosystems are potential contributors to regional development, more attention from policymakers could stimulate the development of additional funding schemes. More funding opportunities would support not only the development of the smallest firms, but have a positive, long-term impact on the entire ecosystem. Moreover, the ecosystem boundaries seen through a competence perspective could indicate a need of policymakers' support in enabling and supporting smart specialization (Foray *et al.*, 2009). The concept originated from a spatial sector, but has recently been raised in a regional context (McCann and Ortega-Argilés, 2013). In order to find the relevant domain and achieve critical mass and a required level of connectedness, open innovation in ecosystems could be thus considered as places where smart specialization could be initiated and developed.

Strong ecosystem identity articulated by its members may be expressed by concerns about further ecosystem development as well as a holistic approach to manpower sourcing not only for single firms, but also for the entire ecosystem. Interestingly, even though some of the interviewees indicated large local firms as socially responsible for supporting the region and its initiatives (Figure 7.4: Company A1, F2), taking into consideration both perception differences as well as power imbalance, larger enterprises were not viewed as natural leaders, which is in contradiction with Moore's (1996) assumptions. This may be a reason why small firms develop an even stronger ecosystem identity and sense of responsibility for solving their shared problems.

An identity evolves over time and is strongly shaped by interactions between ecosystem members (Porac *et al.*, 1995). What strongly supports the development of ecosystem identity in particular are embeddedness as well as interdependencies present within the ecosystem that may in some way indicate the organizational boundaries of the ecosystem, which instead of geographical proximity would focus on cognitive proximity and specialization (Foray *et al.*, 2009). This conceptualization could help in indicating boundary conditions for generalization of this case study. In particular, an ecosystem that we could generalize does not need to be a regional business ecosystem, but could also be an ecosystem that is developed based only on cognitive proximity and a set of specific competencies that could trigger trust and reduce risks related to opportunistic behaviors.

7.5.1. *Managerial implications and further research directions*

Our findings add to the understanding of organizational boundaries both on the firm as well as ecosystem level; therefore, we can derive some interesting managerial implications and directions for future research. From the practical implications point of view, our findings could serve as guidelines both for practitioners as well as policymakers who are part of various types of ecosystems or wish to support the same. More specifically, we propose a set of implications that ecosystem members may take into consideration while establishing new collaborative projects or initiating various types of activities both within and across ecosystem boundaries:

(1) The presence of boundaries of efficiency could serve as a good ground for forming new collaborative initiatives, which may be important for further development and growth of the smallest ecosystem members. However, these also require appropriate ecosystem governance structures (either formally or informally) (Felin and Zenger, 2014), that is, good communication across the ecosystem, driving force of joint activities (leadership/facilitation processes), shared goals, and business objectives.
(2) The presence of boundaries of power serves as a good ground for establishing new initiatives concerning new or further technology, product or processes development, or resource sharing (Wynarczyk *et al.*, 2013). Nevertheless, it is essential to keep into consideration the important role of nonownership mechanisms in the day-to-day ecosystem management, that is, developing cooperative and coopetitive relationships as well as forming friendly ties (Powell *et al.*, 1996). What is more, it would be crucial to pay

attention to the power balance between small and large partners to make sure that despite many interdependencies, which are present between ecosystem members, small firms have a chance to diversify their customer portfolio, so that they do not depend solely on a potential ecosystem orchestrator. This remark is also relevant for the process of establishing new joint initiatives, where the proper power balance could assure long-term commitment and motivation of all (e.g., project) partners.

(3) The presence of boundaries of competencies assures a safe environment to open up for external sources of knowledge, that is, inflows and outflows of knowledge between partners from personal and professional networks as well as through indirect network ties (Stuart, 2000). Monetary resource issues call for strong support from policymakers as well as national and regional governments that could stimulate the ecosystem development by offering tailored funding (Radziwon *et al.*, 2017). These schemes could in the long run target ecosystems that aim at developing smart specialization (Foray *et al.*, 2009), the prerequisites of which would be achieving critical mass and appropriate level of connectedness. This attention point is also in line with further plans to get SMEs and startups more involved in new projects as well as other types of initiatives in the wider European context (Muller *et al.*, 2016).

(4) The presence of boundaries of identity may serve as a supporting tool in solving ecosystem challenges. Strong ecosystem identification by members may act as an important enabler of further joint development and growth of the ecosystem through better alignment of organizational and ecosystem-level goals as well as lowering the costs of communication and coordination (Kogut, 2000). This could be particularly important in the case of manpower sourcing for an entire ecosystem. Attracting new competencies to the region could, in the long run, further stimulate mutual knowledge flows across organizational boundaries and allow its purposeful exchange and recombination between all ecosystem members.

7.5.2. *Limitations and possible extensions*

The main limitations of this study come from the nature of the research methods and in particular the focus on the particular region and type of firms (namely, a mix of high- and low-tech firms). Therefore, it would be interesting to encourage further qualitative research focusing on different regions as well as ecosystems that center around particular competencies, which serve as a bonding force of a shared interest. Exploration of two or three ecosystems could complement the findings of this investigation through the development of better understanding of factors that shape ecosystem dynamics. Such a comparative study could provide a very solid ground for offering implications for policymakers that could support value creation and capturing processes in ecosystems in general (Radziwon *et al.*, 2017; Ritala *et al.*, 2013).

Further research could look closer into the relationships and intersection of organizational boundaries of SMEs and other ecosystem members like research institutions or governmental institutions. In particular, it could take into consideration the context of the triple-helix actors present in the ecosystem as well as community-based enterprises and community of practice (Theodorakopoulos *et al.*, 2012). It would also be interesting to extrapolate the results of this study to both large firms, startups, and research projects. In the case of well-established market players of more complex nature of operational activities, it would be interesting not only to analyze their boundaries and try to extend the conceptualization of Santos and Eisenhardt (2005) by the more dynamic aspect, thus exploring temporal development of their organizational boundaries. As for the latter, despite the initial research of open innovation in startups (Eftekhari and Bogers, 2015) and ecosystem-based interorganizational collaboration centered around research projects (Brem and Radziwon, 2017), our understanding of the process of forming and shifting organizational boundaries as well as the role of open innovation practices in this process in startups and research projects is still relatively low. Exploring this area would enable us to provide better support to these firms (or firms-to-be), thus fostering entrepreneurship and strengthening collaboration not only in the regional but also in the European context (Muller *et al.*, 2016). Finally, additional promising research outreach would be to develop tools that would allow for the efficient mapping of boundaries of ecosystems along with all its members. Our research findings could be the first step in this direction, which if further developed could serve academics, practitioners, and policymakers in the process of exploring an ecosystem's potential and contribution to the economy.

References

Adner, R. (2006). Match your innovation strategy to your innovation ecosystem, *Harvard Business Review*, 84, pp. 98–107.

Adner, R. and Kapoor, R. (2010). Value creation in innovation ecosystems: how the structure of technological interdependence affects firm performance in new technology generations, *Strategic Management Journal*, 31, pp. 306–333.

Agrawal, A., Cockburn, I. and McHale, J. (2006). Gone but not forgotten: knowledge flows, labor mobility, and enduring social relationships, *Journal of Economic Geography*, 6, pp. 571–591.

Aldrich, H. (1999). *Organizations Evolving* (Sage Publications, Thousand Oaks, CA).

Asheim, B. T. and Coenen, L. (2005). Knowledge bases and regional innovation systems: comparing Nordic clusters, *Research Policy*, 34, pp. 1173–1190.

Autio, E., Kenney, M., Mustar, P., Siegel, D. and Wright, M. (2014). Entrepreneurial innovation: the importance of context, *Research Policy*, 43, pp. 1097–1108.

Blomqvist, K., Hurmelinna, P. and Seppänen, R. (2005). Playing the collaboration game right — balancing trust and contracting, *Technovation*, 25, pp. 497–504.

Bogers, M., Zobel, A.-K., Afuah, A., Almirall, E., Brunswicker, S., Dahlander, L., Frederiksen, L., Gawer, A., Gruber, M., Haefliger, S. and others. (2017). The open innovation research landscape: established perspectives and emerging themes across different levels of analysis, *Industry and Innovation*, 24, pp. 8–40.

Brem, A. and Radziwon, A. (2017). Efficient triple-helix collaboration fostering local niche innovation projects: a case from Denmark, *Technological Forecasting and Social Change*, 123, pp. 130–141.

Brunswicker, S. and van de Vrande, V. (2014). *Exploring Open Innovation in Small and Medium-Sized Enterprises*, eds. Chesbrough, H., Vanhaverbeke, W. and West, J., New Frontiers in Open Innovation (Oxford University Press, Oxford), pp. 135–156.

Chesbrough, H. (2003). *Open Innovation: The New Imperative for Creating and Profiting from Technology* (Harvard Business School Press, Boston, MA).

Chesbrough, H. (2006). *Open Innovation: The New Imperative for Creating and Profiting from Technology* (Harvard Business School Press, Boston, MA).

Chesbrough, H. and Bogers, M. (2014). *Explicating Open Innovation: Clarifying an Emerging Paradigm for Understanding Innovation*, eds. Chesbrough, H. Vanhaverbeke W. and West J., New Frontiers in Open Innovation (Oxford University Press, Oxford), pp. 3–28.

Conner, K. R. and Prahalad, C. K. (1996). A resource-based theory of the firm: knowledge versus opportunism. *Organization Science*, 7, pp. 477–501.

Corley, K. G. and Gioia, D. A. (2011). Building theory about theory building: what constitutes a theoretical contribution? *Academy of Management Review*, 36, pp. 12–32.

Davis, J. P. and Eisenhardt, K. M. (2011). Rotating leadership and collaborative innovation recombination processes in symbiotic relationships. *Administrative Science Quarterly*, 56, pp. 159–201.

Eftekhari, N. and Bogers, M. (2015). Open for entrepreneurship: how open innovation can foster new venture creation, *Creativity and Innovation Management*. 24(4), pp. 574–584.

Eisenhardt, K. M. (1989). Building theories from case study research, *Academy of Management Review*, 14, pp. 532–550.

Eisenhardt, K. M. and Galunic, D. C. (2000). Coevolving: at last, a way to make synergies work, *Harvard Business Review*, 78, pp. 91–102.

Eisenhardt, K. M. and Graebner, M. E. (2007). Theory building from cases: opportunities and challenges, *Academy of Management Journal*, 50, pp. 25–32.

Etzkowitz, H. (2003). Research groups as "quasi-firms": the invention of the entrepreneurial university, *Research Policy*, 32, pp. 109–121.

Feijóo, C., Maghiros, I., Abadie, F. and Gómez-Barroso, J.-L. (2009). Exploring a heterogeneous and fragmented digital ecosystem: mobile content, *Telematics and Informatics*, 26, pp. 282–292.

Felin, T. and Zenger, T. R. (2014). Closed or open innovation? Problem solving and the governance choice, *Research Policy*, 43, pp. 914–925.

Foray, D., David, P. A. and Hall, B. (2009). Smart specialisation-the concept, *Knowledge Economists Policy Brief*, 9, pp. 100.

Freel, M. S. (2003). Sectoral patterns of small firm innovation, networking and proximity, *Research Policy*, 32, pp. 751–770.

Gastaldi, L., Appio, F. P., Martini, A. and Corso, M. (2015). Academics as orchestrators of continuous innovation ecosystems: towards a fourth generation of CI initiatives, *International Journal of Technology Management*, 68, pp. 1–20.

Glückler, J., Lazega, E. and Hammer, I. (2017). *Knowledge and Networks*, (Springer, Cham).

Gulati, R. (1995). Does familiarity breed trust? The implications of repeated ties for contractual choice in alliances, *Academy of Management Journal*, 38, pp. 85–112.

Gulati, R., Lavie, D. and Singh, H. (2009). The nature of partnering experience and the gains from alliances, *Strategic Management Journal*, 30, pp. 1213–1233.

Gulati, R., Puranam, P. and Tushman, M. (2012). Meta-organization design: rethinking design in interorganizational and community contexts, *Strategic Management Journal*, 33, pp. 571–586.

Iansiti, M. and Levien, R. (2004). Strategy as ecology, *Harvard Business Review*, 82, pp. 68–81.

Kapoor, R. and Lee, J. M. (2013). Coordinating and competing in ecosystems: how organizational forms shape new technology investments, *Strategic Management Journal*, 34, pp. 274–296.

Kogut, B. (2000). The network as knowledge: generative rules and the emergence of structure, *Strategic Management Journal*, 21, pp. 405–425.

Laursen, K. and Salter, A. (2006). Open for innovation: the role of openness in explaining innovation performance among UK manufacturing firms, *Strategic Management Journal*, 27, pp. 131–150.

Leroi-Werelds, S., Pop, O. and Roijakkers, N. (2017). *Understanding Value Creation In Alliance Ecosystems: Insights From Marketing*, eds. Das, T. K., Managing Alliance Portfolios and Networks (Information Age Publishing, New York), pp. 1–31.

Locke, K. (2001). *Grounded Theory in Management Research* (Sage Publications, Thousand Oaks, CA).

Lusch, R. F. and Vargo, S. L. (2014). *Service-Dominant Logic: Premises, Perspectives, Possibilities* (Cambridge University Press, Cambridge, MA).

Malmberg, A. and Maskell, P. (2002). The elusive concept of localization economies: towards a knowledge-based theory of spatial clustering, *Environment and Planning A*, 34, pp. 429–450.

McCann, P. and Ortega-Argilés, R. (2013). Smart specialization, regional growth and applications to European Union cohesion policy, *Regional Studies*, 49(8), pp. 1291–1302.

Moore, J. F. (1993). Predators and prey: a new ecology of competition, *Harvard Business Review*, 71, pp. 75–86.

Moore, J. F. (1996). *The Death of Competition: Leadership and Strategy in the Age of Business Ecosystems* (Harper Business, New York).

Muller, P., Caliandro, C., Peycheva, V., Gagliardi, D., Marzocchi, C., Ramlogan, R. and Cox, D. (2016). *Annual Report on European SMEs*, 2015/2016.

Mäkinen, S. J., Kanniainen, J. and Peltola, I. (2014). Investigating adoption of free beta applications in a platform-based business ecosystem, *Journal of Product Innovation Management*, 31, pp. 451–465.

Nambisan, S. and Baron, R. A. (2013). Entrepreneurship in innovation ecosystems: entrepreneurs' self-regulatory processes and their implications for new venture success, *Entrepreneurship Theory and Practice*, 37, pp. 983–1247.

Nickerson, J. A. and Zenger, T. R. (2004). A knowledge-based theory of the firm — the problem-solving perspective, *Organization Science*, 15, pp. 617–632.

Oh, D.-S., Phillips, F., Park, S. and Lee, E. (2016). Innovation ecosystems: a critical examination, *Technovation*, 54, pp. 1–6.

Pfeffer, J. and Salancik, G. R. (1978). *The External Control of Organizations: A Resource Dependence Approach* (Harper and Row Publishers, New York, NY).

Porac, J. F., Thomas, H., Wilson, F., Paton, D. and Kanfer, A. (1995). Rivalry and the industry model of Scottish knitwear producers, *Administrative Science Quarterly*, 40, pp. 203–227.

Powell, W. W., Koput, K. W. and Smith-Doerr, L. (1996). Interorganizational collaboration and the locus of innovation: networks of learning in biotechnology, *Administrative Science Quarterly*, 41(1), pp. 116–145.

Prahalad, C. K. (2005). *The Fortune at the Bottom of the Pyramid* (Pearson Education, India).

Prahalad, C. K. and Bettis, R. A. (1986). The dominant logic: a new linkage between diversity and performance, *Strategic Management Journal*, 7, pp. 485–501.

Radziwon, A., Bogers, M. and Bilberg, A. (2017). Creating and capturing value in a regional innovation ecosystem: a study of how manufacturing SMEs develop collaborative solutions, *International Journal of Technology Management*, 75(1–4), pp. 73–96.

Ritala, P., Agouridas, V., Assimakopoulos, D. and Gies, O. (2013). Value creation and capture mechanisms in innovation ecosystems: a comparative case study, *International Journal of Technology Management*, 63, pp. 244–267.

Ritala, P. and Almpanopoulou, A. (2017). In defense of "eco" in innovation ecosystem, *Technovation*, 60–61, pp. 39–42.

Rohrbeck, R., Hoelzle, K. and Gemünden, H.G. (2009). Opening up for competitive advantage — how Deutsche Telekom creates an open innovation ecosystem, *R&D Management*, 39, pp. 420–430.

Rothwell, R. and Dodgson, M. (1991). External linkages and innovation in small and medium-sized enterprises, *R&D Management*, 21, pp. 125–138.

Santos, F. M. and Eisenhardt, K. M. (2005). Organizational boundaries and theories of organization, *Organization Science*, 16, pp. 491–508.

Shan, W., Walker, G. and Kogut, B. (1994). Interfirm cooperation and startup innovation in the biotechnology industry, *Strategic Management Journal*, 15, pp. 387–394.

Sjödin, D., Eriksson, P. E. and Frishammar, J. (2011). Open innovation in process industries: a lifecycle perspective on development of process equipment, *International Journal of Technology Management*, 56, pp. 225–240.

Sternberg, R. (1999). Innovative linkages and proximity: empirical results from recent surveys of small and medium sized firms in German regions, *Regional Studies*, 33, pp. 529–540.

Stuart, T. E. (2000). Interorganizational alliances and the performance of firms: a study of growth and innovation rates in a high-technology industry, *Strategic Management Journal,* 21, pp. 791–811.

Theodorakopoulos, N., Preciado, D. J. S. and Bennett, D. (2012). Transferring technology from university to rural industry within a developing economy context: the case for nurturing communities of practice, *Technovation*, 32, pp. 550–559.

Tödtling, F. and Kaufmann, A. (2001). The role of the region for innovation activities of SMEs, *European Urban and Regional Studies*, 8, pp. 203–215.

Valkokari, K. (2015). Business, innovation, and knowledge ecosystems: how they differ and how to survive and thrive within them, *Technology Innovation Management Review*, 5.

van der Borgh, M., Cloodt, M. and Romme, A. (2012). Value creation by knowledge-based ecosystems: evidence from a field study, *R&D Management*, 42, pp. 150–169.

van de Vrande, V., De Jong, J. P., Vanhaverbeke, W. and De Rochemont, M. (2009). Open innovation in SMEs: trends, motives and management challenges, *Technovation*, 29, pp. 423–437.

Vanhaverbeke, W. (2017). *Managing Open Innovation in SMEs* (Cambridge University Press, Cambridge).

Vargo, S. L. and Lusch, R. F. (2011). It's all B2B … and beyond: toward a systems perspective of the market, *Industrial Marketing Management*, 40, pp. 181–187.

Weick, K. E. (1995). *Sensemaking in Organizations* (Sage Publications, London, UK).

West, J. and Bogers, M. (2014). Leveraging external sources of innovation: a review of research on open innovation, *Journal of Product Innovation Management*, 31, pp. 814–831.

Williamson, O. E. (1981). The economics of organization: the transaction cost approach, *American Journal of Sociology*, pp. 548–577.

Wynarczyk, P., Piperopoulos, P. and McAdam, M. (2013). Open innovation in small and medium-sized enterprises: an overview, *International Small Business Journal,* 31(3), pp. 240–255.

Yin, R. K. (2009). *Case Study Research: Design and Methods* (Sage Publications, Thousand Oaks, CA).

Zahra, S. and Nambisan, S. (2012). Entrepreneurship and strategic thinking in business ecosystems, *Business Horizons*, 55, pp. 219–229.

Chapter 8

Open Innovation Within the Low-Technology SME Sector

Lawrence Dooley and David O'Sullivan*

* *l.dooley@ucc.ie*

Abstract

Innovation studies have focused in the main on so called high technology (HT) firms that demonstrate a culture of continuous innovation evidenced by steady growth rates and profitability. HT firms invest significantly in R&D. Less explored are the significant number of so-called low-to-medium technology (LMT) firms and in particular those defined as small-to-medium sized enterprises (SMEs). Despite their low investment in R&D and relatively small size, LMT–SME firms continuously adapt and innovate to maintain a measure of growth and profitability. These firms that make up the majority of firms in developed economies have developed their own very different routines and practices that match their own particular needs and adapt to their own particular competitive environment. These routines include agile and continuous innovation and a focus on processes and services. These practices when studied and shared

* Corresponding author.

provide learning opportunities within the sector as a whole. This chapter presents the routines and practices of innovation within a number of case studies from the LMT–SME sector and in particular their agile approaches towards open innovation.

8.1. Introduction

For decades, innovation studies have focused on the behaviors of high-technology (HT) firms whose growth rates are lauded by researchers as evidence of exemplary creativity and exploitation. The behavior of large enterprises (LE) now routinely epitomize the concept of innovation for researchers and commentators. Yet, while HT–LE firms do generate value for significant numbers of customers, they represent only one perspective of organizational innovation routines and practice. Less well-researched are the behaviors of low-and medium-low-technology (LMT) firms, which represent a significant portion of all economic activities and are largely small to medium-sized enterprises (SMEs) in scale. The prevailing innovation perspective is one of scientific and technological R&D, as the origins of innovation capability and this bias ignore both the importance and innovativeness of those LMT firms not active in traditional R&D (Hirsch-Kreinsen and Jacobson, 2008; Santamaria *et al.*, 2009; Som and Kirner, 2015). The LMT and SME sectors have contextual objectives and challenges that are different to their HT or larger-scale counterparts and which will influence the innovation practices and strategies they develop. Challenges such as resource constraints, commoditized products, and limited management capability can act as drivers for LMT firms to look beyond organizational boundaries in their struggle for innovation development. This chapter explores the innovation routines and practices of mature SMEs within the LMT sectors and the nature of "open" innovation (OI) practices being adopted to support their innovation efforts. In deepening our knowledge of lower-tech SME innovation practices, we strive to better understand how to nurture and enhance the innovation capacity in this "forgotten sector" (Hirsch-Kreinsen, 2008) and optimize the value of OI for SMEs by understanding how it manifests and creates value in this unique context.

8.2. Innovation, Low-Tech Industries, and SMEs

Despite the significant research undertaken within innovation studies over the last half century, the lower-tech industries (Hirsch-Kreinsen and Jacobson, 2008; Robertson *et al.*, 2009; Som and Kirner, 2015) and the SME sector (Hoffman *et al.*, 2008; van de Vrande *et al.*, 2009; Gassmann *et al.*, 2010) have been largely overlooked. The Organisation for Economic Cooperation and Development (OECD) classification identifies four distinct categories (low-tech, medium-low tech, medium-low tech, and high-tech) based on relative R&D intensities (direct and indirect) of sector and classification of the products produced (Hatzichronoglou, 1997; OECD, 2011). Discussing the OECD classification, the research of Hirsch-Kreinsen and Jacobson (2008) categorizes industries which are not R&D-intensive (LMT) as those with expenditure less than 3% of turnover, while from a firm-level perspective, Som and Kirner (2015) categorize non-R&D-intensive and non-R&D-performing firms as having percentage R&D expenditure on sales of less than 2.5%. The LMT sector accounts for 40%–60% of industrial value-added and more than half of industry-based employees in the European Union (Hirsch-Krensen, 2008; Som and Kirner, 2015). Such firms are highly embedded in regional and national supply chains and local markets (Som and Kirner, 2015). Many of these firms within the LMT sector are not only mature and indigenous to their country of origin, but are also SMEs (Pavitt, 1984; Nouman *et al.*, 2011; Spithoven *et al.*, 2011), and as such will have to overcome the challenges of innovation that all SMEs face.

The importance of SMEs for economic growth and innovation has been well established (Rothwell and Zegveld, 1986; Rothwell and Dogson, 1991; Gibb and Li, 2003; Radas and Bozic, 2009; Lee *et al.*, 2010). A review of UK literature on the state of knowledge about SMEs and innovation (Hoffmann *et al.*, 1998) stresses that SMEs are generally innovative, but also acknowledges that not all SMEs in all sectors innovate. Major SME innovation challenges relate to shortage of resources in terms of technological knowledge, infrastructure, and financial and human capital, which constrain the scope of innovation activity. These pressures can drive firms to look beyond their internal capabilities to consider collaborative innovation with partner organizations to leverage external knowledge to support innovation activity (Rothwell and Dogson, 1991; Narula, 2004). Research highlights there are no significant structural impediments to SMEs engaging in collaborative innovation since "collaboration is a critical input for the innovation processes of SMEs" (Nieto and Santamaria, 2010, p. 61) and the sectors "heavily draw on their network to find missing resources" (van de Vrande *et al.*, 2009, p. 426).

A firm's uniqueness and competitiveness are provided by the heterogeneity of its resource configuration (Penrose, 1959), based on skills developed, and fit with external environment requirements of the firm's capabilities. The "capability" view of the firm promotes that there are organizational traits and embedded ways of working, developed over time in response to organizational problem-solving (Dosi *et al.*, 2008), which underpin core competencies and offer advantage in pursuit of future opportunities (Teece *et al.*, 1997; Chatenier *et al.*, 2010). While LMT-SMEs have characteristics common to all SMEs, they are also likely to have differences given their nonreliance on traditional R&D. LMT firms often personify a culture that is informal, unstructured, and adaptive in response to the evolving context (Pullen *et al.*, 2009). Hirsch-Kreinsen (2008) suggests that the LMT sector's capabilities are largely based on management and unskilled workers, a factor contributing to lower levels of knowledge appropriability (Vale and Caldeira, 2008) by such firms. Many of the innovations undertaken are driven by necessity of market forces or the emergence of an entrepreneurial opportunity and result in LMT firm's innovation efforts being more prone toward incremental innovation, concentrated within their core business (Hirsch-Kreinsen, 2008; Hervas-Oliver *et al.*, 2011). In fact, although promoting the need for LMT innovation activity, Hervas-Oliver *et al.* (2011) suggest that over emphasis on (cutting-edge) R&D may damage the sector's innate innovation capability, which often originates from "design, the use of advanced machinery and training" (Santamaria *et al.*, 2009, p. 514). Hirsch-Kreinsen (2015) identifies three innovation strategies common to LMT firms: (1) continuous, step-by-step development of existing products; (2) enhancing market position through increased customer-oriented products; and (3) introduction of increased process specialization. While firms from lower-tech industries may often lag behind their HT counterparts regarding radicalness of product innovation produced, they report similar performance with respect to process innovation, where R&D capability is not as central (Santamaria *et al.*, 2009; Hervas-Oliver *et al.*, 2011). Process innovation is advantageous for LMT-SMEs not only in improving overall costs and responsiveness but also in providing enhanced opportunities for future product innovation from the resulting capabilities developed.

In highlighting the sector's innovation capability, Hervas-Oliver *et al.* (2011, p. 438) view it as one "characterized by process, organizational and marketing innovations, weak internal innovative capabilities ... and strong dependencies on the external provision of inputs...." Two key determinants of LMT firms' innovative output are their problem-solving/experimentation and entrepreneurial orientation of the firm to respond to market turbulence (von Tunzelmann and Acha, 2005; Hirsch-Kreinsen 2008; Santamaria *et al.*, 2009). When reflecting on the LMT firms, Bender (2008, p. 39) views their "ability to bridge the gap between different knowledge domains creatively and to create new bonds between different actors and/or processes and/or components" as core to their success. Santamaria *et al.* (2009) also emphasize this collaborative innovative capability of LMT firms, stressing their importance within the innovation ecosystem, not only with respect to innovating in their own right but also as adopters of advanced technology, developed by HT firms. Thus, OI practices may be advantageous for LMT-SMEs' innovation activities, especially given their natural collaborative capabilities and constrained resources.

8.3. Open Innovation

An important innovation practice highlighted by research of the past decade is the leverage of external resources, not controlled by the firm, to support innovation development. Much of this practice has been discussed as the OI paradigm that promotes that firms that source or commercialize innovations through leveraging external resources may benefit in terms of costs and risk reduction, pooling of complementary competencies, and realization of organizational synergies that enhance innovation process efficiency (Chesbrough, 2003; Gassmann, 2006; Vanhaverbeke, 2012). Gassmann *et al.* (2010) highlight that OI research has been primarily focused on large-scale enterprises, with a focus on SME adoption largely absent to date. In discussing the nature of OI practice, Gassmann and Enkel (2004) have identified: (1) the outside-in process (inbound) of acquiring and integrating external knowledge resources to support commercialization; (2) inside-out process (outbound) of leveraging external knowledge sources for the external exploitation of internally generated ideas or IP; and finally (3) coupled process of linking outside-in and inside-out processes in more strategic and interactive engagements between complementary partners as the three core processes in OI. Research to date has been mainly focused on the outside-in process mode, with the inside-out and coupled OI process modes attracting significantly less attention (Chesbrough and Bogers, 2014). A firm's leveraging of outside-in innovation practices often contributes to improving "exploration" capabilities, introducing nonredundant knowledge for later synthesis and application, whereas inside-out innovation practices relate more to "exploitation" capabilities, as the firm leverages its existing knowledge and technology resources for increased value capture (He and Wong, 2004). Gassmann (2006) suggests that industry sectors characterized by globalization, technology intensity, technology fusion, and new business models are more inclined toward OI adoption, indicating lower-tech SMEs may be less supportive of OI practices. Yet, harnessing complementary external resources to support the constrained and often myopic internal innovation capability of LMT-SMEs' innovation routines and practices may be beneficial not only in enhancing the competitive advantage of the sector but also in revealing an alternative manifestation of OI adoption given the differing LMT context and firm capabilities when compared with large-scale enterprises, which have been the focus of the majority of the OI research to date.

The nature of OI practice is often categorized in terms of its breadth and depth (Laursen and Salter, 2006), with innovation breadth relating to the diversity of external sources relied upon in innovative activities and innovation depth relating to the depth of interaction and importance for the firm of that external source engagement. Similar to the OI mode adopted, the nature of the OI practice may result in nurturing differing phases of the innovation process, with greater breadth supporting opportunity scanning and creative capabilities at the earlier process stages, while greater prevalence of deeper engagements may support more significant innovation implementation activities as greater synthesis of complementary resources is achieved in a "coupled" mode. Examining the nature of the external partners engaged with, Heindereich (2009) concludes that relative to HT firms, LMT firms are less likely to be engaged in innovation-focused partnerships or the leveraging of external knowledge sources. However, Hervas-Oliver *et al.* (2011) find that LMT firms have strong dependencies on external provision of inputs and that the most important source of this knowledge is from suppliers. External sources leveraged by firms in the LMT sectors can include the expertise of nonequity alliances (product innovation related), consultants (process-related), the hiring of better staff (absorptive capacity and capability), and the leveraging of other complementary knowledge sources as enablers of their innovation process. While leveraging external knowledge may be advantageous to the LMT-SME innovation capability, much of their external knowledge flows occur via informal transfer routines from sources within their existing networks (Nouman *et al.*, 2011).

The literature acknowledges the validity of further study of innovation both within lower-tech sectors and the SME sector, given their importance in economies, the unique challenges they face, and the relative underrepresentation to date when compared with large-scale firms and HT sectors. As a step in restoring the imbalance, our research examines the innovation of mature SMEs within two industries of the LMT sectors and explores the nature of OI practices being adopted as part of their innovation effort. The research seeks to explore two research questions; first, how LMT-SMEs are engaging in innovation activity, and second, are these firms leveraging OI practices to support their innovation efforts. In deepening our knowledge of lower-tech SME innovation practices, we strive to better understand how to nurture and enhance the innovation capacity in this "forgotten sector" (Hirsch-Kreinsen, 2008) and ways in which OI is optimized within such SMEs so that they are not at a competitive disadvantage relative to larger, better resourced firms.

8.4. Case Studies

Four case studies are now presented of LMT-SMEs in the manufacturing sector. For the purpose of this research, the scope of the cases studied was constrained to firms that are of small or medium size (10–249 employees), manufacturing-based, which have been in existence for five years or more, operate in an industry sector that is defined as either low-tech or medium low-tech as per the OECD (1997) classification, and where the firm is recognized as being innovation-active. In each case study, the Managing Director and/or founder completed a short innovation-related questionnaire followed by a number of interviews exploring how the firm undertook innovation. A total of 15 semistructured interviews were undertaken across the four case sites. Multiple interviews were undertaken within each firm, not only to facilitate triangulation but also to provide multiple opportunities for the researcher and interviewee to collectively "unpeel the reality" of innovation practice as it applies within the SME case. This was necessary since SME innovation activity is often *ad hoc* and reactive in nature, and SME management can struggle to conceptualize the reality of innovation routines and practices at the first attempt, given it is just one of the multiple and interrelated roles they undertake daily. While all firms were forthcoming with information, the reiterative nature of multiple interviews was beneficial because the space between interviews allowed for reflection and deeper understanding of the innovation context since the interviewees had time to ponder the routines/practices by which they innovate and the researcher was afforded the opportunity to probe unresolved issues from previous interview rounds.

In analysis, each case was examined, not only in terms of its innovation routines across the innovation process but also the specific nature of the projects/practice progressing along this process.

Given the absence of explicit OI strategies within the cases, OI practice is only evident from a project-level, bottom-up perspective. For analysis purposes, the scope of projects studied was limited to those occurring in the previous two years to ensure comparability, relevance to the current strategic direction of the firm, and reliability of the interview data provided. An innovative practice was deemed *open* when external knowledge flows (either outside-in or inside-out) were leveraged to support part of the action. Where actions did leverage OI, then the type of partner engaged with by the firm (as an indicator of breadth) and the reliance on that partner relative to that achievement of the specific action (as an indicator of depth) were explored. Following individual analysis of the cases, a cross-case analysis was conducted to enhance generalizability and also to allow deeper explanation of the research question (Miles *et al.*, 2014).

A single industry category from the two relevant OECD categorizations of the LMT sector was taken. Classified as low-tech, both firm A and firm B produce animal nutrition products for the agri-industry. Classified as medium-low tech, firm C and firm D produce plastic extrusion finished products for multiple markets. All four firms are mature firms, have been established for a number of decades, and have a regional reputation for being innovative and growth-orientated. All four firms still have an entrepreneurial founder engaged in the business operations and have family members active within the management team. Table 8.1 provides a brief synopsis of the four cases.

8.4.1. *Innovation practices*

In terms of the innovation practices of LMT-SMEs, the findings highlight that all cases demonstrate evidence of innovation, with specific firms highlighting examples of product, process, market, and organizational innovation activity. The culture of all cases was one best described as *market-orientated, entrepreneurial-adhocracies*, with the entrepreneur/managing director central to the decisions taken. This is something which not only enhanced firm agility and responsiveness but also constrained the diversity of knowledge and

Table 8.1. Case studies — Industry profiles.

Industry profile	Case studies			
	Tivio (A)	Newcast (B)	Clon (C)	KCF (D)
OECD classification	Low-tech	Low-tech	Medium-low tech	Medium-low tech
Industry	Animal nutrition	Animal nutrition	Plastics	Plastics
Disruptive product technology	Low	Very low	Low	Low
Customer	B2C	B2C	B2C	B2C
Foundation year	1985	1982	1977	1987
Employees[a]	80	50	50	220
Market focus	National	Regional	European	European
Management	Family/ professional	Family	Family	Family/ professional
Innovation types	Product/process/ market niche	Product/ process	Product/ process	Product/ process/ market niche
Design capability	Yes	Yes	Yes	Yes

Note: [a]Rounded.

experience underpinning decision-making. The firm size of the four cases varied significantly, and this factor influenced their resource constraints, geographical spread of markets served, and professionalism of the management team. The underpinning technology of each case's market offering was low, which makes it challenging to identify the product value-added compared with competitors and increases commodity pressures on the firm. In examining the LMT innovation strategies (Hirsch-Kreinsen, 2015) pursued by the four cases, cases B and C reported the expected low investment in R&D (<3% of turnover) and showed evidence of a step-by-step product development of existing products, while cases A and D reported above the expected investment in R&D (>3% of turnover) and, with respect to product innovation, showed evidence of greater customer orientation through focus on market segmentation and enhanced design capability. Process innovation was something highly evident in all cases, with process knowledge perceived by both cases A and D as a key source of advantage for growth, whereas cases B and C highlighted their agility to respond to customer requests and solve customer problems as underpinning their success (responsiveness was also highlighted in cases A and D, but not to same degree).

When examining the nature of process innovations introduced within the cases, activities of cases A and D had more technologically advanced process specialization, which may underpin the development of core competence for the firms, especially given that both identified process knowledge as a source of primary advantage for the firm and were increasingly focused on addressing more demanding customer needs in niche markets. Like all SMEs, resources were an issue for all four cases, but constraints were more extreme for cases B and C, partially due to the relative scale of the firms. Case B made the point with respect to innovation investment that "we have to make do...and live within our means with the little resource we have available," whereas due to a "healthier" order book, case D referenced introduction of process equipment representing the firm's largest capital investment to date. While cost reduction was a common focus for all cases, so was recognition of the need to increase the novelty and customer value within the product range. Cases A and D highlighted innovation activity with respect to market innovation, with both firms having significant international ambitions for the immediate future through the leveraging of their process knowledge and design capabilities. However, cases B and C were more focused on incremental step-by-step product and process innovation as a strategy for growth. While innovation routines and practices of all cases were *ad hoc* and strongly influenced by the entrepreneur/managing director, again cases A and D showed evidence of increased efforts toward systemization of their innovation routines and management practices when compared with cases B and D. While all cases exhibited good technical capabilities, there was a tendency across all cases toward leveraging less formal mechanisms such as reputational responsiveness and internal knowledge as opposed to more formal patents as a strategy toward protecting their position. All cases had exemplars of innovation practices that harnessed external resources to support innovation within the two-year window studied, but such practices were often reactive in nature and not viewed by the firms as a strategic thrust of their innovation efforts. A summary of innovation practices is presented in Table 8.2.

8.4.2. *Open innovation practices*

In terms of OI, all cases leveraged both internal and external knowledge resources to support their innovation activities, having evidence of at least one OI project occurring in the previous two years. In the majority of scenarios, the "openness" related to process innovations and involved the inward flow of external knowledge into the organization to support the innovation activity. When discussing the openness of specific product innovations, all cases identified knowledge flows from customers and suppliers that aided the activity but down played their importance, stressing they occurred primarily through informal channels and that it occurred at only one or two specific stages of the development. In fact, all cases exemplified strong preference toward leverage of internal resources and only sought out external resources when driven by necessity or the perception of emerging opportunity beyond their internal scope. Examining the nature of interaction, there was an obvious lack of more formal, strategic collaborations between the LMT-SMEs and external entities, and where collaborative activities did deepen, then relationship avoided scenarios of joint ownership of the output. For example, case D highlighted the informal reciprocity of innovation collaboration with external partners, stating "they benefit because if it's successful then they will sell more widgets [partner component included in assembly of product innovation] and we get into a new, higher margin market for our product."

In terms of the external knowledge sources leveraged, while all cases highlighted the importance of knowledge originating from supplier and customer, cases A and D highlighted a wider breadth

Table 8.2. Case studies — Innovation practices.

	Case studies			
Innovation practices	Tivio (A)	NewCast (B)	Clon (C)	KCF (D)
R&D intensity	>3%	<3%	<3%	>5%
Appropriability — formal	Trademarks	Trademarks	Trademarks	Trademarks
Appropriability — informal	Market knowledge, process know-how	Market knowledge, process know-how	Market knowledge, process know-how	Market knowledge, process know-how
Innovation management	Evolving to formal	Informal	Informal	Evolving to formal
Innovation culture	*Ad hoc*	*Ad hoc*	*Ad hoc*	Medium
Innovation order winner	Cost reduction and value creation	Cost reduction and responsiveness	Cost reduction and responsiveness	Cost reduction and value creation
Innovation focus	Process and product	Process	Process	Process and product

of the external knowledge sources searched in terms of partner heterogeneity. Cases A and D had experience of engagement with public research labs, universities, and other commercial firms for the purpose of supporting the innovation process, whereas the more traditional LMT cases (B and C) primarily engaged with cognitively and geographically closer complementors. Irrespective of the breadth of knowledge sources accessed, a common theme of the four cases was that such external engagements originated primarily from sources within established and existing networks, and such initiatives were under pressure to deliver results for the SME firm in the short to medium term. Case D highlighted a reluctance to re-enter research projects such as H2020, given they were "too blue-sky and far away from what the [current] market needs." This requirement for immediacy of innovation impact for the firm corresponds with a general trait of the LMT cases toward incremental innovation, where risks are less since knowledge is easier to absorb and where commercial return is more likely, if also more humble.

Given the reluctance of the four cases to engage in equity-sharing with external partners with respect to innovation collaborations, unsurprisingly, the majority of OI interactions were lacking in depth as measured by strategic importance to the firm and the longevity of the engagement. While the majority of the cases leveraged OI in the outward-in mode and showed little evidence of depth with respect to innovation partnerships, there was one exception, namely a single project within case A, which was more "coupled" and long-term. Case A had begun engaging with a foreign-based firm, serving a similar market in their home country, and while the early engagement was highly transactional, it has led to further initiatives, which are perceived as of strategic potential for the future growth of the firm. Case A describes its current project as phase 2 of an evolving three-phase relationship, with each subsequent phase deepening the relationship between the partners and also providing immediate commercial return and organizational validation of the collaboration. Phase 3 (at some future point) would involve both parties jointly exploiting new commercial opportunities across both markets. No grand plan exists for this strategy, nor were there any contractual articles between the parties of this long-term vision. Instead, each project phase was driven by entrepreneurial recognition of an opportunity and synergy/desire existing between the partners to exploit knowledge arising from the collaboration. The success and relational understanding of earlier phases of the interaction have proven advantageous, since they provide the necessary foundation for "deeper" interactions to occur through enhancement of organizational collaborative capabilities. In examining the development trajectory of this project, it is moving from one born in exploitative innovation, now transitioning toward more explorative focused innovation for the future. Within this "deeper" OI innovation, there is recognition by the case that potential opportunities must be underpinned by relational and operational understanding between the parties. The risk management practiced in this project also underpins the unstable nature of alliances and the scarcity of SME resources, since value has to be appropriated and the strategic potential of any future opportunities has to be balanced against the reality of commercial return and the current context. The experience of this case (and the other three also) indicates the emergent and operational nature of OI practice within the LMT-SME context where its adoption is primarily viewed as a problem solution and rarely in terms of its strategic potential for the firm in terms of exploring new trajectories for the business.

Despite limited internal capabilities constraining innovation capability, OI practices represented only a small proportion of the LMT innovation portfolio, and although they contribute to firm innovative capability, they are viewed as low priority. The relatively low strategic importance of OI, given the resource constraints evident across the cases, may relate to the absence of formal appropriation mechanisms, which reduce their ability to capture value with more distant collaborative partners, the challenge in attracting complementary partners if the internal resource base is not perceived as valuable, and rare by potential collaborators or weakness in the firm's human resource to absorb and apply the external knowledge exchanged. Likewise, it may relate to the lower strategic planning capability evident across the cases, and thus one might

Table 8.3. Case studies — OI practices.

Open innovation practices	Case studies			
	Tivio (A)	Newcas (B)	Clon (C)	KCF (D)
OI incidents	2	2	2	4
OI main focus	Process	Process	Process	Process
Collaborating partners heterogeneity	Low-medium	Low	Low	Low-medium
Collaborative breadth	Low	Low	Low	Low
Collaborative depth	Medium	Low	Low	Low
Outside-in	Yes	Yes	Yes	Yes
Inside-out	Yes	No	No	No
Primary mode of knowledge acquisition	Informal	Informal	Informal	Informal
Importance of OI	Low	Low	Low	Low

conclude that the full potential of OI practice for LMT-SMEs is not currently being leveraged. Table 8.3 presents a summary of OI practices in the four LMT-SME case studies.

8.5. Reflection

The research results from the four LMT cases studied indicate that SMEs are innovating and that their range of innovation is not only curtailed to product and process-related innovation but also can include market and organizational innovations. Process innovation is core to organizational sustainability, not only for cost control but also for enhanced internal capabilities that support future product development. The research supports earlier research findings that identify the innovative potential of the LMT sectors (Hirsch-Kreinsen *et al.*, 2006; Santamaria *et al.*, 2009; Hervas-Oliver *et al.*, 2011; Spithoven *et al.*, 2011; Som and Kirner, 2015). While forces of competitive necessity and emerging opportunity drive innovative practice, the influence of the entrepreneur/managing director on its realization is central, since their hunches and beliefs determine the firm's course of action. Such behavior aligns with the upper echelon theory perspective of Hambrick and Mason (1984) and reinforces the entrepreneur's importance in SME actions and strategy determination of the firm. The research finds that there is significant variation in the percentage R&D expenditure of LMT-SMEs, even within the same industry sector and suggests that factors other than the industry sector such as firm size, distance of markets served, and management experience/capability all influence innovation behavior of the firm. Controlling for such factors within sample selection of future LMT-SME research may further enhance our understanding of capabilities that nurture innovation practice. When examining the nature of innovation, the research highlights the dominance of process innovation for competitive advantage and that product innovation is primarily incremental and conservative in nature, even within more ambitious cases (those of higher innovation investment). This reinforces the limitation of resource constraints on the innovation practices of LMT-SMEs and the related pressure of immediacy of return from such actions, which can create an innovation "glass ceiling" skewing focus toward exploitative as opposed to more explorative innovation, and consequently constraining the novelty and value added that such innovation activity can deliver.

In examining the adoption of OI within LMT-SMEs, the research identifies the importance of studying OI practice as part of the wider organizational innovation practice since it is only within this context that the pressures and capabilities that drive OI decision-making can be understood. The cases show that LMT-SMEs are leveraging OI practices to support internal innovation capabilities, but highlight that this harnessing of external resources primarily occurs due to operational necessity of realizing the project rather than any strategic intent for the business. Serendipity plays a role in all innovation, but the lower strategic planning capability evident within the LMT cases and the *ad hoc* nature of their innovation process negatively impact the ability to harness the full benefit of OI when practiced. While all cases were constrained by resource limitations, the OI adoption was limited as all cases had strong preference toward internal capabilities for innovation development. This preference toward internal capability development may relate to the desire to grow the scale of the firm but also may indicate lack of self-efficacy within LMT management regarding their ability to capture value and sustain long-term engagement with external collaborative partners, given their size and novelty of internal assets. Given that OI practice is only possible when firms can attract a suitably synergistic partner to collaborate with, then LMT-SMEs are at a disadvantage compared to more technologically intense counterparts in terms of "partner attractiveness" and thus may need to enhance the value and novelty of their internal capabilities for prospective partners and future OI engagement.

Examining the OI project shows strong bias of the LMT-SMEs toward outward-in knowledge flows to facilitate exploitative-focused innovation, and this impacts both the diversity of partner breadth engaged with and the longevity and depth of such interorganizational interaction. The findings indicate that OI activity is primarily focused toward engagements with external knowledge sources in quadrant 1 (see Figure 8.1), with significantly less activity present in quadrant 2 and quadrant 3, and trivial levels of activity within quadrant 4. This aligns with the earlier findings of Dooley *et al.* (2016),

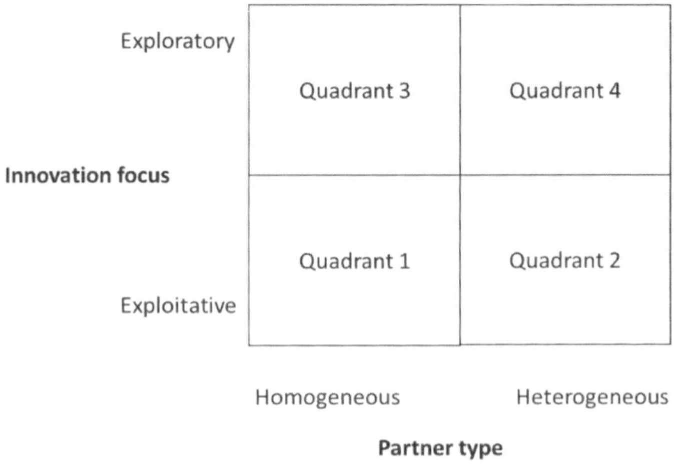

Figure 8.1. Open innovation activity map.

which show that SMEs have bias toward exploitative-focused engagements with geographical and cognitively close partners. Thus, while such collaborations may enhance the ability to absorb and apply knowledge exchange for existing problem solutions, it also implies LMT-SMEs are failing to realize OI practices' full potential. Increased focus on OI for more exploratory-focused innovation, where risks might be shared across partners, could offer increased strategic potential and a trajectory to break through the LMT innovation glass ceiling that maintains the low value added of market offering and the pressure of "commodity hell." There is opportunity for policy initiatives to nurture innovation activity of the LMT sector toward more exploratory-focused innovation activities with more distant, yet synergistic partners that will enhance competitive advantage of the firm into the future. Such an endeavor would reduce the reactive nature of the LMT sector, encourage management to lengthen its planning horizon, and ultimately be more proactive in their actions. Finally, an advantageous consequence of increases in OI practices by any firm is that it enhances collaborative capability, widens exposure to new knowledge and market trends, and offers the potential to deepen interorganizational interaction with specific partners as experience and trust develops over time. OI adoption can be advantageous for the LMT-SME, but more practice is needed since it takes time to develop the organizational capabilities and experience to effectively capture this value. Currently, OI adoption is primarily occurring in a more emergent, speculative, and bottom-up approach than would be the norm in larger-scale enterprises, and this may imply that LMT-SMEs are forgoing some of the transformational advantages of OI engagement and thus eroding their competitive advantage, especially given the OI propensity of larger-scale enterprise.

8.6. Practical Implications

The implications of this chapter from a managerial and policy dimension are as follows. First, it reinforces the importance of the LMT sector for national economies and highlights that despite the sector undertaking significant innovation, it does not readily align with the scientific and technological R&D perspective promoted by policymakers. Greater emphasis by policymakers on process innovation development and market innovation would better align with the innovation focus of the sector and facilitate capability development that may lead to subsequent development of increased product novelty. Second, it reinforces the need for education of the LMT sector management to improve their strategic ability to manage the challenges of balancing exploitative and explorative-focused innovation, enhance the core competence of their enterprise, and optimize the leveraging of OI practice in terms of potential breadth and depth. Such upskilling of management would also reduce the firm's reliance on the entrepreneur with respect to innovation decisions and nurture greater collective strategic capability as a consequence of enhanced knowledge diversity within the management team. Third, the findings highlight the need for the sector to engage in more diverse innovation collaborations since this would allow the development of collaboration experience to nurture collaborative capability and increased exposure to nonredundant knowledge and potential transformative trajectories. This may enhance the SME competitive potential, allowing them to embrace the full potential of innovations of fifth generation (Rothwell, 1992, 1994) and reinforce their natural collaborative competence that has allowed them to achieve innovative output comparable with larger, better resourced firms (Nieto and Santamaria, 2010).

References

Bender, G. (2008). How to grasp innovativeness of organizations: outline of a conceptual tool, Eds. Hirsch-Kreinsen, H. and Jacobson, D., *Innovation in Low-Tech Firms and Industries* (Edward Elgar Publishing), pp. 25–44.

Chatenier, E. D., Verstegen, J. A., Biemans, H. J., Mulder, M. and Omta, O. S. (2010). Identification of competencies for professionals in open innovation teams, *R&D Management*, 40, pp. 271–280.

Chesbrough, H. W. (2003). *Open Innovation: The New Imperative for Creating and Profiting from Technology* (Harvard Business Press, Boston, MA).

Chesbrough, H. and Bogers, M. (2014). Explicating open innovation: clarifying an emerging paradigm for understanding innovation, Eds. Chesbrough, H., Vanhaverbeke, W. and West, J., *New Frontiers in Open Innovation* (Oxford University Press, Oxford), pp. 3–28.

Dooley, L., Kenny, B. and Cronin, M. (2016). Interorganizational innovation across geographic and cognitive boundaries: does firm size matter? *R&D Management*, 46, pp. 227–243.

Dosi, G., Faillo, M. and Marengo, L. (2008). Organizational capabilities, patterns of knowledge accumulation and governance structures in business firms: an introduction, *Organization Studies*, 29, pp. 1165–1185.

Gassmann, O. (2006). Opening up the innovation process: towards an agenda, *R&D Management*, 36, pp. 223–228.

Gassmann, O. and Enkel, E. (2004). Towards a theory of open innovation: three core process archetypes, *R&D Management Conference*, pp. 1–18.

Gassmann, O., Enkel, E. and Chesbrough, H. (2010). The future of open innovation, *R&D Management*, 40, pp. 213–221.

Gibb, A. and Li, J. (2003). Organizing for enterprise in China: what can we learn from the Chinese micro, small, and medium enterprise development experience, *Futures*, 35(4), pp. 403–421.

Hambrick, D. and Mason, P. (1984). Upper Echelons: the organization as a reflection of its top managers, *The Academy of Management Review*, 9(2), pp. 193–206.

Hatzichronoglou, T. (1997). *Revision of the High-Technology Sector and Product Classification* (OECD), Science, Technology and Industry Working Papers, No. 1997/02, OECD Publishing, Paris.

He, Z. L. and Wong, P. K. (2004). Exploration vs. exploitation: an empirical test of the ambidexterity hypothesis, *Organization Science*, 15, pp. 481–494.

Heidenreich, M. (2009). Innovation patterns and location of European low- and medium-technology industries, *Research Policy*, 38, pp. 483–494.

Hervas-Oliver, J. L., Garrigos, J. A. and Gill-Pechuan, I. (2011). Making sense of innovation by R&D and non R&D innovators in low technology contexts: a forgotten lesson for policymakers, *Technovation*, 31, pp. 427–446.

Hirsch-Kreinsen, H. (2008). "Low-technology": a forgotten sector in innovation policy, *Journal of Technology Management & Innovation*, 3(3), pp. 11–20.

Hirsch-Kreinsen, H. (2015). The power of the practical, *University of Auckland Business Review*, 18(1), p. 8.

Hirsch-Kreinsen, H. and Jacobson, D. (2008). *Innovation in Low-Tech Firms and Industries*. Edward Elgar Publishing, Cheltenham, p. 320.

Hirsch-Kreinsen, H., Jacobson, D. and Robertson, P. L. (2006). "Low-tech" industries: innovativeness and development perspectives — a summary of a European research project, *Prometheus*, 24(1), pp. 3–21.

Hoffmann, K., Parejo, M., Bessant, J. and Perren, L. (1998). Small firms, R&D, technology and innovation in the UK: a literature review, *Technovation*, 18, pp. 39–55.

Laursen, K. and Salter, A. (2006). Open for innovation: the role of openness in explaining innovation performance among U.K. manufacturing firms, *Strategic Management Journal*, 27, pp. 131–150.

Lee, S., Park, G., Yoon, B. and Park, J. (2010). Open innovation in SMEs — an intermediated network model, *Research Policy*, 39(2), pp. 290–300.

Miles, M. B., Huberman, A. M. and Saldaña, J. (2014). Qualitative data analysis: a methods sourcebook, SAGE Publications, Thousand-Oaks, California, p. 373.

Narula, R. (2004). R&D collaboration by SMEs: new opportunities and limitations in the face of globalisation, *Technovation*, 24, pp. 153–161.

Nieto, M. J. and Santamaria, L. (2010). Technological collaboration: bridging the innovation gap between small and large firms, *Journal of Small Business Management*, 48(1), pp. 44–69.

Nouman, M., Warren, L. and Thomas, S. R. (2011). Researching the "forgotten sector": low and medium tech (LMT) innovation — present light on future trends, *Proceedings of the 1st International Technology Management Conference, ITMC* 2011, DOI: 10.1109/ITMC.2011.5995934.

OECD (2011). Rev, I.S.I.C., 3. Technology intensity definition. http://www.oecd.org/sti/ind/48350231.pdf. (Accessed: June 2017).

Pavitt, K. (1984). Sectoral patterns of technical change: towards a taxonomy and a theory, *Research Policy*, 13(6), pp. 343–373.

Penrose, E (1959). *The Theory of the Firm* (John Wiley & Sons, New York, NY).

Pullen, A., Weerd-Nederhof, D., Groen, A., Song, M. and Fisscher, O. (2009). Successful patterns of internal SME characteristics leading to high overall innovation performance, *Creativity and Innovation Management*, 18(3), pp. 209–223.

Radas, S. and Božić, L. (2009). The antecedents of SME innovativeness in an emerging transition economy, *Technovation*, 29(6), pp. 438–450.

Rammer, C., Czarnitzki, D. and Spielkamp, A. (2009). Innovation success of non-R&D-performers: substituting technology by management in SMEs, *Small Business Economics*, 33(1), pp. 35–58.

Robertson, P., Smith, K. and von Tunzelmann, N. (2009). Innovation in low- and medium-technology industries, *Research Policy*, 38, pp. 441–446.

Rothwell, R. (1992). Successful industrial innovation: critical factors for the 1990s, *R&D Management*, 22(3), pp. 221–240.

Rothwell, R. (1994). Towards the fifth-generation innovation process, *International Marketing Review*, 11(1), pp. 7–31.

Rothwell, R. and Dodgson, M. (1991). External linkages and innovation is small and medium-sized enterprises, *R&D Management*, 21, pp. 125–138.

Rothwell, R. and Zegveld, W. (1986). *Innovation and the Small and Medium Sized Firm* (Francis Pinter, London).

Santamaria, L., Nieto, M. J. and Barge-Gill, A. (2009). Beyond formal R&D: taking advantage of other sources of innovation in low-and medium-technology industries, *Research Policy*, 38, pp. 507–517.

Som, O. and Kirner, E., eds. (2015). *Low-Tech Innovation: Competitiveness of the German Manufacturing Sector*. Springer, Berlin.

Spithoven, A., Clarysse, B. and Knockaert, M. (2011). Building absorptive capacity to organise inbound open innovation in traditional industries, *Technovation*, 31(1), pp. 10–21.

Teece, D. J., Pisano, G. and Shuen, A. (1997). Dynamic capabilities and strategic management, *Strategic Management Journal*, 18(7), pp. 509–533.

Vale, M. and Caldeira, J. (2008). Fashion and the governance of knowledge in a traditional industry: the case of the footwear sectoral innovation system in the northern region of Portugal, *Economics of Innovation and New Technology*, 17(1–2), pp. 61–78.

Van de Vrande, V., de Jong, J. P. J., Vanhaverbeke, W. and de Rochemont, M. (2009). Open innovation in SMEs: trends, motives and management challenges. *Technovation*, 29(6), pp. 423–437.

Vanhaverbeke, W., Vermeersch, I., De Zutter, S. (2012). Open innovation in SMEs: How can small companies and start-ups benefit from open innovation strategies? *Research Report*. Flanders DC: Vlerick Leuven Gent Management School.

Von Tunzelmann, N. and Acha, V., (2005). Innovation in 'low-tech' industries, Eds. Jan Fagerberg and David C. Mowery, *The Oxford Handbook of Innovation* (Oxford Handbooks, Oxford, U.K.), pp. 407–432.

Yin, R. K. (2009). *Case Study Research: Design and methods* (Sage publications), p. 312.

Part IV

Sectoral Patterns of Open Innovation in SMEs

Chapter 9

Open Innovation Practices of Clustered SMEs: The Intermediate Role of Cluster Governance

Anne Berthinier-Poncet[*,‡],
Simona Grama-Vigouroux[†,§] *and Sana Saidi*[†,¶]

[*] *Conservatoire National des Arts et Métiers*
Laboratoire LIRSA, Paris, France
[†] *South Champagne Business School,*
Troyes, France
[‡] *anne.berthinier-poncet@lecnam.net*
[§] *simona.grama@get-mail.fr*
[¶] *sana.saidi@get-mail.fr*

Abstract

Little is known about open innovation (OI) of small and medium-sized enterprises (SMEs) colocated in clusters, although these specific agglomerations seem to offer SMEs some fruitful opportunities for networking and open innovation. This chapter aims to fill this

[*] Corresponding author.

gap and contribute to a better understanding of the intermediate role of cluster governance in supporting open innovation practices of clustered SMEs. Following the framework of Lee *et al.* (2010) on open innovation intermediaries, we examine a single case study of the Technopole de l'Aube in the Champagne Region of France. We carried out 26 face-to-face semistructured interviews with technopolitan firms' chief executive officers and governance's members, between June 2014 and January 2015. Our findings highlight direct and indirect support for OI projects of the technopolitan SMEs. However, our case study also suggests that cluster governance has to pay more attention to the support of collaborative projects and to allocate additional resources for developing an organizational climate for OI.

9.1. Introduction

Innovation is regarded as a major source of performance improvement for both large firms and small and medium-sized enterprises (SMEs). Innovation strategies and processes progressively have opened up, with firms increasingly relying on collaboration to acquire or exploit external knowledge (Brunswicker and Vanhaverbeke, 2015; Laursen and Salter, 2006; Verbano *et al.*, 2015; Brunswicker and Vanhaverbeke, 2015; Vanhaverbeke, 2017) to develop new products, services and processes, or to commercialize them faster and more successfully (Huizingh, 2011; Verbano *et al.*, 2015). Using external knowledge to improve innovation processes and developing collaboration to find sources of external commercialization is a relatively new strategy (Huizingh, 2011). This general trend of opening up the innovation process from a traditional, closed model to an open one has been popularized by Chesbrough under the concept of "open innovation" (OI) (Chesbrough, 2003), which can be defined as "the use of purposive inflows and outflows of knowledge to accelerate internal innovation and expand the markets for external use of innovation, respectively" (Chesbrough, 2003, p. 1).

Despite the increasing interest of academics and practitioners in the concept of OI, most studies have focused on large, technology-based companies, excluding SMEs from the mainstream discussion, even though scholars agree on their growing role in innovation (Brunswicker and Vanhaverbeke, 2015; Freel, 2005; Lee *et al.*, 2010; Spithoven *et al.*, 2011; Van de Vrande *et al.*, 2009). In a special issue of *R&D Management* on the "Future of open innovation," Gassmann *et al.* (2010) looked at the main trends in OI research and pointed out that even if SMEs are opening up their innovation process as a means of overcoming their "liability of smallness," lack of resources and catching up with the OI concept, the field remains under-researched.

Parallel to this literature on OI, the network and cluster literature highlights the impact of geographical proximity and networking on the innovation performance of small and medium-sized companies. It suggests that colocating in a cluster might be an interesting solution to overcome challenges for SMEs in terms of resources or size (Becattini, 2002; Iansiti and Levien, 2004; Porter, 1998). Clustering has effectively received constant attention as a means to boost SMEs' innovation and competitiveness (Baptista and Swann, 1998; Carpinetti *et al.*, 2007; McCann and Folta, 2011).

However, very few studies link OI, SMEs and clustering (Huang and Rice, 2013; Lee *et al.*, 2010; Vanhaverbeke, 2006), and fewer discuss the role of the management teams as intermediaries in supporting OI for SMEs (Lee *et al.*, 2010). Mutambi (2011), for example, looks at business incubators as OI intermediaries and suitable spaces that encourage OI for very small firms. This analysis is supported by other studies that underline the role of management teams in supporting networking and a culture of openness in business incubators (Hansen *et al.*, 2000; Bollingtoft, 2012). Nevertheless, the role played by cluster governance in encouraging OI practices of the clustered SMEs is not investigated yet (Vásquez-Urriago *et al.*, 2016).

In this chapter, we therefore specifically investigate the intermediate role of the governance of a French technopole in enabling SMEs' collaboration and OI practices. Following the work of Lee *et al.*,

(2010) that points to the importance of an intermediated network as an effective model to enable OI for SMEs, we conduct a qualitative study in a French technopole, the Technopole Aube-en-Champagne (TAC). We analyze the practices of the technopole's governance to enhance the adoption and implementation of OI practices for its SMEs members. Our empirical study is based on the interviews of 26 members of the TAC, 19 chief executive officers (CEOs) of firms belonging to the TAC and 7 governance members.

9.2. The Role of Cluster Governance on Open Innovation Practices

9.2.1. *Open innovation practices of SMEs and the role of clusters*

Although OI has been proposed as a new paradigm for innovation management (Van de Vrande *et al.*, 2009; Chesbrough, 2003), relatively little data has been collected on the practices of OI for SMEs (Lichtenthaler, 2008). The reasons are numerous but are mostly due to the limited ability of SMEs to access external resources and the smaller number of technological assets they can exchange (Brunswicker and Vanhaverbeke, 2015; Lee *et al.*, 2010; Vanhaverbeke, 2017). Compared to larger firms that have the resources to scan their environment and find potential partners (Lichtenthaler and Lichtenthaler, 2009), SMEs lack the financial resources to gather vital information and knowledge and decide with whom to collaborate (Lee *et al.*, 2010), making it very difficult to develop a network (Simard and West, 2006). They also lack access to larger knowledge stocks in order to develop higher absorptive capacity that will benefit their innovation performance (Cohen and Levinthal, 1990; Spithoven *et al.*, 2011). Absorptive capacity is understood indeed as the firm's ability to identify, assimilate and exploit knowledge from the environment (Cohen and Levinthal, 1990).

One possible means for SMEs to overcome these constraints and develop OI practices might be to be located in or join innovation clusters (Carpinetti *et al.*, 2007; Karaev *et al.*, 2007; McCann and

Folta, 2011). Clusters are geographical concentrations of interconnected companies, specialized suppliers, service providers, and associated institutions (e.g., universities, standard agencies, trade associations) in a particular field, which both compete and cooperate (Porter, 1998). They have an impact on the formation of a critical mass of SMEs and start-ups and the creation of OI opportunities (Karaev *et al.*, 2007), facilitating the "connection and development" considered as the operational philosophy of OI (Huang and Rice, 2013).

Moreover, the geographical and cognitive proximity of clustered firms facilitate explicit and tacit knowledge transfer as well as the development of their knowledge base, enhancing therefore the absorptive capacity of SMEs (McCann and Folta, 2011). Thus, it facilitates the recognition of innovation opportunities in the case of collaborative projects and is regarded as an essential parameter for developing OI practices (Muscio, 2007; Vanhaverbeke *et al.*, 2008; Arikan, 2009; Spithoven *et al.*, 2011).

However, these studies do not tell us anything about *how* cluster governance can facilitate collaboration between cluster members and the development of OI practices. A stream of literature has recently questioned the role of cluster governance in the creation of innovation practices to foster a specific institutional environment conducive to more collaboration and innovation (Bahlmann and Huysman, 2008; Berthinier-Poncet *et al.*, 2011; Arikan, 2009; Bocquet and Mothe, 2010) and calls for more research into this relevant topic.

9.2.2. *The role of technopoles for open and collaborative innovation*

For Jung and Andrew (2014), an institutional collective action is necessary to create an environment favorable to building R&D collaboration between SMEs, identifying key actors, potential risks, and incentives to form collaborative activities. Technopoles, defined as a concentration of innovative companies (often high-tech), public and private research units, and higher educational institutions in a given geographical area (Berthinier-Poncet *et al.*, 2015), play an important role in establishing this favorable culture for a regional sustainable innovation (Zeng *et al.*, 2010a). Similar to Italian and Spanish science and technology parks, French technopoles were created by the 1980s. Centrally planned and established as a part of a regional development policy, technopoles are mainly located in urban environments, aim at fostering synergies between R&D activities (mainly with local universities) and industry, and focus on the creation of new firms (Vásquez-Urriago *et al.*, 2016).

Vásquez-Urriago *et al.* (2016) recently provided empirical evidence that location in a technopole increases the likelihood of cooperation for innovation and positively affects the intangible results of cooperation (i.e. increased ability to formulate strategies, enhanced human resources, better management of information and relationships). These positive effects of technopoles on OI have to be found in the creation of "an environment that spurs the creation and development of technological and non-technological ecosystems" (Lopez-Vega and Vanhaverbeke, 2010, p. 21).

The issue of cluster governance is crucial in that it can sustain the development of this specific environment. Cluster governance is here understood as a management and coordination structure that plays a distinctive role in creating an institutional environment conducive to open and collaborative innovation practices (Berthinier-Poncet *et al.*, 2011). An effective management of the technopole can thus facilitate access to flows of information and knowledge such as technological change, sources of technical assistance, market requirements, and stronger competitive edge (Inkpen and Tsang, 2005; Lee *et al.*, 2010). Crespin-Mazet *et al.* (2013) confirm the role of a third party in facilitating OI processes between firms in the cluster by initiating contacts between development partners and key customers, and organizing knowledge transfer.

Therefore, technopoles should not only provide physical settings for innovating firms (as it has been often criticized (Asheim, 2007)), but they should play an "orchestrating" role (Lopez-Vega and Vanhaverbeke, 2010) by providing access to a large face-to-face collaboration network, sharing research facilities, and therefore improving connectedness within the ecosystem, as well as an "animateur" role (Howells, 2006) of creating new possibilities and dynamism within the innovation ecosystem.

These different activities draw the outlines of the intermediate role of the technopole's governance in developing OI practices for its firms' members.

9.2.3. *The intermediate role of cluster governance for OI practices*

According to Lopez-Vega and Vanhaverbeke (2010), innovation intermediaries facilitate the development of OI practices for SMEs through three main functions: (1) connection, (2) collaboration and support, and (3) provision of technological services. Technopoles act as innovation intermediaries in so far as they facilitate technology transfer between firms and universities, act as knowledge hub (easier access to scientific and technological resources for innovation), and provide a privileged access to a large face-to-face collaboration network (Lopez-Vega and Vanhaverbeke, 2010).

In order to support SMEs' ability to construct an open and collaborative innovation network, Lee *et al.* (2010) suggest a collaboration model that emphasizes the intermediate role of cluster governance. This conceptual model highlights both *direct* and *indirect* support activities. Direct activities encompass (1) the network database, (2) the network construction, and (3) the network management.

First, the main purpose of the *network database* is to identify suitable partners by collecting information on technologies, markets, and competitors. By facilitating the access to and the appropriation of new knowledge, the network builds the absorptive capacity of its members, an essential lever for developing OI practices. In innovation clusters, the higher the absorptive capacity, the more SMEs benefit in return from agglomeration — "local buzz" (Bathelt *et al.*, 2004) — building specific and combinative capabilities (McCann and Folta, 2011), and from external and global pipelines (Wolfe and Gertler, 2004).

Second, the cluster governance can impact directly on the *network construction* by offering an adapted structure that facilitates matching and networking between the potential partners and encourages geographical clustering. Following Lee *et al.* (2010), the cluster governance acts as a third-party warranty to support technology transfer and facilitate cooperation between potential partners who might otherwise be reluctant to disclose detailed R&D information or to cooperate on a collaborative project. Moreover, normative practices such as defining a global strategy and an identity specific to the cluster, creating innovation communities, facilitating the formalization of exchanges, and developing collaborative projects will facilitate the adhesion of clustered SMEs to the network (Berthinier-Poncet *et al.*, 2015).

Finally, the *network management* is useful not only as it can solve potential conflicts between partners, but mainly because it offers a regulative framework and consulting services (general management, information on tax, law, process innovation, etc.). Berthinier-Poncet *et al.* (2015) highlight the importance of regulatory and constitutive practices of the cluster governance in order to limit the risks of opportunism within the cluster and strengthen membership cohesion by defining the scope of the cluster cooperative framework.

Two sets of indirect activities complete the framework of the intermediate role of cluster governance: (1) the *development of a culture of collaboration* and (2) the *facilitation of collaboration* (Lee *et al.*, 2010). Previous studies on science parks and technopoles highlighted the controversial fact that firms were locating in technopoles mainly for reputation-building purposes and for creating an innovative firm image rather than for networking and knowledge exchanges (Chan *et al.*, 2010; Huber, 2012; Asheim, 2007). To enhance SMEs' OI practices, cluster governance must therefore foster culture that encourages innovation, entrepreneurship, collaboration, and a proper environment conducive to the free flow of innovation, ideas, and know-how (Zeng *et al.*, 2010b).

Table 9.1 summarizes the framework we will use to analyze our case study.

Table 9.1. Analytical framework for the intermediate role of cluster governance.

Cluster governance activities for sustaining OI practices		Details of the practices
Direct	Network database	• Identification, codification, and dissemination of new and external knowledge between cluster members (common knowledge database) • Collect information on technologies, markets, and competitors; strategic innovation watch • Create a detailed database on cluster members
	Network construction	• Define a global strategy and a specific identity for the cluster • Develop networking events (formal and informal) • Develop collaborative projects for innovation (act as an innovation intermediary) • Enhance technology transfer between university and firms within the cluster (closer relationships between the two communities) • Help formalize collaborative projects (consortia agreements)
	Network management	• Build a strategic and an operational governance body to manage the cluster • Define rules for selection and recruitment of cluster members • Implement regulation rules (formal and informal disciplinary mechanisms) and define respective roles of cluster members • Offer consulting services (tax, law, finance of innovation projects) and general services
Indirect	Culture of collaboration	• Collaboration needs analysis • Develop innovation communities • Disseminate an innovative culture oriented on openness
	Facilitation of collaboration	• Collaboration process analysis • Collaboration benchmarking • Develop links with institutional actors of the innovation ecosystems.

Source: Adapted from Lee *et al.* (2010) and Berthinier *et al.* (2015).

9.3. Methodology

After describing the peculiarity of our single case study, we detail our data collection and analysis and highlight the main OI characteristics of the SMEs belonging to our sample.

9.3.1. *Description of the case study*

Empirical research focused on the OI practices adopted by SMEs is scarcely developed (Lee *et al.*, 2010). There is very little evidence related to SMEs operating in a geographically bounded cluster. To fill this gap, we chose to study the OI practices of SMEs located in a "technopole." More precisely, we chose the Technopole Aube-en Champagne (TAC) to identify the OI practices of cluster governance for numerous reasons. First, the TAC is a member of the International Association of Science Parks network, certified CEEI since 2002 and labeled French Tech since 2015. Second, within the TAC, the majority of firms are SMEs, mostly very small firms (from one to ten employees), working in the service sector (knowledge-intensive business services, called KIBS; see Table 9.2). Third, the easiness to access to data also plays an important role to conduct our empirical research, as the authors have privileged professional links with the TAC. Finally, we also complete previous empirical research on French technopoles through this analysis.

The TAC was set up in a region with a strong textile tradition that was hit by a major crisis between 1973 and 1996. In view of this crisis, the President of the Aube General Council supported the creation of the University of Technology of Troyes (UTT) in 1997, followed by the TAC in 1998 at the same site. The TAC was aimed initially at UTT students to make it easier for basic research to be transformed into profitable businesses. Today, the TAC is a technology park of 70 hectares, housing 45 firms, a student and a business incubator, Troyes Business School, and the UTT (Technical University of Troyes). It should be noted that there are no real sectorial themes in the choice of location of companies within the TAC.

The governance of the TAC is ensured by an autonomous organization, the SEMTAC (Society of Mixed Economy of the Technopole of Aube in Champagne), which has its elected representatives and its own budget. The SEMTAC votes all the decisions of the TAC. It is important to note that the Regional Council is both the shareholder and principal agent of the TAC: It owns all the premises and is responsible for its management and maintenance. Within the SEMTAC, there are two distinct governance bodies: strategic governance and operational governance. At the strategic level, we find the main bodies: a president (mayor of the commune), an executive director (external and currently the director of the Troyes Business School), an assistant director, and a board of directors who validate the budget for the TAC at the end of the year and the annual general meeting of shareholders (General Meeting) which takes place once a year in June. It should be pointed out that strategic governance members are exclusively institutional; so, no TAC companies are part of it. At the operational level, the TAC is managed by eight employees (including the assistant director) who are under the authority of the general manager, who has been in charge since the TAC was created. The assistant director is responsible for four operational divisions in order to better focus the governance action on the innovation needs of the member companies and to facilitate the synergies between research and entrepreneurship: (1) prospection, animation, communication; (2) project incubation; (3) business services and (4) student entrepreneurship (in association with higher education institutions, UTT, and ESC Troyes). That is why the operational governance is made up of eight permanent members to manage the day-to-day business of the technopole. To sum up, the TAC counts 12 strategic and operational governance members (see Table 9.2).

9.3.2. *Data collection and analysis*

Following the recommendations of Edmondson and McManus (2007), we adopted a qualitative analysis based on a single case study in order to identify the practices implemented and to evaluate their effectiveness according to the technopole actors. We carried out 26 face-to-face semistructured interviews between June 2014 and January 2015. These interviews were mostly conducted with 19 CEOs from different companies and 7 governance members.

The choice of these companies was approved by the director of the technopole. This choice is based on the diversity of the actors in order to have a representative sample extracted from the initial population which is made up of 45 SMEs operating in various sectors of activities (manufacturing industries, construction, wholesale and retail trade, information and communication, financial and insurance activities, specialized, scientific and technical activities, business support activity, and associative sector). From the initial population of SMESs operating as KIBS (75.6%) and non-KIBS (24.4%), we identified a sample of firms composed by 15 KIBS and 4 non-KIBS companies. Moreover, we were able to interview most of the strategic and operational governance members within the TAC.

We also collected additional data by means of direct observation and participation (Plug and Start events), and external (website, institutional brochures) and internal (statutes, activity reports, members' details, etc.) documentation. This allowed us to triangulate our data in order to increase its analytical robustness.

The characteristics of the TAC's companies and its governance's members can be summarized as in Table 9.2.

Data analysis was performed in two stages. First, the interviews were triple coded separately, after retranscription, to ensure the validity and quality of data processing. The coding is based on a grid developed from the analytical framework resulting from our review on OI practices for SMEs and the role of cluster governance for developing these practices.

Once the coding was completed, we proceeded to the second stage of our analysis by examining data extracted from the interviews. In this way, we carried out four types of special work. First, we analyzed the two series of interviews taken separately. We began with vertical interview-by-interview analyses to find out what was answered by an individual for all given themes. Horizontal analyses for each of the 18 OI practices were conducted for both enterprises (noted E for the verbatims in the results part) and members of governance (noted G for the verbatims). Then, we proceeded to syntheses (both horizontal and vertical) in order to have a global vision related to the studied theme.

Table 9.2. The characteristics of TAC's companies and governance's members.

<table>
<tr><td></td><td colspan="3">TAC' s initial population characteristics</td><td colspan="2">Selected sample</td></tr>
<tr><td>Companies (100% SMEs)</td><td>Sectors of activity</td><td>Number of SMEs</td><td>KIBS/non-KIBS SMEs</td><td>Number of SMEs</td><td>KIBS/non-KIBS SMEs</td></tr>
<tr><td rowspan="8"></td><td>Manufacturing industries</td><td>6</td><td rowspan="4">11 non-KIBS (24.4%)</td><td>3</td><td rowspan="4">4 non-KIBS</td></tr>
<tr><td>Construction</td><td>1</td><td>0</td></tr>
<tr><td>Wholesale and retail trade</td><td>3</td><td>0</td></tr>
<tr><td>Associative sector</td><td>1</td><td>1</td></tr>
<tr><td>Information and communication</td><td>19</td><td rowspan="4">34 KIBS (75.6%)</td><td>8</td><td rowspan="4">15 KIBS</td></tr>
<tr><td>Financial and insurance activities</td><td>1</td><td>0</td></tr>
<tr><td>Specialized, scientific and technical activities</td><td>11</td><td>6</td></tr>
<tr><td>Business support activity</td><td>3</td><td>1</td></tr>
<tr><td colspan="2">Total</td><td colspan="2">45</td><td colspan="2">19</td></tr>
<tr><td rowspan="4">Governance's members</td><td colspan="3">Initial population characteristics</td><td colspan="2" rowspan="2">Selected sample</td></tr>
<tr><td>Position characteristics</td><td colspan="2">Number of members</td></tr>
<tr><td>Strategic governance</td><td colspan="2">4</td><td colspan="2">2</td></tr>
<tr><td>Operational governance</td><td colspan="2">8</td><td colspan="2">5</td></tr>
<tr><td colspan="2">Total</td><td colspan="2">12</td><td colspan="2">7</td></tr>
</table>

These analyses clarified the individual and collective points of view of the interviewees regarding OI practices and the role of cluster governance within the TAC. Table 9.3 reproduces the logic of analysis and synthesis pursued.

Table 9.3. Analysis of interviews.

	Interview 1	*Interview 2*	*Interview 3*	*Interview 4*		*Horizontal synthesis*
Theme 1						
Theme 2						
	Vertical synthesis	Vertical synthesis	Vertical synthesis	Vertical synthesis		

9.3.3. *OI characteristics of TAC's SMEs*

9.3.3.1. *Collaboration Patterns of Clustered SMEs*

Technopolitan SMEs choose to integrate TAC for many reasons: finding *internal* partners (firms, research labs, institutions, all members of the technopole) as well as *external* partners (not belonging to the TAC but potentially related to the technopole's members); obtaining privileged access to financers, other clusters, academics and researchers, innovation support institutions, and other entrepreneurs' support and advisory services; being part of a favorable environment for innovation, proximity with other entrepreneurs. Technopolitan SMEs collaborate with research centers and universities, in particular with the laboratory of the University of Marseille, and nonprofit and government organizations, for example, the Regional Incubator CARINNA. Concerning collaboration with firms, technopolitan SMEs cooperate with internal and external firms. They mainly work with clients and suppliers to develop their innovation (see also Vanhaverbeke, 2017).

9.3.3.2. *Open Innovation Activities of Clustered SMEs*

SMEs' CEOs are involved in OI projects especially in the format of *outside-in* activities, absorbing external knowledge from outside partners in multiple forms such as buying licenses or new software, or using technologies available in open source. However, very few technopolitan SMEs are engaged in *inside-out* activities like opening up their API terminals or finding new ways to market their technologies such as using a FREEMIUM model. A small number of the firms practice the coupled process of OI through a joint property patent or the cocreation of a platform. These coupled processes are also performed with professionals involved in the joint development of the product, and also with close partners.

9.3.3.3. *Open Innovation Barriers for Clustered SMEs*

SMEs belonging to the TAC encounter some difficulties regarding the OI projects related to many areas: human resources, finance, market, R&D management, and difficulties associated with OI projects. In human resources management, SMEs' CEOs complain about the shortage of appropriate skills. This means firms are obliged to search for new competencies outside. They also complain about the difficulties related to the organizational culture and those related to the instauration of a more participative leadership approach, which is appropriate for the OI context. The majority of SMEs' CEOs also face financial problems in pursuing their innovation projects. Entrepreneurs understand that they have to pay a lot of money in order to attract the best talent with the right skills to the firm. In the marketing field, entrepreneurs from technopolitan SMEs are aware of the constraints linked to the time to market. Either the market changes too quickly or they need to constantly adapt. They also complain about the uncertainty linked to innovative products, claiming that it is difficult to find the right time to conquer a new market or to launch a new product. Concerning the R&D management challenges, SMEs encounter two major barriers. The first is related to the difficulty of finding appropriate R&D partners due to their small size or geographical location. The second barrier is linked to the technological knowledge management to find either an appropriate empirical application or the right technologies needed for the development of the products and services. Concerning the difficulties associated with OI projects, some entrepreneurs point to the wide diversity of partners' profiles, which sometimes impacts negatively on collaboration.

9.4. Results

Our findings relate to the support given by the TAC governance organization for SMEs' collaborative projects. There are two categories linked to the direct and indirect support of cluster governance on OI practices.

9.4.1. *Direct support of cluster governance*

Concerning direct support, there are three aspects to take into account: the network database, network construction, and network management.

9.4.1.1. *Network Database*

Network database refers to how governance members manage their knowledge of partners and the technopole's environment (e.g., market, trends and technologies). Governance's bodies are convinced that they are making good efforts to improve the database of TAC with innovative projects and ideas (G3: "We're working on the local ecosystem to identify innovative business ideas and to do all we can to develop them"; G6: "For us, innovation is based on getting more and better quality business ideas"). However, both SMEs and TAC members report that knowledge management inside the technopole needs to be improved suggesting the need for a platform so as to facilitate the exchange of information on TAC's firms' profile and external partners (E_8: "a platform where you can internally value all experiences, skills and business expertise capitalized"; E_6: "It seems to be a topic that can unify and make the technopolitan enterprises work together").

At the same time, entrepreneurs regret that TAC's governance members do not use all their knowledge about their partners to facilitate collaboration. In this purpose, they also suggest that the TAC's operational governance makes an in-depth analysis of technopolitan SMEs' profile that could promote matching and collaboration (E7: "Typically, if the experts are also in these firms [technopolitan SMEs], why not have advisors by business field... These advisors could act as an intermediary for topics related to this field").

9.4.1.2. *Network Construction*

Network construction concerns governance practices to support the collaboration between members. Technopolitan SMEs are aware of the formal and informal procedures implemented by the governance's bodies that support collaboration such as via different events, for example: Plug and Start competition, breakfasts, and barbecues (E8: "We had some insights, let's say, from these events, moments, breakfasts and others, we discussed it and said: 'look, it would be interesting to meet each other to see if there are any synergies'; and then we met but there are no synergies so far"; E13: "The Plug & Start ... is a time to get together, to meet each other. There's also the annual barbecue, which can be a friendly moment. There are the breakfast mornings, when a new business arrives and we're invited to meet it [the new contacts breakfast]"; "G3: "barbecue, it works well (...) it is a friendly event and it helps to talk with families too that are coming, that's important").

Despite these events implemented by the cluster governance and aimed at facilitating collaboration, entrepreneurs need to collaborate more with other technopolitan firms, but they feel that they are not enough supported (I6: "... the governance's team of TAC could better help us to know each other and to meet more regularly, more often."). On their side, governance's members explain that collaborative projects are interesting but they are very risky and costly for the SMEs' CEOs (G3: "Collaborative projects need a different approach and, for some entrepreneurs, they need to learn to be an entrepreneur first and can worry about collaborative projects later"; G1: "it's dangerous, it's true that the ease of work together, it is immediate, it is well, it is funny, it is friendly, but it does not solve my cash flow problems").

9.4.1.3. *Network Management*

Network management is related to all the roles and services delivered by the governance. Technopolitan SMEs understand that governance plays an important role in their innovation projects through the help given in terms of infrastructure and logistics (E14: "The direct support for innovation is not in their remit, but what is important for them is the contribution they offer you in terms of infrastructure and organization. For example, we count on them to organize meetings with important people, and that's great").

However, entrepreneurs suggest to the governance the need for a physical space that foster collaboration (E6: "And then they have to try to recreate a place where we must always come across the others, at least fairly regularly"). Governance's bodies are also offering consulting and training services (E18: "It [the governance] can give advice. It can help us with specific issues...I hear them sometimes when they talk about start-ups; they attract the attention of the entrepreneur by saying to him: watch out because maybe that's not the right solution"; E_{12}: "Yes, the management team play a real role, they're available every day for questions, giving advice"). They also supervise the start-ups' evolution and search for appropriate solution in their network to help TAC's firms to exceed their difficulties (E7: "... the panel of experts (mobilized by TAC), which is the only way to see the progress in a project ..."; G1: "When we feel that there's been a substantial negative impact, we search in our network for the right people to solve the situation (for TAC's firms) if possible"). All technopolitan SMEs have to demonstrate their evolution by presenting regularly some indicators (G1: "... there is also a certain rhythm imposed for reports, indicators and support").

The companies feel that TAC's governance pays close attention to the companies' operations (E_{17}: "There are very regular contacts in the initial stages of locating here for businesses..."), while the governance body claims the opposite (G7: "For our part, we monitor from a distance to ensure that the company is not heading for disaster").

Governance's members are also involved in resolving conflicts in the case of collaboration, playing the role of a mediator (I8: "I think it's more the mediator, so I think that if they're the meditator, the project needs to have a strong local presence, and the technopole has to be known and recognized in the project"; G5: "the best for us is to play a mediator role but we will not interfere in such conflicts").

9.4.2. *Indirect support of cluster governance*

Concerning the indirect support given by the governance, there are two dimensions to consider: the culture of collaboration and the facilitation of future collaboration.

In terms of collaboration culture, the interviewees declared that they felt isolated and not very close to other technopolitan enterprises (I3: "The technopole is something quite closed. Everyone has his office ...It's pretty isolated, so people don't meet"). This feeling is not highlighted by cluster governance, which feels quite optimistic about the organizational climate that they implemented inside of TAC (G7: "we managed to create a spirit that makes people enjoy to meet, discuss and eventually create things together").

Some enterprises also consider that the governance body organizes some events to encourage future collaboration, such as the Plug and Start event for instance which offers good contacts for the business (I7: "...What's good here [Plug and Start] is that we meet the CEOs of other companies: it's with them that we acquire experience in recruiting, on everything related to organizational methods, working contracts, commercial networks and so on"). The governance's members and SMEs also propose some common ways to facilitate future collaborations inside of the TAC, as for example the realization of a road map for success for all entrepreneurs or a digital FAB LAB. Moreover, technopolitan SMEs propose some meetings with the other technopolitan firms' owners that could encourage fruitful exchanges of information and knowledge and that could facilitate the OI projects (E6: "Why not organize a meeting of one and a half to two hours per quarter between companies' directors, between relevant persons").

The results of our study are summarized in Table 9.4. For each type of support given by cluster governance for OI practices of technopolitan SMEs, we show the existent practices and the ones recommended for the future.

9.5. Discussion

This chapter focuses on the support given by the governance's organization for the OI projects of the technopolitan SMEs. Our findings represented by the testimonials of firms' owners and managers of the governance's organization helped us to understand the intermediate role played by the TAC's governance in enhancing OI

Table 9.4. Summary of results.

TAC Governance's Support	Practices	Details of the practices
Direct support	Network data base	**Existent:** • TAC's governance exploits regularly the local ecosystem to develop new business ideas • TAC's governance recruits a big diversity of innovative projects **Recommended for the future:** • TAC's governance needs to implement a tool to facilitate the exchange of information on TAC's members and external partners and that could facilitate the collaborative projects
	Network construction	**Existent:** • TAC's governance organizes some events that encourage networking (Plug and Start) **Recommended for the future:** • TAC's governance has to give more attention and resources for the collaborative projects between technopolitan firms
	Network management	**Existent:** • TAC's governance is offering satisfactory infrastructure for offices and meeting • TAC's governance if offering good consultancy and training services • TAC's governance is a good mediator in collaborative projects **Recommended for the future:** • TAC's governance must involve more efforts in creating a common space to facilitate interactions

(Continued)

Table 9.4. (*Continued*)

TAC Governance's Support	Practices	Details of the practices
Indirect support	Culture for collaboration	**Existent:** • TAC's governance is not aware that technopolitan SMEs feel isolated **Recommended for the future:** • TAC's governance need to create an organizational climate more focused on openness to foster collaborations
	Future collaborations	**Existent:** • TAC's governance realizes some events as Plug and Start that facilitate future collaborations **Recommended for the future:** • TAC's governance must put in place a digital FAB LAB, that we will make firms to collaborate and use common resources • TAC's governance must create a map for success for TAC's business owners, underlying the role of collaborations as a determinant for innovation performance • TAC's governance must establish regular meetings between technopolitan SMEs to foster future collaborations

practices for SMEs in clusters. We first discuss the results concerning the direct support activities, and second the indirect support activities of the cluster governance. We then highlight the theoretical and managerial implications of our findings.

9.5.1. *Direct support of cluster governance on OI practices*

Concerning the direct support the governance body provides for the innovative project, three aspects have to be taken into consideration: network database, network construction, and network management (Lee *et al.*, 2010).

Concerning the network database, entrepreneurs report the need for a tool that promotes easier collaboration and a better matching between SMEs. TAC's governance needs to put in more efforts to target appropriate partners in order to foster collaborative projects. Previous findings showed that the number, the diversity, and the use of partners are positively linked to performance (Lee *et al.*, 2010). This tool could also facilitate the communication between TAC's members. Communication is very important, especially in OI projects, since it facilitates knowledge of the project's objective, as well as information linked to each member's profile (Du Chatenier *et al.*, 2009; Vanhaverbeke, 2017). Effective communication also serves to direct, coordinate, and regulate the organization members' activities.

Concerning the network construction, the TAC governance body organizes events to facilitate collaboration such as Plug and Start, an annual barbecue, or morning breakfasts where the startups can meet new firms. TAC organizes some networking events (Plug and Start, breakfasts, barbecues, etc.), but the current efforts are insufficient to result in strategic alliances. These strategic alliances demand specific support activities and resources in order to be maintained (Das and Teng, 2000). Previous studies have shown that the development of collaborative projects foster performance in terms of innovation (Berthinier-Poncet *et al.*, 2011; Bocquet and Mothe, 2010). Involvement in collaborative projects helps SMEs to cover their resource needs in terms of knowledge, finance, and skills (Lee *et al.*,

2010). Since governance members fail to encourage more collaboration, this can have a negative impact on the innovative performance. Previous studies have shown that collaborative projects which are not adequately supported by the clusters impact negatively the performance (Zeng *et al.*, 2010b). The TAC governance's reluctance to support the collaborative projects could be explained by the difficulty to match the big diversity of profiles existent inside the technopole. As we have seen in the methodological part, TAC is a mixed technopole, which recruits a large variety of innovative projects. Consistent with the findings of Du Chatenier *et al.* (2009), too much diversity inside the OI projects can harm the innovative performance. Also, OI projects could encourage the clandestine behavior of a partner who does not have the same involvement in the project as the other members (du Chatenier *et al.*, 2009). This fact leads to conflict between the OI partners. Cluster governance could be reluctant to manage such a conflict because it involves many risks and costs.

Concerning the network management, the governance of the TAC is providing a satisfying infrastructure (IT services, offices, catering) and business support services. These services are important in order to cover the gaps of competencies, knowledge, and also financial resources (Lee *et al.*, 2010). While the technopolitan infrastructure is very important for entrepreneurs, they feel the need for a physical space where they could interact with each other in order to avoid isolation. OI projects require a balance between working in proximity with the OI partners and working on distance (Du Chatenier *et al.*, 2009). Also, in spite of the efforts of the governance to manage relative to the physical space layout (Cooke, 2001; Doloreux, 1999), a solution still needs to be found to build a common space where entrepreneurs can regularly meet.

9.5.2. *Indirect support of cluster governance on OI projects*

Concerning indirect support, two aspects need to be considered: the technopole's culture and the governance organization's actions that encourage future collaboration. Concerning the technopole culture, the governance of the TAC puts the focus on building trust between its members and tries to offer a suitable environment for innovative ideas that are facilitated by exchanges with other entrepreneurs (Lee *et al.*, 2010). A culture based on trust is associated with high performance outcomes (Dovey, 2009). Building trust between members of an OI project also positively influences exchanges and innovative performance (Du Chatenier *et al.*, 2009; Vanhaverbeke, 2017). While the technopole attempts to facilitate trust via the organization of certain events such as Plug and Start, the annual barbecue, and breakfasts, some entrepreneurs complain about a feeling of isolation due to the physical space, and also the lack of information and communication about the other technopolitan members, which can harm innovative performance (Ismail and Abdmajid, 2007). We communicated these results (feeling of isolation, need of a better knowledge of the TAC's internal network of partners) to the governance. Although agreeing on the fact that they could do more to reinforce networking and the spirit of the openness, they argued that they did not have enough human and financial resources to deal with these issues. However, they took our proposals such as organizing joint activities and regular meetings to avoid isolation and develop openness (Conrad and Phillips, 1995, Hansen *et al.*, 2000) into account.

Concerning the actions that encourage future collaboration, the cluster governance uses *Plug and Start*; this is not only a tool to encourage network construction, but also constitutes an opportunity for future collaboration. It is not enough, however, because the entrepreneurs complain and ask for more actions. Cluster governance needs to extend the number and quality of its partners in order to boost the innovation results of the technopolitan SMEs and future collaboration. The number of innovation partners and the degree to which firms use these partners in their innovation activities are highly correlated with innovative performance (Laursen and Salter, 2006).

9.5.3. *Theoretical and managerial implications*

These results have implications for both OI and the cluster governance literature. Concerning the OI literature, we came with some insights regarding OI of clustered SMEs (Gassmann *et al.*, 2010), and we have seen that our findings are consistent with the literature on SMEs (Van de Vrande *et al.*, 2009; Gassmann *et al.*, 2010). We have shown that TAC's SMEs collaborate with research centers and universities, either for technology purchasing or for strategic alliances. Besides these partners, technopolitan SMEs also collaborate with internal and external, and also with nonprofit organizations and governmental institutions. The main motives that encourage TAC's SMEs to enter into collaboration processes are linked to a better fit with market developments, client demand, as well as increased growth or market share. Consistent with previous literature (Van de Vrande *et al.*, 2009), SMEs in the TAC also engage in outside-in strategies made concrete in terms of technology purchase and the use of open-source technologies. They also adopt fewer inside-out strategies or coupled OI practices such as strategic alliances or codevelopment processes with professionals and local partners.

Technopolitan firms are confronted with the same OI barriers as the other SMEs examined in previous literature (Van de Vrande *et al.*, 2009; Lee *et al.*, 2010). Entrepreneurs from the TAC encounter human resource challenges, especially in finding appropriate competencies inside their firms. They also have marketing challenges especially in finding the perfect market launch time of their products or services. They are confronted also with some financial difficulties related to the high costs of technology or skills.

Regarding the cluster governance literature, little is known about OI practices proposed by cluster governance (Berthinier-Poncet, 2014). To cover this gap, our study highlights some governance practices that encourage or limit collaboration in the context of OI. Following the framework proposed by Lee *et al.* (2010), we completed the literature by showing the intermediated role of cluster governance in supporting OI projects. Our results are consistent with the literature (Lee *et al.*, 2010), but came with some new insights by examining both parties of the technopole's ecosystem represented by the governance's members and firms' CEOs. We have detected that there are some common views between the governance's and technopolitan firms, as for example the services provided or the networking events, which facilitate collaborations. There are also some points where the governance's members still need to improve their practices, as for example the support of collaborative projects and the creation of an organizational climate that favor openness.

Our results also have some practical implications for both governance's members and firms' CEOs. Governance's organizations could be encouraged by our results to detect and classify their practices in terms of direct and indirect support for OI projects. They could be motivated to pay more attention to the support of collaborative projects, which are determinant for innovation (Berthinier-Poncet, 2014). Also, they could engage more resources for the foundation of an institutional and organizational climate focused on openness, which determines a willingness to innovate and to involve in collaborative projects (du Chatenier *et al.*, 2009). Our results could help firms' owners to have a better understanding of how cluster's governance is organizing its support for OI projects. They can also use this understanding to convince governance's members to get more focus for the support of OI projects. At the same time, our findings could help entrepreneurs to realize the importance of the involvement in OI projects in order to improve their innovative performance.

9.6. Conclusions

The objective of this study was to show the intermediate role of cluster governance in supporting the OI projects of technopolitan SMEs. In order to fill this objective, we performed a single case study in a French technopole, Technopole de l'Aube en Champagne (TAC). Our principal results highlight a direct and indirect support given by the cluster's governance for SMEs' OI practices. These results have implications for both OI and the cluster governance literature. We fill the gaps on OI in SMEs (Gassmann *et al.*, 2010) and show how SMEs in technopoles adapt to this specific OI context. Regarding the governance literature, we come with new insights regarding the intermediate role of cluster governance in supporting OI projects (Berthinier-Poncet, 2014).

Our findings are limited by the fact that we focus on a single case study. In order to generalize the results, future studies should focus on a range of technopoles, adopting a more appropriate quantitative approach. Future studies should investigate the link between OI strategies used by entrepreneurs, governance practices, and innovative performance. The results of our study may encourage more entrepreneurs to join collaborative projects in order to boost their internal innovation and enhance their value creation. They could also influence members of cluster governance to adjust their practices and focus more on how collaboration can be encouraged.

References

Arikan, A. (2009). Interfirm knowledge exchanges and the knowledge creation capability of clusters. *Academy of Management Review* 34(4), pp. 658–676.

Asheim, B. T. (2007). Differentiated knowledge bases and varieties of regional innovation systems. *Innovation: The European Journal of Social Sciences* 20(3), pp. 223–241.

Bahlmann, M. D. and Huysman, M. H. (2008). The emergence of a knowledge-based view of clusters and its implications for cluster governance. *The Information Society* 24(5), pp. 304–318.

Baptista, R. and Swann, P. (1998). Do firms in clusters innovate more? *Research Policy* 27(5), pp. 525–540.

Bathelt, H., Malmberg, A. and Maskell, P. (2004). Clusters and knowledge: Local buzz, global pipelines and the process of knowledge creation. *Progress in Human Geography* 28(1), pp. 31–56.

Becattini, G. (2002). Industrial sectors and industrial districts: Tools for industrial analysis. *European Planning Studies* 10(4), pp. 483–493.

Berthinier-Poncet, A. (2014). Gouvernance et dynamiques d'innovation au sein d'un technopôle: Une analyse par les pratiques institutionnelles d'innovation. *Management International* 19(1), pp. 94–112.

Berthinier-Poncet, A., Bocquet, R., Brion, S. and Mothe, C. (2011). Institutional innovation practices in technopoles: An example in France. In: *Technological, Managerial and Organizational Core Competencies: Dynamic Innovation and Sustainable Advantage*, Eds. Farley Nobre, David Walker, and Robert Harris, IGI Global, USA, pp. 450–473.

Berthinier-Poncet, A., Grama, S. and Saidi, S. (2015). Perceptual distance between firms and governance of a technopole. What impact on innovation? Paper presented at the *R&D Management Conference*, Pisa, Italy.

Bocquet, R. and Mothe, C. (2010). Knowledge governance within clusters: The case of small firms. *Knowledge Management Research & Practice* 8(3), pp. 229–239.

Bøllingtoft, A. (2012). The bottom-up business incubator: Leverage to networking and cooperation practices in a self-generated, entrepreneurial-enabled environment. *Technovation* 32(5), pp. 304–315.

Brunswicker, S. and Vanhaverbeke, W. (2015). Open innovation in small and medium-sized enterprises (SMEs): External knowledge sourcing strategies and internal organizational facilitators. *Journal of Small Business Management* 53(4), pp. 1241–1263.

Carpinetti, L., Ribeiro, C., Gerolamo, M. and Cardoza Galdámez, E. (2007). Continuous innovation and performance management of SME clusters. *Creativity and Innovation Management* 16(4), pp. 376–385.

Castells, M. and Hall, P. (1994). *Technopoles of the World: The Making of 21st Century Industrial Complexes*. Abingdon, Oxon: Routledge.

Chan, K. A., Oerlemans, L. and Pretorius, M. W. (2010). Knowledge exchange behaviours of science park firms: The innovation hub case. *Technology Analysis & Strategic Management* 22(2), pp. 207–228.

Chesbrough, H. W. (2003). *Open Innovation*. Boston, MA: Harvard Business School Press.

Cohen, W. M. and Levinthal, D. A. (1990). Absorptive capacity: A new perspective on learning and innovation. *Administrative Science Quarterly* 35(1), pp. 128–152.

Conrad, L. and Phillips, E. M. (1995). From isolation to collaboration: A positive change for postgraduate women? *Higher Education* 30(3), pp. 313–322.

Cooke, P. (2001). From technopoles to regional innovation systems: The evolution of localised technology development policy. *Canadian Journal of Regional Science* 24(1), pp. 21–40.

Crespin-Mazet, F., Goglio-Primard, K. and Scheid, F. (2013). Open innovation processes within clusters — the role of tertius iugens. *Management Decision* 51(8), pp. 1701–1715.

Das, T. K. and Teng, B. (2000). Instabilities of strategic alliances: An internal tensions perspective. *Organization Science* 11(1), pp. 77–101.

Doloreux, D. (1999). Technopoles et trajectoires stratégiques: Le cas de la ville de Laval (Québec). *Cahiers de Géographie du Québec* 43(119), pp. 211–235.

Dovey, K. (2009). The role of trust in innovation. *The Learning Organization* 16(4), pp. 311–325.

Du Chatenier, E., Verstegen, J., Biemans, H., Mulder, M. and Omta, O. (2009). The challenges of collaborative knowledge creation in open innovation teams. *Human Resource Development Review* 8(3), pp. 350–381.

Edmondson, A. C. and McManus, S. E. (2007). Methodological fit in management field research. *Academy of Management Review* 32, pp. 1155–1179.

Freel, M. (2005). Patterns of innovation and skills in small firms. *Technovation* 25(2), pp. 123–134.

Gassmann, O., Enkel, E. and Chesbrough, H. (2010). The future of open innovation. *R&D Management* 40(3), pp. 213–221.

Hansen, M., Chesbrough, H., Nohria, N. and Sull, D. (2000). Networked incubators. *Harvard Business Review* 5(78), pp. 74–84.

Howells, J. (2006). Intermediation and the role of intermediaries in innovation. *Research Policy* 35(5), pp. 715–728.

Huang, F. and Rice, J. (2013). Does open innovation work better in regional clusters? *Australasian Journal of Regional Studies* 19(1), pp. 85–120.

Huber, F. (2012). Do clusters really matter for innovation practices in information technology? Questioning the significance of technological knowledge spillovers. *Journal of Economic Geography* 12(1), pp. 107–126.

Huizingh, E. (2011). Open innovation: State of the art and future perspectives. *Technovation* 31(1), pp. 2–9.

Iansiti, M. and Levien, R. (2004). *The Keystone Advantage: What the New Dynamics of Business Ecosystems Mean for Strategy, Innovation and Sustainability*. Boston, MA: Harvard Business School Press.

Inkpen, A. C. and Tsang, E. W. K. (2005). Social capital, networks, and knowledge transfer. *Academy of Management Review* 30(1), pp. 146–165.

Ismail, W. and Abdmajid, R. (2007). Framework of the culture of innovation: A revisit. *Jurnal Kemanusiaan* 9, pp. 366–377.

Jung, K. and Andrew, S. (2014). Building R&D collaboration between university-research institutes and small medium-sized enterprises. *International Journal of Social Economics* 41(12), pp. 1174–1193.

Karaev, A., Lenny Koh, S. C. and Szamosi, L. T. (2007). The cluster approach and SME competitiveness: A review. *Journal of Manufacturing Technology Management* 18(7), pp. 818–835.

Laursen, K. and Salter, A. (2006). Open for innovation: The role of openness in explaining innovation performance among UK manufacturing firms. *Strategic Management Journal* 27(2), pp. 131–150.

Lee, S., Park, G., Yoon, B. and Park, J. (2010). Open innovation in SMEs — an intermediated network model. *Research Policy* 39(2), pp. 290–300.

Lichtenthaler, U. (2008). Open innovation in practice: An analysis of strategic approaches to technology transactions. *IEEE Transactions on Engineering Management* 55(1), pp. 148–157.

Lichtenthaler, U. and Lichtenthaler, E. (2009). A capability-based framework for open innovation: Complementing absorptive capacity. *Journal of Management Studies* 46(8), pp. 1315–1338.

Lopez-Vega, H. and Vanhaverbeke, W. (2010). Connecting open and closed innovation markets: A Typology Intermediaries. MPRA Munich. https://mpra.ub.uni-muenchen.de/id/eprint/27017.

McCann, B. T. and Folta, T. B. (2011). Performance differentials within geographic clusters. *Journal of Business Venturing* 26(1), pp. 104–123.

Muscio, A. (2007). The impact of absorptive capacity on SMEs collaboration. *Economics of Innovation and New Technology* 16(8), pp. 653–668.

Mutambi, J. (2011). *Stimulating industrial development in Uganda through open innovation business incubators*. PhD dissertation. Karlskrona, Sweden: School of Planning and Media Design, Blekinge Institute of Technology.

Porter, M. E. (1998). Clusters and the new economics of competition. *Harvard Business Review* 76(6), pp. 77–90.

Simard, C. and West, J. (2006). Knowledge networks and locus of innovation, In: *Open Innovation: Researching a New Paradigm*, Eds. Chesbrough, H., Vanhaverbeke, W. and West, J., Oxford University Press, New York, pp. 220–240.

Spithoven, A., Clarysse, B. and Knockaert, M. (2011). Building absorptive capacity to organize inbound open innovation in traditional industries. *Technovation* 31(2), pp. 10–21.

Van de Vrande, V., De Jong, J. P. G., Vanhaverbeke, W. and De Rochemont, M. (2009). Open innovation in SMEs: Trends, motives and management challenges. *Technovation* 29(6), pp. 423–437.

Vanhaverbeke, W. (2006). The inter-organizational context of open innovation. In: *Open Innovation: Researching a New Paradigm*, Eds. Henry W. Chesbrough, Wim Vanhaverbeke, and Joel West. Oxford: Oxford University Press.

Vanhaverbeke, W. (2017). *Managing Open Innovation in SMEs*. Cambridge, UK: Cambridge University Press.

Vanhaverbeke, W., Van de Vrande, V. and Cloodt, M. (2008). Connecting absorptive capacity and open innovation. *SSRN Electronic Journal,* p. 22. https://doi.org/10.2139/ssrn.1091265.

Vásquez-Urriago, Á. R., Barge-Gil, A. and Modrego Rico, A. (2016). Science and technology parks and cooperation for innovation: Empirical evidence from spain. *Research Policy* 45(1), pp. 137–147.

Verbano, C., Crema, M. and Venturini, K. (2015). The identification and characterization of open innovation profiles in italian small and medium-sized enterprises. *Journal of Small Business Management* 53(4), pp. 1052–1075.

Westhead, P. and Batstone, S. (1998). Independent technology-based firms: The perceived benefits of a science park location. *Urban Studies* 35(12), pp. 2197–2219.

Wolfe, D. A. and Gertler, M. S. (2004). Clusters from the inside and out: Local dynamics and global linkages. *Urban Studies* 41(5–6), pp. 1071–1093.
Zeng, S., Tam, C. and Xie, X. (2010a). Evaluating innovation capabilities for science parks: A system model. *Technological and Economic Development of Economy* 3, pp. 397–413.
Zeng, S., Xie, X. and Tam, C. (2010b). Relationship between cooperation networks and innovation performance of SMEs. *Technovation* 30(3), pp. 181–194.

Chapter 10

Exploring Collaborative Practices Between SMEs and Designers in the Italian Furniture Industry

Claudio Dell'Era, Stefano Magistretti and Roberto Verganti*

* *claudio.dellera@polimi.it*

Abstract

Design has recently gained substantial attention among practitioners and scholars as a source of innovation. In contemporary business and academia, design is increasingly viewed as an important strategic asset. Moreover, the role of design and designers in generating value is growing in relevance to the point that they can be considered part of inbound open innovation strategies. Indeed, managing designers through open innovation processes is becoming particularly important for companies in several industries. This chapter aims to investigate the influence of the practices adopted by small and medium-sized enterprises (SMEs) in collaboration with external designers on the relationship between design investments

* Corresponding author.

and competitive performance. In doing that, the research focuses on the Italian furniture industry as empirical setting. The reason for this selection is twofold. First, because more than 95% of Italian furniture companies are small and medium-sized. Second, many Italian furniture companies collaborate with several external designers to the point that over 80% of the product portfolio is on average conceived by external designers. Analyzing the results coming from a survey conducted on SMEs of the furniture industry, different collaborative practices have been explored. In particular, the effect that three different design management approaches, named Styling, Thinking, and Meaning, have on SME performance as a result of collaborations with external designers is highlighted.

10.1. Introduction

Although technological innovation has captured the greatest attention in the last decades, especially in relation to radical technological change, practitioners and scholars increasingly view design as a strategic asset and a source of innovation. This has led to growing interest in design management and organization studies over the last decade. Management scholars initially contributed to the debate through specialized journals (e.g., Dumas and Mintzberg, 1989, 1991; Verganti, 2003), while more recently, research in this field has also been published in journals with a wider scope (e.g., Hargadon and Sutton, 1997; Boland and Collopy, 2004; Rindova and Petkova, 2007; Verganti, 2006, 2009, 2011; Brown, 2008; Brown and Wyatt, 2015; Michlewski, 2008). The growing attention to design has led scholars and managers to investigate the links between design, innovation, and competitive advantage as well as the impacts of design management and design practices on firm performance (Hertenstein *et al.*, 2005; Veryzer, 2005; Gemser and Leenders, 2001).

Several scholars agree on the relevance of design as a strategic tool able to generate competitive advantage (Hertenstein *et al.*, 2005; Abecassis-Moedas and Mahmoud-Jouini, 2008). Furthermore, it has been found to be especially important for small and medium-sized enterprises (SMEs) since it can improve communication and product development, and can become a source of support for innovation

(Brazier, 2004; Acklin, 2010). Managing design through open innovation processes is becoming an important option for firms (Utterback *et al.*, 2006), and especially SMEs (Berends *et al.*, 2011; Vanhaverbeke, 2017) because experienced designers can mitigate the lack of internal firm design skills and create solutions that no SME on its own could ever achieve. At the same time, studies that analyze the relationship between investment in design and competitive performance show two main limitations. First, they adopt different definitions of design, rendering the comparison of empirical results difficult (Swan and Luchs, 2011). Second, they mainly show a significant relationship between investment in design and competitive performance (Candi, 2010; Swink, 2000; Wallace, 2001). Few studies demonstrate that this relationship can be influenced by several factors (Chiva and Alegre, 2009; Gemser and Leenders, 2001; Candi and Gemser, 2010), such as the managerial processes and practices adopted to foster innovation (as also noted by Perez-Freije and Enkel, 2007; Ravasi and Stigliani, 2012).

To address the first limitation, we consider three different interpretations of design:

- **Styling:** The first approach looks at design from a particular perspective. Design is associated with the form of products, often in juxtaposition with the product function. Indeed, when reduced to its real essence, many believe that design essentially deals with form. Modernists in particular questioned the predominance of form, from American architect Louis Sullivan's claim that "form follows function" at the dawn of the 20th century to the aphorism "less is more" of Ludwig Mies van der Rohe, one of the Bauhaus directors (the art and architecture school active in Germany between 1919 and 1933).
- **Thinking:** The second approach considers design in its broader interpretation encompassing creative activities and professions that produce a change in the environment, associating design with a "better way of thinking" (usually referred to as design thinking) to address any type of problem (Boland and Collopy, 2004; Brown, 2008, 2009; Martin, 2009).

- **Meaning:** The third approach leverages on the definition of John Heskett, a major design theorist: "Design, can be defined as the human capacity to shape and make our environment in ways without precedent in nature, to serve our needs and give meaning to our lives" (Heskett, 2002). He clearly points to a specific characteristic of design, namely, making things more meaningful. Design is the activity through which we innovate the meaning of things.

To overcome the second limitation, we focus on the practices adopted by companies in collaboration with external designers and design consultancies. In the literature about inbound open innovation, designers and design consultancies can be interpreted as a specific category of intermediaries and sources of knowledge (Dell'Era *et al.*, 2008; Dell'Era and Verganti, 2009; Dell'Era and Verganti, 2010; Abecassis-Moedas *et al.*, 2012). Case studies of companies such as Alessi, Apple, Bang & Olufsen, Kartell, Philips, Sony, and Swatch show that designers are becoming key actors in terms of product innovation and strategic renewal (Ravasi and Lojacono, 2005); as a consequence, firms are looking for new practices that allow to effectively collaborate with them (Hargadon and Sutton, 1997; Hargadon, 1998; Chiaroni *et al.*, 2008; Abecassis-Moedas and Ben Mahmoud Jouini, 2008; Colombo *et al.*, 2011). Designers and design consultancies have been traditionally depicted as "knowledge brokers" (Hargadon, 2003; Sutton, 2002), i.e., "firms that span multiple markets and technology domains and innovate by brokering knowledge from where it is known to where it is not" (Hargadon, 1998, p. 210). They play a critical role in knowledge transfer and exchange processes (Hargadon, 1998; Hargadon and Sutton, 1997).

This chapter aims to investigate the influence of the practices adopted by SMEs in collaboration with designers on the relationship between design investments and competitive performance. The chapter is structured as follows. Section 10.2 analyzes the primary contributions to literature. Thereafter, the methodology used in the analysis is presented. The subsequent sections describe and discuss the empirical results. The conclusions and avenues for future research are outlined in the last section.

10.2. Literature Review

Open innovation has been one of the most debated topics in technology management research in the last decades (Chesbrough, 2003; Gassmann, 2006). It has been proposed as a new paradigm for industrial innovation management, according to which firms use "purposive inflows and outflows of knowledge to accelerate innovation, and to expand the markets for external use of innovation, respectively" (Chesbrough *et al.*, 2006). In this chapter, we focus on the collaboration practices between SMEs and designers, interpreted as inbound open innovation. The literature review is presented in three subsections. First, the role of design in SMEs is described. Second, we describe the three aforementioned design interpretations. Thereafter, we summarize the principal contributions in literature on the relationship between design and competitive advantage.

10.2.1. *Design in SMEs*

Design is recognized by researchers as a key element in sustaining firm' performances (Abecassis-Moedas and Mahmoud-Jouini, 2008; Hertenstein *et al.*, 2005). Its primary role in competitiveness is even greater in SMEs. This is because it can help companies in improving communication and new product development, allowing them to foster innovation (Acklin, 2010; Brazier, 2004). Given the centrality of design, a proper way to manage it is mandatory (Fernández-Mesa *et al.*, 2013). Indeed, Walsh (1996) clearly reported the fact that design is crucial in innovation because it helps SMEs in putting ideas into material form. This convergent process is quite linear in large firms but it takes place under several constraints in the SMEs environment. Indeed, the limited amount of resources create problems in managing innovation. The main barriers to design innovations in

SMEs are: lack of human resources, financial issues, less formality in new product development process, poor design understandings, and difficulties in accessing design resources (Fueglistaller, 2004; Cox, 2005; Moultrie, 2007; Acklin, 2013). In addition to them, the complexity of the market, the technology, and the interaction with external partners can influence as well the process. As it is clear, the first set of them are internal barriers while the others are external barriers (D'Este *et al.*, 2012). Researchers have shown that in order to cope with this problem SMEs can leverage on different strategies. On one hand, they can empower and reinforce internal resources, and this normally leads to a better performing process (Belso-martinez *et al.*, 2011). On the other, they can leverage on inbound open innovation and outsource the design capabilities in order to mitigate and overcome the lack of internal skills (Utterback *et al.*, 2006; Berends *et al.*, 2011). Especially this second approach is favorable for SMEs because it can help in reaching and creating solutions that on its own they cannot even imagine. This approach toward design in SMEs is not easy to be implemented. Indeed, good design does not emerge by chance (Chiva and Alegre, 2009), but it requires a proper and effective management of the external (e.g., external designers) and internal resources (e.g., research and development (R&D) employees). So, this brief overview of the literature among design and SMEs stresses the crucial role that design management approaches have in these types of companies in order to sustain their competitiveness and in overcoming the traditional barriers that they daily encounter. Considering the current economic context and the environment in which SMEs compete, a proper way to manage the design capabilities is mandatory (Fernández-Mesa *et al.*, 2013). Indeed, this is one of the aim of the chapter.

10.2.2. *Design management approaches*

In this subsection, we introduce the three approaches that differently interpret the opportunities that design offers: Styling, Thinking, and Meaning.

10.2.2.1. *Styling: Design as the Form of Things*

Historically, design is associated with the form of products, often in juxtaposition with the product function. Indeed, as previously mentioned, when reduced to its real essence, many believe that design essentially deals with form. If engineers make products function through technology, designers make things beautiful. At the beginning of the 20th century, the "form follows function" paradigm became the dominant approach in the US and was later reinterpreted and refined by the German Bauhaus and the Ulm School. However, partly in the 1930s and above all in the 1970s, new approaches shifted the interest to the aesthetic dimension of products. Modernists in particular questioned the predominance of form, from the claim of American architect Louis Sullivan that "form follows function" to the aphorism "Less is more" of Ludwig Mies van der Rohe. The classic dialectic of "function versus form" frequently leads to confining the interpretation of the term "form" to the aesthetic appearance of products. However, a product can bring messages to the market in several ways, and styling is merely one of the ways. While the functionalities of a product aim to satisfy the operational needs of customers, product meanings aim to satisfy their emotional and sociocultural needs (Csikszentmihalyi and Rochberg-Halton, 1981; Margolin and Buchanan, 1995). Yet, if we look at the most frequent applications of design in business in the past century, these are infused with a prevailing attention to form, complying more with the "ugliness does not sell" dogma coined by Raymond Loewy, the French-American designer and founding father of "styling."

10.2.2.2. *Thinking: Design as Creative Problem-Solving*

The successes in the last decades of major design firms such as IDEO (Kelley, 2001) or Continuum (Lojacono and Zaccai, 2004) introduced a new interpretation of design as "as a discipline that uses the designer's sensibility and methods to match people's needs with what is technologically feasible and what a viable business strategy can convert into customer value and market opportunity" (Brown, 2008). This implies that product development should start from a deep analysis of user needs. The underlying assumption is that a firm may infer unique insights to inform product innovation by asking users about their needs or more effectively by observing users as they use the products and track their behavior in the consumption process. Investigations of user-driven design and analyses of successful cases have provided a deeper and more valuable interpretation of design as an organizational process, a process to get closer to users and their actual needs (Stein and Iansiti, 1995; Leonard and Rayport, 1997; Seybold, 2001; Thomke and Von Hippel, 2002; Chayutsahakij and Poggenpohl, 2002). Interest in design thinking grew significantly at the start of the 21st century when the press began popularizing the term. Design thinking is a formal method for the practical and creative resolution of problems and the generation of solutions with the intent of improving future results (Brown, 2008, 2009; Martin, 2009). In 2005, Stanford University's d.school began teach engineering students design thinking as a formal method. In 2007, the Hasso Plattner Institute for IT Systems Engineering in Potsdam, Germany, established a design-thinking program.

10.2.2.3. *Meaning: Design as Making Sense of Things*

According to Heskett (2002), the third approach interprets design driven innovation as the innovation of the purpose and reason why people buy and use things, in other words, focusing on the innovation of meanings. Customers seldom anticipate possible radical changes in product meanings. The sociocultural context in which they are currently immersed inclines them to interpretations in line with what is happening today. Radical changes in meanings instead require radically new interpretations of what a product is meant for, which can only be understood (and affected) by looking at things with a broader perspective (Dell'Era *et al.*, 2008; Verganti and Oberg, 2013; Verganti and Dell'Era, 2013). The role of market factors in the design-driven approach is assumed to change according to the level of novelty in product meanings. In the case of incremental innovation of product meanings, market drivers are key. In other words, incremental adaptations of product meanings are determined by the continuous and natural evolution of the explicit cultural models that customers adopt. Conversely, market factors lose importance in the radical innovation of product meanings, where innovations originate from a cultural scenario developed through collaborations between companies and designers. In other words, radical innovations of product meanings require understanding the possible or latent social dynamics that can influence consumer lifestyles and behaviors (Norman and Verganti, 2013).

10.2.3. *Design and competitive advantage*

As previously mentioned, in the contemporary business and academic arenas, design is increasingly viewed as an important strategic resource. The last decade has witnessed an explosion of contributions in business and research literature from scholars and companies alike continuously attempting to link design and competitive advantage. A series of case studies demonstrates that companies apply design to better understand user needs (Patnaik and Becker, 1999; Sutton, 2001; Whitney and Kumar, 2003; Rosenthal and Capper, 2006; Bloch, 2011; Moreau, 2011). Design is recognized as a strategic resource, and hence numerous scholars study the link between design and competitive advantage (Gemser and Leenders, 2001; Platt *et al.*, 2001; Borja de Mozota, 2003; Boland and Collopy, 2004; Creusen and Schoormans, 2005; Hertenstein *et al.*, 2005; Veryzer, 2005). In 2005, the *Journal of Product Innovation Management* dedicated two special issues to *Marketing Meets Design* (*JPIM*, 2005a, 2005b), while in 2011, the same journal dedicated a special issue to *Product Design Research and Practice* (*JPIM*, 2011). Several management and innovation scholars have attempted to quantify the contribution of design to the bottom line by providing interesting results and suggestions, albeit often struggling with the lack of a common language and a unifying theory. Two separate and complementary research streams focus on the economic and financial results deriving from a design-driven strategy.

The first research stream investigates the relationship between investments in design and competitive performance with the aim of identifying a type of Return on Investment (ROI) for design (Wallace, 2001; Zec and Jacob, 2010). Companies with larger design budgets are generally in a better financial position than their competitors (Candi, 2010; Gemser and Leenders, 2001; Swedish Industrial Design Foundation, 2004). However, the results derived from design investments are significantly influenced by the managerial practices adopted (Danish Design Centre, 2003; Chiva and Alegre, 2009). Some authors note that financial indicators alone cannot capture the full value of design. The impact of design on a company's performance is multifaceted and mediated by various other factors, rendering it difficult to measure by exclusively analyzing financial data (Borja de Mozota, 2006; Gabrielsen *et al.*, 2007). Indeed, the effects of design can be tangible when they generate direct financial returns, but they may also be intangible when they contribute to the future performance of a company by affecting difficult-to-quantify factors such as the company's cultural and strategic assets (Inns, 2002). Extant literature shows that investments in design not only allow enhancing financial performance (Gemser and Leenders, 2001), but also foster indirect benefits such as acquiring design management skills (Roy, 1994). Figure 10.1 illustrates the conceptual framework generally adopted by authors contributing to the first research stream, focusing mainly on the direct impact of design investments on competitive performance,

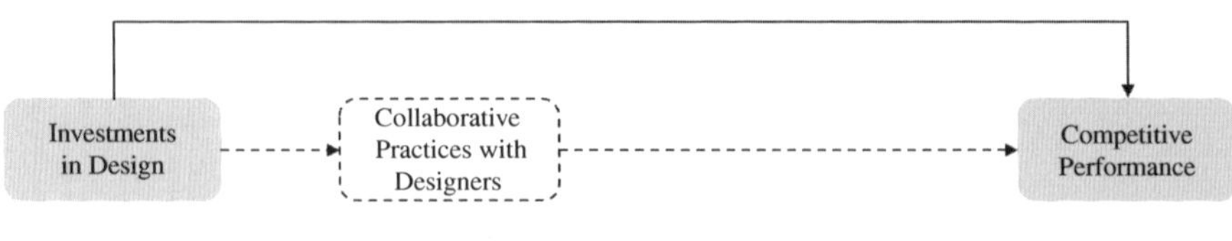

Figure 10.1. Literature review — conceptual framework adopted in the first research stream.

although some also consider the mediating role of practices adopted in the design process.

The second research stream investigates the relationship between design-based innovations and competitive performance. More precisely, this research stream aims to explore the relationship between the design innovation outcome, interpreted as the goodness of new designs, and competitive performance. Several scholars approximate the capability to launch design-driven innovations by considering the number of design awards or asking design professionals to rate the skills and achievements of companies (Goodrich, 1994; Walsh *et al.*, 1988; Platt *et al.*, 2001; Hertenstein *et al.*, 2005; Talke *et al.*, 2009). Alternatively, customer evaluations are used to assess the design innovation outcomes (Veryzer and Hutchinson, 1998; Bloch *et al.*, 2003; Creusen and Schoormans, 2005). Firms characterized by the successful use of design consistently outperform their competitors on a wide variety of financial indicators (Roy, 1994; Trueman and Jobber, 1998; Swan *et al.*, 2005; Bedford *et al.*, 2006; Gemser *et al.*, 2011) and even in overall stock performance (Hertenstein *et al.*, 2001; Rich, 2004). Additionally, good design would appear to have long-term effects that are persistent and stable over time (Roy, 1994; Hertenstein *et al.*, 2005) while the effects of technical newness on sales would seem to decrease as a product reaches its maturity. Innovative design is a significant driver of sales over the entire product life cycle (Talke *et al.*, 2009). Figure 10.2 illustrates that contributions to this second research stream mainly focus on the direct impact of design innovation outcomes on competitive

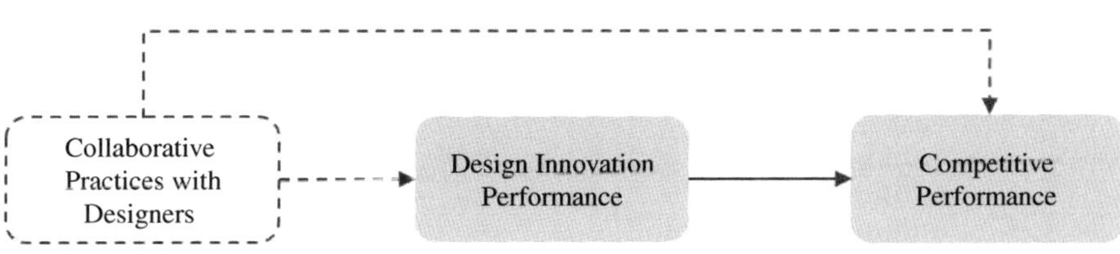

Figure 10.2. Literature review — conceptual framework adopted in the second research stream.

performance, although some authors also consider the contribution of practices adopted in the design process.

As previously mentioned, this chapter explores the practices adopted by SMEs in collaborating with external designers for innovation purposes. More specifically, it aims to investigate the influence of the practices adopted by SMEs in collaboration with designers on the relationship between design investments and competitive performance.

10.3. Research Methodology

In view of the aim of this chapter, survey research is considered the most appropriate methodology (Forza, 2002), and the questionnaire design (for more details, see Appendix A) is a central issue to ensure its correct interpretation. To collect the information, different types of questions were formulated including a multiple-choice question (for one data point), 5-point Likert-type scale (for 13 data points), and open questions (for a further 10 data-points). The different types of questions were positioned in different parts of the questionnaire according to some considerations, for example, the open questions were positioned at the end of the survey following Heberlein and Baumgartner (1978). Indeed, if respondents have invested time in answering closed questions, they are more likely to provide general information through open questions to complete the questionnaire. In more detail, the questionnaire is divided into four sections: (i) design management approaches, (ii) investments in design, (iii) collaborative practices with designers, and (iv) competitive performance. These four sections are indicated in gray in Figure 10.3 referring to the information collected through the questionnaire, while the white box (design innovation performance) refers to information gathered from secondary sources.

The first section of the questionnaire consists of a multiple-choice question to enable understanding the design management approach adopted by the SMEs. Considering the nonlinear inquiry,

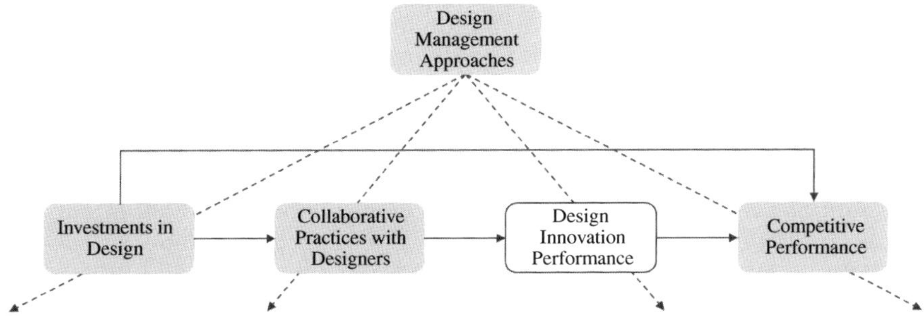

Figure 10.3. Conceptual framework adopted in the research.

the respondents selected the most relevant of the short descriptions reflecting the three approaches reported in the literature review (Styling, Thinking, and Meaning). The second section refers to design investments with open-ended questions on the number of employees and designers involved in new product development. The third set of questions concerns the collaborative practices adopted using a 5-point Likert-type scale to gather information on the strategic and operational collaborative practices where respondents were asked to indicate the percentage of time designers spent in different interactions. The last part of the questionnaire refers to more descriptive aspects, namely, total revenues and those derived from new products. In addition, evidence of awards received was collected through secondary resources (the white box in Figure 10.3), particularly referring to the Compasso D'Oro Awards.[1]

The questionnaire was tested on five firm experts in face-to-face meetings to ensure that it satisfied the readability and completeness requirements of this methodology. Thereafter, the questions were reformulated and rephrased for a more defined inquiry.

From the empirical standpoint, the research focuses on the Italian furniture industry for several reasons: first, the furniture industry is particularly relevant in most industrialized countries,

[1] Compasso D'Oro is the oldest and most influential international design award established in 1954 by the famous Italian designer Gio Ponti.

Table 10.1. Descriptive statistics of the sample.

Average number of employees	69.7
Average number of designers (external collaborators)	4.9
Average revenues	€16.08 million

generally representing between 2% and 4% of the production value of the manufacturing sector; second, more than 95% of Italian furniture companies are small and medium-sized; finally, many Italian furniture companies collaborate with several external designers to the point that over 80% of the product portfolio is on average conceived by external designers. Collaborating with Webmobili (Internet spin-off of Federmobili, Italian Association of Furniture Manufacturers) enabled us to identify the appropriate sample frame and then collect the data on the sample. Indeed, the DesignBest (www.designbest.com) Internet database developed by Webmobili is particularly appropriate for our research objectives. Although the database does not contain the products of all Italian companies, it can be considered representative of the offering of the Italian furniture industry. DesignBest is supported by relevant institutions operating in the Italian furniture industry, including the Italian Association for Industrial Design (ADI),[2] Confcommercio,[3] and Federlegno.[4] Starting from Webmobili, a sample size of 358 furniture SMEs was initially identified. The final sample is composed of 56 SMEs operating in the Italian furniture market (response rate = 16%). Table 10.1 reports some descriptive statistics of the final sample.

[2]The Association for Industrial Design (ADI) founded in 1956 "brings together professionals, researchers, teachers, critics, journalists around the main topics of design: project, energy consumption, recycling and training. ADI has a leading role in the development of the industrial design as a cultural and economic phenomenon." ADI confers the prestigious Compasso D'Oro award.

[3]Confcommercio is the Association of Italian Retailers and has over 700,000 member companies.

[4]Federlegno, founded in 1945, is the association for companies operating in the furniture industry and promotes trade interests in Italy and the world.

10.4. Empirical Results

To interpret and evaluate the responses received, we extrapolate some descriptive statistics. In particular, the results will be presented according to the conceptual framework previously reported (Figure 10.3), starting from the design management approaches, investments in design, collaborative practices with designers, design innovation performance, and finally, competitive performance.

10.4.1. *Design management approaches*

Considering the design management approaches described in the literature review (Styling, Thinking, and Meaning), Figure 10.4 reports the percentage of firms that adopt each approach and features some companies and their products.

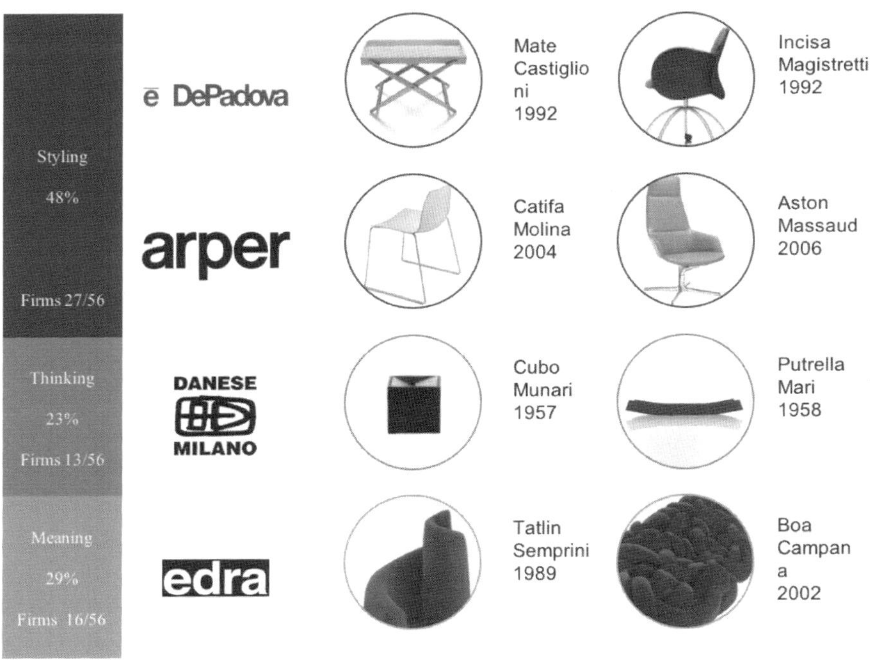

Figure 10.4. SMEs design approaches.

The figure clearly shows that the traditional approach considering design as the form of things is largely adopted by Italian SMEs (almost 50%). Conversely, the emerging consideration of design as creative problem-solving and sense-making account for 23% and 29%, respectively.

10.4.2. *Investments in design*

To understand the design management approaches, the second set of questions intended to gather information on investments in design. Considering the strategic relevance of designers and the characteristics of the Italian furniture industry, we approximate investments in design with the ratio between number of designers and number of employees. More specifically, we focus on collaborations with external designers, which are a particular trait of Italian furniture companies (Dell'Era and Verganti, 2010). On average, each company engages less than two internal designers (who mainly manage operational activities) and more than eight external designers. The ratio between external designers and employees is around 16%.

Emerging from Table 10.2 is that the average number of external designers increases from companies that adopt the Styling approach to those embracing the Meaning approach. Considering the ratio between number of external designers and number of employees, or investments in terms of collaborations, the evidence shows that companies using the Meaning approach invest less than those employing the Thinking approach. This is in line with literature, as companies

Table 10.2. Investments in design.

	Styling	Thinking	Meaning	Total
Average number of employees	63.33	71.31	79.38	69.76
Average number of external designers	7.22	7.38	9.50	7.91
Average ratio between number of external designers and number of employees	16%	20%	14%	16%

Figure 10.5. Focus on two SMEs: Driade and FontanaArte.

operating the Meaning approach focus more on relations and opt for a lower ratio compared to those using the Thinking approach. In this latter approach, the higher ratio is justified by the pursuit of new ideas. Indeed, if we consider two comparable companies, namely, Driade (adopting the Thinking approach) and FontanaArte (adopting the Meaning approach), the different investments in design are evident (Figure 10.5). Driade had 35 employees and 25 designers, while in the same year, FontanaArte had 42 employees and only 6 designers, and the former invested 71% compared to FontanaArte's 14%.

10.4.3. *Collaborative practice with designers*

The third set of data concerns the management practices that different SMEs adopt in managing external collaborations with designers. In particular, this section reports the results of the different collaborative practices investigated. In view of their heterogeneity, three different aspects are considered: the strategic, organizational, and operational practices. The first provides a general understanding of the possible collaborative practices between a company and its external designers and takes into account the strategic activities in which different external designers are involved (i.e., future scenario, lifestyles, customer experience, concepts, technologies, production processes, user needs, brokering solutions, and aesthetic quality).[5] The second set of collaborative practices refers to the organizational aspects of the relationship between the companies and external designers, namely, evaluating the percentage of time external designers spent in the interaction with different business units in the SME (i.e., board, R&D, engineering, and marketing). Finally, the last set of data concerns the different tools the companies adopt to interact with external designers. In this case, the adopted survey method calls for a simplification, as not all the possible tools can be considered and were thus aggregated (i.e., user-centered tools, participative tools, concept generation tools, and visualization tools).[6] As the practices refer to different aspects of the interactions between firms and designers, the metrics adopted differ slightly, and the responses led to creating the graphs in Figure 10.6.

Some interesting evidence emerges from Figure 10.6. Considering the Styling approach, it is clear that aesthetic quality (3.96) is the most important strategic collaborative practice, while future scenario (2.43) is the activity that least involves external designers. In terms of the Thinking approach, the figure shows that the concepts activity is extremely high (4.36), in line with the literature showing that the Thinking approach supports creativity and the ideation of

---

[5] *Future Scenario* is the identification of new business strategy for the SMEs; Life styles is the identification of new emerging sociocultural trends; *Brokering solutions* is the ability of designers in bringing in the SMEs solutions existing in other fields and situations.

[6] *User-centered tools* are those tools used in order to understand user needs and develop solutions starting from them; *Participative tools* are those tools that are used by both developer and users in order to co-create solutions; *Visualization tools* are those tools that allow designers to represent and communicate their envisioned solutions.

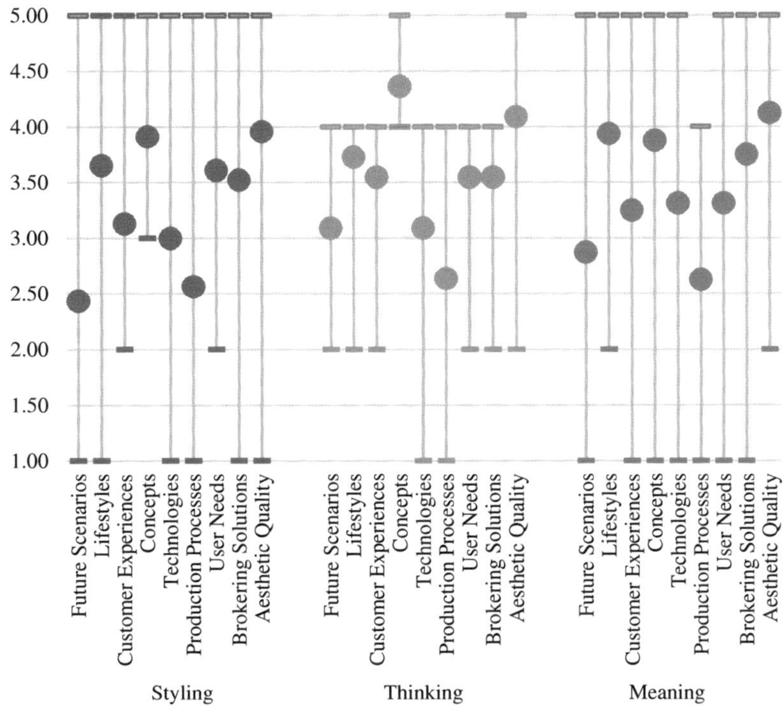

Figure 10.6. Strategic collaborative practices.

new solutions, i.e., brainstorming sessions.[7] Conversely, the impact on the production process is relatively low (2.64), since Thinking pertains more to idea generation than to the implementation process. In terms of the Meaning approach, the high levels of lifestyle (3.94) and brokering solutions (3.75) are unsurprising given the influence that lifestyle has on "why" people use and buy products.

Considering the variance, the most evident difference is in the concepts activity. Indeed, in the Styling approach, the variation is between 3 and 5; in the Thinking approach, between 4 and 5; while in the Meaning approach, it covers the entire range from 1 to 5. This is particularly coherent with the definition of the different

[7] *Brain storming* is the moment where different people envision and create lots of ideas through collaboration, creativity, and openness of mind.

design management approaches. Indeed, in both the Styling and Thinking approaches, the relevance of collaborations with external designers in the concept activity is high since prototypes and products are at the center, while this is not always the case in the Meaning approach.

Looking at the interactions between designers and other business units, in the Styling approach, on average, almost half the time (43.65%) is spent interacting with R&D (Figure 10.7). This is coherent with the fact that R&D is an essential part of the product engineering process to give form to things. In terms of the Thinking approach, the time spent with the engineering unit is relatively high and on average a quarter of the entire work period (25.27%), since engineers also provide stimuli in the ideation phase. With regard to the Meaning approach, interactions between external designers and the different business units is relatively high for the board of directors (29%). This

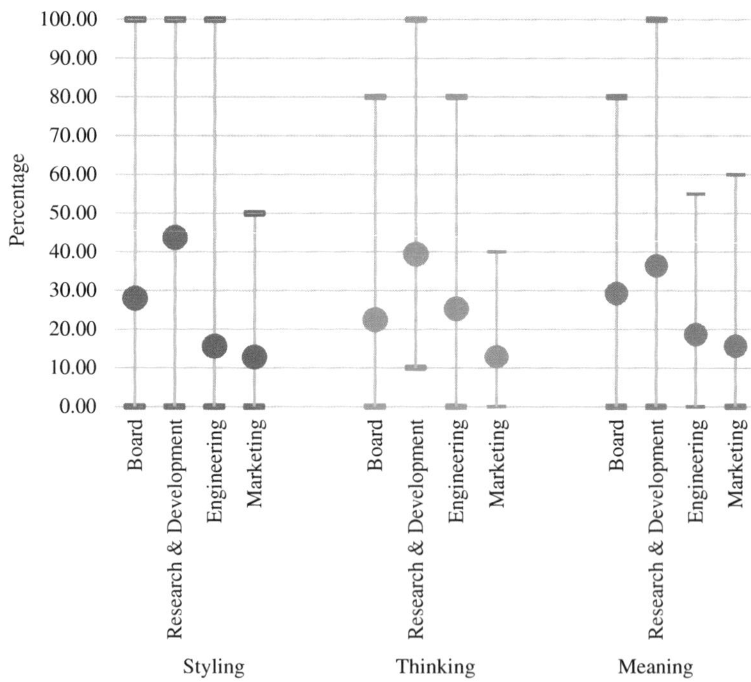

Figure 10.7. Organizational collaborative practices.

also finds support in literature as the strategic aspects of the innovation of meanings reshape the reason why people buy products.

Considering the variance in terms of collaboration time, the most significant is in marketing interactions. Indeed, if in the Styling approach this varies from 0% to 50%, in the Thinking approach it decreases to the 0–40% range, while in the Meaning approach it even reaches 60%. This is particularly interesting, as it shows that the Meaning approach incentivizes collaborations with a more varied set of business units and for longer periods as the relations and interactions can be a source of different understanding of the meanings of products.

The tool most used in the Styling approach relates to concept generation. This is in line with the overarching aim of this approach where the definition of the concept in terms of shape is essential. In the Thinking approach, visualization is the most frequently used tool (3.45) as this enables conveying new ideas to the board (Figure 10.8).

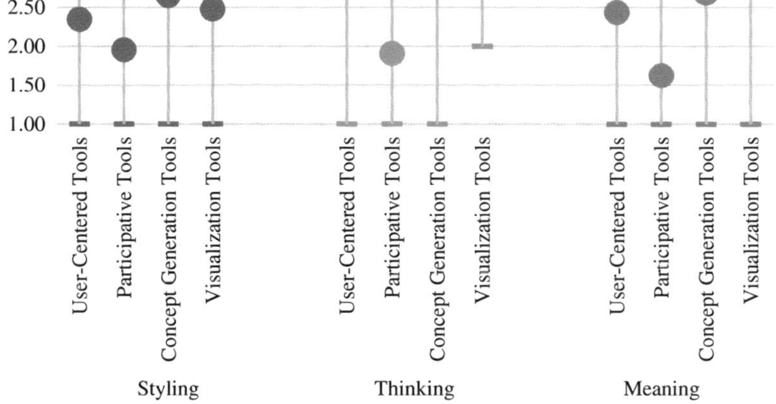

Figure 10.8. Operational collaborative practices.

Hence, external designers widely use moodboards[8] and other similar tools. In the Meaning approach, the most relevant finding is that participative design shows low adoption (1.63). Indeed, the Meaning approach leverages more on making proposals to the market rather than codesigned products and services with external designers. Coherently, visualization is relevant due to the need to clearly frame and embody the new meaning that designers want to propose.

In terms of variance, the metrics show a high fluctuation in participative tools. Indeed, if in the Styling approach these cover the entire range from 1 to 5, in the Thinking approach they reach at most level 4, while in the Meaning approach, the variation can be from 1 to 3. This shows that directly involving external designers in the generation of new products is more relevant in the Styling and Thinking approaches than in the Meaning approach. Considering that the meaning paradigm is defined as an inside-out approach, this result is confirmed and corroborated by literature.

10.4.4. *Design innovation performances*

In terms of awards, only the ADI Award is considered, because this is the most significant design award for all Italian enterprises and not only SMEs.

On average, 27% of SMEs received an award, but as Table 10.3 shows, this ratio increases for companies adopting the Meaning approach and decreases when adopting a Styling approach. In more

Table 10.3. Design innovation performances.

	Styling	Thinking	Meaning	Total
Firms that received the ADI award in 2012 or 2013	6	4	5	15
Ration between awarded and total number of firms	22%	31%	31%	27%

[8] *Mood boards* are collage of photos, or images that through juxta position convey emotions and feelings to the people that looks at them.

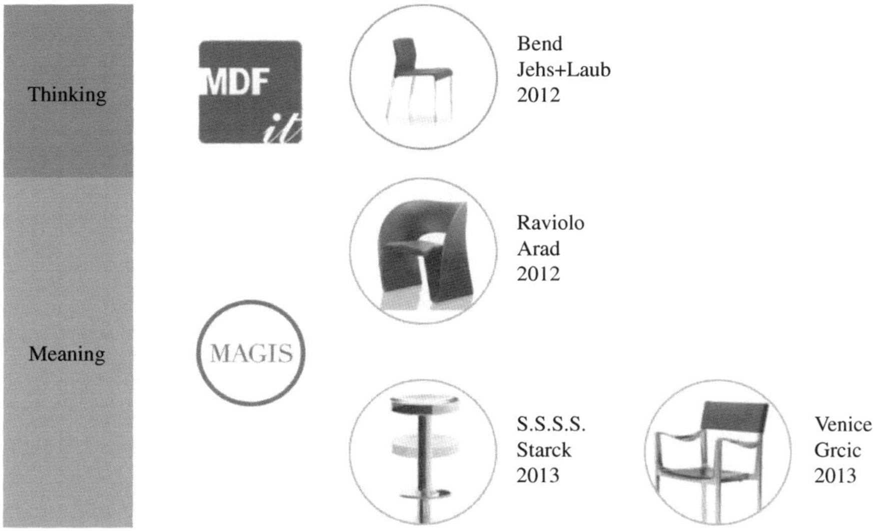

Figure 10.9. ADI awards received in the period 2012–2013.

detail, on average, 22% of the companies adopting the Styling approach received an award in the years 2012–2013. Companies adopting the Thinking or the Meaning approach show a higher ratio, where one out of three received an award in the same period. In particular, if we compare two companies, such as MDF Italia and Magis, some differences emerge. MDF Italia has 40 employees and uses the Thinking approach, and received only one award in the period 2012–2013 as can be seen in Figure 10.9. Conversely, Magis adopts the Meaning approach and received three ADI awards in the same period and leverages on 15 designers.

10.4.5. *Competitive performance*

The last results concern the revenues in the years 2012 and 2013. The main aim is to understand the efficacy of each design management approach. The detailed analysis of competitive performances can be seen in Table 10.4.

Analyzing the results, it would seem that companies adopting the Styling approach on average have lower revenue (13.99 mln) than those employing Thinking and Meaning approaches (18.25 mln and

Table 10.4. Competitive performances.

	Styling	Thinking	Meaning	Total
Total revenues (2013)	13.99 mln	18.25 mln	18.78 mln	16.35 mln
Revenues from new products (2013)	2.17 mln	2.36 mln	2.18 mln	2.22 mln
Percentage of revenues from new products (2013)	7.9%	6.2%	7.5%	7.4%
Average percentage growth of revenues (2012–2013)	5.1%	0.8%	10.9%	5.8%

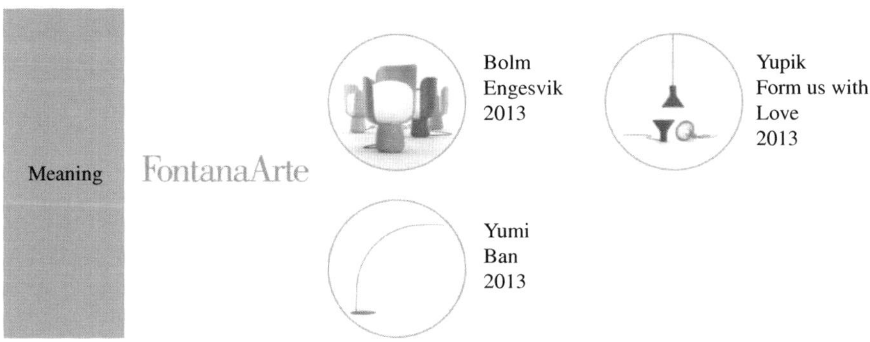

Figure 10.10. FontanaArte products launched in 2013.

18.5 mln). Looking at the average growth of revenues, the Thinking approach is more stable, while the Styling and Meaning approaches on average show an increase in annual revenues. In particular, the Meaning approach records double the rate of growth in terms of revenues compared to the Styling approach. SMEs adopting the Styling or Thinking approach show lower variance compared to those adopting a Meaning approach, for example, FontanaArte, which moved from 13 mln to 29 mln (123% growth) (Figure 10.10).

10.5. Discussion

As previously mentioned, the open innovation paradigm, defined as a distributed innovation process that leverages the flow of knowledge crossing company boundaries (Chesbrough and Bogers, 2014), is based on both inflows and outflows that are respectively able to accelerate the internal development process and the external exploitation process (Chesbrough, 2006). In design-intensive industries such as the furniture one, the collaboration with external designers represents one of the crucial inflows to the point that companies are looking for new practices in order to engage and involve designers. Table 10.5 reports the most significant findings of this study.

As claimed by several scholars, design is a strategic tool able to generate competitive advantage. The investments in design shown by the furniture companies belonging to our sample demonstrate the significant effort allocated in the collaboration with designers: in fact, designers represent a significant part of the external organization, if compared with the number of employees (about 16%). Managing design through open innovation processes is becoming particularly important for SMEs to the point that designers not only contribute to the aesthetic qualities of the products, but they also participate in the generation of new concepts and in the identification of emerging lifestyles. Even if the R&D function represents the main interface with external designers, the strategic role played by designers is demonstrated by the fact they directly interact with the Board.

Table 10.5 shows that the investments in design, calculated as the ratio between external designers and employees, differ across the three approaches (Styling, Thinking, and Meaning). The high investments in the Thinking approach (20%) are coherent with the view sustaining that this approach should generate a large number of new solutions (Brown, 2008). The increase in the number of solutions proposed is directly linked to the increase in the source of ideas and consequently in the number of collaborations with designers. In addition, and coherently with literature, the meaning approach shows the lowest level of investments in design. In this peculiar approach, what really matters is the strength and duration of the relations with external designers rather than merely the number of collaborations (Dell'Era and Verganti, 2007; Verganti, 2009). The Styling approach lies between the two other approaches, requiring a greater number of designers compared to the Meaning approach, but fewer than the Thinking approach.

Table 10.5. Main research findings.

	Styling 16%		Thinking 20%		Meaning 14%	
Investments in design: Ratio between designers and employees						
Strategic Collaborative Practices (Main contributions provided by designers)	Aesthetic quality	3.96	Concepts	4.36	Aesthetic quality	4.13
	Concepts	3.91	Aesthetic quality	4.09	Lifestyle	3.94
	Lifestyle	3.65	Lifestyle	3.73	Concepts	3.88
	User needs	3.61	Customer experience	3.55	Brokering solutions	3.75
Organizational Collaborative Practices (Main organizational units collaborating with designers)	R&D	43.65%	R&D	39.36%	R&D	36.38%
	Board	27.96%	Engineering	25.27%	Board	29.25%
Operational Collaborative Practices (Main tools adopted in the collaboration with designers)	Concept tools	2.65	Visualization tools	3.45	Visualization tools	3.13
	Visualization tools	2.48	User-centered tools	3	Concept tools	2.69
Design Innovation performance: Ratio between awarded and total number of firms	22%		31%		31%	
Competitive Performance: Percentage of revenues from new products (2013)	7.9%		6.2%		7.5%	
Competitive Performance: Average percentage growth of revenues (2012–2013)	5.1%		0.8%		10.9%	

The strategic contributions provided by designers to companies adopting different approaches are quite aligned: generation of new concepts, improvement of the aesthetic quality, and identification of emerging lifestyles represent the most diffused strategic contributions across all three approaches (Styling, Thinking, and Meaning). The foremost diffused strategic contribution differs across the three approaches: analysis of the users' needs in the case of the Styling approach, design of the customer experience in the case of the Thinking approach, and finally reinterpretation of existing solutions in the case of the Meaning approach. Considering that the Styling approach refers to the aesthetic and symbolic values of products, the presence of user needs is reasonable. Indeed, the main goal of furniture companies using a Styling approach is to fulfill user needs with shapes and attractive products. Moving to the Thinking approach, the customer journey is an essential aspect as it is only through this that designers can solve user-centered problems and propose new solutions (Martin, 2009). Finally, brokering solutions represents a key aspect of the Meaning approach; cross-boundary contamination is one of the most inspiring ways to propose new meanings to the market (Verganti, 2009). Considering the organizational aspects of the collaborative practices instead, the most important finding relates to the unequal time external designers dedicate to interactions with the business units. On one hand, designers that collaborate with companies adopting the Styling approach mainly deal with the R&D department (almost 50%). On the other hand, designers that interact with firms adopting the Meaning approach collaborate with all the different business units in a more balanced way (30% with almost all business units). Interactions in the Thinking approach fall between the two approaches. Concluding with the tools, the Thinking approach relies more on tools than the other two approaches. This is in line with literature reporting that this relates to structuring the creative process into a set of working tools (Brown and Wyatt, 2015).

About the relationship between investments in design and competitive performances, interesting results emerge. Even if companies adopting the Thinking approach show the highest average investment in design (ratio between designers and employees = 20%), they do not show the best competitive performances both in terms of percentage of revenues from new products (6.2%) and average percentage growth of revenues (0.8%). Vice versa, firms adopting the Meaning approach show the lowest average investment in design (ratio between designers and employees = 14%), but better performance in terms of average percentage growth of revenues (10.9%). To some extent, the empirical results suggest to consider the mediating role played by the collaborative practices (Strategic, Organizational, and Operational) in influencing the relationship between investments in design and competitive performances. The high level of investments in the Thinking approach can be connected to the fact that the tools used and the number of new ideas generated are higher than in other approaches. Nevertheless, in today's world where the problem is more in the choice than in the creation of ideas, this approach alone cannot guarantee higher performance. The Styling approach where the level of tools used is lower and the interactions mainly take place with the R&D department ensures a more traditional strategy where products are more easily accepted by end-users. Finally, the Meaning approach shows that with lower investments (14%), companies can attain higher performance both in the competitive area (revenue growth 10.8%) and in terms of design performance (31%), thanks to the identification of future scenarios and the reasons why people should use and buy products through more accurately managing collaborations with all the possible partners involved.

10.6. Conclusions

Given the growing number of studies on design as a source of innovation, this chapter attempts to briefly define the effect that three different design management approaches (Styling, Thinking, and Meaning) have on SME performance as a result of collaborations with external designers. This research has implications for both researchers and practitioners:

(1) Indeed, from a theoretical perspective it enriches the knowledge around inbound open innovation (Chesbrough, 2006). As a

matter of fact, it sheds lights on the particular influence that external sources of innovation, e.g., designers, have on SMEs collaboration with external actors.

(2) Moreover, it improves the consciousness of researchers on the fact that design can sustain firm performances (Abecassis-Moedas and Mahmoud-Jouini, 2008). This is by highlighting how different design management approaches can generate different performances such as awards or revenues.

Focusing on the practical implication of the study the empirical results show these three approaches significantly differ in terms of (a) contributions from collaboration with designers, (b) time spent in interactions between designers and business units, (c) and tools used. From a managerial standpoint, the study provides managers with a greater understanding of the possible collaborative practices to leverage with external designers and how this could affect the results. In particular, the inquiry conducted shows how the different approaches toward design management are related to different strategic, organizational, and operational practices. So, managers leveraging on these can understand which levers they have in their hands in order to adopt a different design management approach. This study is based on a single nation sample, Italy, and on a particular sector, the furniture one. To ensure that generalization of the findings is possible, further researches can be explored. First of all, an extension of the firm's sample within the industry should be pursued. This can be managed in two directions, preserving the focus on SMEs or enlarging it also to large enterprises. In the second case, an additional analysis can be managed in order to compare the evidence of the studies on SMEs and the one emerging from the large companies. Second, considering the fact that collaboration with external partners such as designers is not limited to the furniture industry, a broader sample of market segments can be analyzed (see also Vanhaverbeke, 2017). Especially, similar industries, such as fashion or car industry where external collaborations are relevant, can be taken into consideration. This is in order to enrich the evidences on the strategies companies can adopt in collaborating with external sources of innovation. In addition to the previous line of inquiries to reinforce the generalization of the study, a wider set of nations can be considered. Indeed, the national boundaries can limit the findings of the research. Moreover, considering the fact that the national culture impacts the research, as clearly reported in several researches (Steensma *et al.*, 2000; Hardy *et al.*, 2003), the study can be repeated on a broader scale involving firms belonging to different nations. This can outline different approaches or different nuances in the approaches, Styling, Thinking, and Meaning, previously reported. Finally, the robustness of the findings can be enriched by adopting a more structured and statistical relevant analysis of the data such as a structural equation modeling analysis. These are just few of the future researches that, leveraging on the investigation reported above, can enrich knowledge on collaborative practices between company and external sources of innovation such as designers.

References

Abecassis-Moedas, C. and Ben Mahmoud-Jouini, S. (2008). Absorptive capacity and source-recipient complementarity in designing new products: an empirically derived framework. *Journal of Product Innovation Management*, 25, pp. 473–480.

Abecassis-Moedas, C., Mahmoud-Jouini, S. B., Dell'Era, C., Manceau, D. and Verganti, R. (2012). Key resources and internationalization modes of creative knowledge-intensive business services: the case of design consultancies. *Creativity and Innovation Management*, 21(3), pp. 315–331.

Acklin, C. (2010). Design-driven innovation process model. *Design Management Journal*, 5(1), pp. 50–60.

Acklin, C. (2013). Design management absorption model: a framework to describe and measure the absorption process of design knowledge by SMEs with little or no prior design experience. *Creativity and Innovation management*, 22(2), pp. 147–160.

Bedford, C., Hertenstein, J. H., Plat, M. B. and Desbarats, G. (2006). Profiting by design. *Design Management Review*, 17(2), pp. 54–59.

Belso-Martinez, J., Molina-Morales, F. and Mas-Verdu, F. (2011). Clustering and internal resources: moderation and mediation effects. *Journal of Knowledge Management*, 15(5), p. 738.

Berends, H., Reymen, I., Stultiens, R. G. L. and Peutz, M. (2011). External designers in product design processes of small manufacturing firms. *Design Studies*, 32(1), pp. 86–108.

Bloch, P. H. (2011). Product design and marketing: reflections after fifteen years. *Journal of Product Innovation Management*, 28(3), pp. 378–380.

Bloch, P. H., Frederic, F. B. and Todd, J. A. (2003). Individual differences in the centrality of visual product aesthetics: concept and measurement. *Journal of Consumer Research*, 29(March), pp. 551–565.

Boland, R. J. and Collopy, F. (2004). *Managing as Designing* (Stanford, CA: Stanford University Press).

Borja De Mozota, B. (2003). *Design Management: Using Design to Build Brand Value and Corporate Innovation* (New York: Allworth Press).

Borja De Mozota, B. (2006). The four powers of design: a value model in design management. *Design Management Review*, 17(2), pp. 44–53.

Brazier, S. (2004). Walking backward into design: support for the SME. *Design Management Review*, 15(4), p. 61.

Brown, T. (2008). Design thinking. *Harvard Business Review*, 86(6), pp. 84–88.

Brown, T. (2009). *Change by Design: How Design Thinking Transforms Organizations and Inspires Innovation* (HarperCollins, New York).

Brown, T. and Wyatt, J. (2015). Design thinking for social innovation. *Annual Review of Policy Design*, 3(1), pp. 1–10.

Candi, M. (2010). Benefits of aesthetic design as an element of new service development. *Journal of Product Innovation Management*, 27(7), pp. 1047–1064.

Candi, M. and Gemser, G. (2010). An agenda for research on the relationships between industrial design and performance. *International Journal of Design*, 4(3), pp. 67–77.

Chayutsahakij, P. and Poggenpohl, S. (2002). User-centered innovation: the interplay between user-research and design innovation. In: *Proceedings of The European Academy of Management 2nd Annual Conf. on Innovative Research in Management*. EURAM, Stockholm, Sweden.

Chesbrough, H. (2003) *Open Innovation: The New Imperative for Creating and Profiting From Technology* (Boston: Harvard Business School Press).

Chesbrough, H. (2006). *Open Business Models: How Thrive in the New Innovation Landscape* (Harvard, MA: Harvard Business Press).

Chesbrough, H. and Bogers, M. (2014). Explicating open innovation: clarifying an emerging paradigm for understanding industrial innovation, in H. Chesbrough, W. Vanhaverbeke and J. West (eds.), *New Frontiers in Open Innovation.* (Oxford: Oxford University Press), pp. 3–28.

Chesbrough, H. and Crowther, A. K. (2006). Beyond high-tech: early adopters of open innovation in other industries. *R&D Management*, 36(3), pp. 229–236.

Chiaroni, D., Chiesa, V., De Massis, A. and Frattini, F. (2008). The knowledge bridging role of technical and scientific services in knowledge-intensive industries. *International Journal of Technology Management*, 41(3/4), pp. 249–272.

Chiva, R. and Alegre, J. (2009). Investment in design and firm performance: the mediating role of design management. *Journal of Product Innovation Management*, 26, pp. 424–440.

Colombo, G., Dell'Era, C. and Frattini, F. (2011). New product development (npd) service suppliers in open innovation practices: processes and organization for knowledge exchange and integration. *International Journal of Innovation Management*, 15(1), pp. 165–204.

Cox, G. (2005). *Cox Review of Creativity in Business: Building on the UK's Strengths* (London: Design Council).

Creusen, M. E. H. and Schoormans, J. P. L. (2005). The different roles of product appearance in consumer choice. *Journal of Product Innovation Management*, 22, pp. 63–81.

Csikszentmihalyi, M. and Rochberg-Halton, E. (1981). *The Meaning of Things: Domestic Symbols and the Self* (Cambridge: Cambridge University Press).

Danish Design Centre, National Agency for Enterprise and Housing (2003). *The Economic Effects of Design*. Copenhagen.

Dell'Era, C. and Verganti, R. (2007). Strategies of innovation and imitation of product languages. *Journal of Product Innovation Management*, 24(6), pp. 580–599.

Dell'Era, C. and Verganti, R. (2009). The impact of international designers on firm innovation capability and consumer interest. *International Journal of Operations & Production Management*, 29(9), pp. 870–893.

Dell'Era, C. and Verganti, R. (2010). Collaborative strategies in design-intensive industries: knowledge diversity and innovation. *Long Range Planning*, 43, pp. 123–141.

Dell'Era, C., Marchesi, A. and Verganti, R. (2008). Linguistic network configurations: management of innovation in design-intensive firms. *International Journal of Innovation Management*, 12(1), pp. 1–19.

D'Este, P., Iammarino, S., Savona, M. and von Tunzelmann, N. (2012). What hampers innovation? Revealed barriers versus deterring barriers. *Research Policy*, 41(2), pp. 482–488.

Dumas, A. and Mintzberg, H. (1989). Managing design/designing management. *Design Management Journal*, 1(1), pp. 37–43.

Dumas, A. and Mintzberg, H. (1991). Managing the form, function and fit of design. *Design Management Journal*, 2, pp. 26–31.

Fernández-Mesa, A., Alegre-Vidal, J., Chiva-Gómez, R. and Gutiérrez-Gracia, A. (2013). Design management capability and product innovation in SMEs. *Management Decision*, 51(3), pp. 547–565.

Forza, C. (2002). Survey research in operations management: a process-based perspective. *International Journal of Operations & Production Management*, 22(2), pp. 152–194.

Fueglistaller, U. (2004). Charakteristik und Entwicklung von Klein- und Mittelunternehmen (St Gallen: KMU Verlag HSG).

Gabrielsen, G., Gronhaug, K., Kahle, L., Kristensen, T., Plenborg, T. and Wilke, R. (2007). *Is Good Design Good Business?* In: Rusten, G. and Bryson, J. R. (eds.), Industrial Design, Competition and Globalization (Palgrave Macmillan, Basingstoke), pp. 220–42.

Gassmann, O. (2006). Opening up the innovation process: towards an agenda. *R&D Management*, 36(3), pp. 223–228.

Gemser, G. and Leenders, M. (2001). How integrating industrial design in the product development process impacts on company performance. *Journal of Product Innovation Management*, 18, pp. 28–38.

Gemser, G., Candi, M. and van den Ende, J. (2011). How design can improve firm performance. *Design Management Review*, 22(2), pp. 72–77

Goodrich, K. (1994). The designs of the decade: quantifying design impact over ten years. *Design Management Journal*, 5(2), pp. 47–55.

Hardy, C., Phillips, N. and Lawrence, T. B. (2003). Resources, knowledge and influence: the organizational effects of interorganizational collaboration. *Journal of Management Studies*, 40(2), pp. 321–347.

Hargadon, A. B. (1998). Firms as knowledge brokers: lessons in pursuing continuous innovation. *California Management Review*, 40(3), pp. 209–227.

Hargadon, A. (2003). *How Breakthroughs Happen? The Surprising Truth about How Companies Innovate* (Boston, MA: Harvard Business School Press).

Hargadon, A. and Sutton, R. I. (1997). Technology brokering and innovation in a product development firm. *Administrative Science Quarterly*, 42, pp. 716–749.

Heberlein, T. and Baumgartner, R. (1978). Factors affecting response rates to mailed questionnaires: a quantitative analysis of the published literature. *American Sociological Review*, 43(4), pp. 447–462.

Hertenstein, J. N., Platt, M. B. and Brown, D. R. (2001). Valuing design: enhancing corporate performance through design effectiveness. *Design Management Journal*, 12(3), pp. 10–19.

Hertenstein, J. H., Platt, M. B. and Veryzer, R. W. (2005). The impact of industrial design effectiveness on corporate financial performance. *Journal of Product Innovation Management*, 22, pp. 3–21.

Heskett, J. (2002). *Design, A Very Short Introduction* (Oxford: Oxford University Press).

Hipp, C. (2000). Information flows and knowledge creation in knowledge-intensive business services: scheme for a conceptualization, in J.S. Metcalfe and I. Miles (eds.), *Innovation Systems in the Service Economy* (Boston, MA: Kluwer Academic Publishers).

Inns, T. (2002). *The Tangible and Intangible Impact of Design Activity* (Dundee: Duncan of Jordanstone College of Art & Design).

JPIM (2005a). *Journal of Product Innovation Management*, 22(1).

JPIM (2005b). *Journal of Product Innovation Management*, 22(2).

JPIM (2011). *Journal of Product Innovation Management*, 28(3).

Kelley, T. (2001). *The Art of Innovation* (New York: Currency).

Lawrence, P. and McAllister, L. (2005a). Marketing Meets Design: Core Necessities for Successful New Product Development from the Special Issue Guest Editors. *Journal of Product Innovation Management*, 22(1), 1–1.

Lawrence, P. and McAllister, L. (200b5). Marketing meets design: Core necessities for successful new product development. *Journal of Product Innovation Management*, 22(2), pp. 107–108.

Leonard, D. and Rayport, J. F. (1997). Spark innovation through empathic design. *Harvard Business Review*. Available at: https://hbr.org/1997/11/spark-innovation-through-empathic-design.

Lojacono, G. and Zaccai, G. (2004). The evolution of the design-inspired enterprise. *Sloan Management Review*, Spring, pp. 75–79.

Margolin, V. and Buchanen, R. (eds.) (1995). *The Idea of Design: A Design Issues Reader* (Cambridge: MIT Press).

Martin, R. (2009). *The Design of Business: Why Design Thinking is the Next Competitive Advantage* (Harvard Business School Press, Boston, MA).

Michlewski, K. (2008). Uncovering design attitude: inside the culture of designers. *Organization Studies*, 29, pp. 373–392.

Moultrie, J., Clarkson, P. J. and Probert, D. (2007). Development of a design audit tool for SMEs. *Journal of Product Innovation Management*, 24(4), pp. 335–368.

Moreau, C. P. (2011). Inviting the amateurs into the studio: understanding how customer engagement in product design creates value. *Journal of Product Innovation Management*, 28(3), pp. 409–410.

Norman, D. and Verganti, R. (2013). Incremental and radical innovation: design research vs technology and meaning change. *Design Issues*, 30(1), pp. 78–96.

Patnaik, D. and Becker, R. (1999). Need finding: the way and how of uncovering people's needs. *Design Management Journal*, 2, pp. 37–43.

Perez-Freije, J. and Enkel, E. (2007). Creative tension in the innovation process: how to support the right capabilities. *European Management Journal*, 25(1), pp. 11–24.

Platt, M. B., Hertenstein, J. H. and Brown, D. R. (2001). Valuing design: enhancing corporate performance through design effectiveness. *Design Management Journal*, 12(3), pp. 10–19.

Ramlau, U. H. (2004). In Denmark, design tops the agenda. *Design Management Review*, 15(4), 48–54.

Ravasi, D. and Lojacono, G. (2005). Managing design and designers for strategic renewal. *Long Range Planning*, 35, pp. 51–77.

Ravasi, D. and Stigliani, I. (2012). Product design: a review and research agenda for management studies. *International Journal of Management Reviews*, 14(4), pp. 464–488.

Rich, H. (2004). Proving the practical power of design. *Design Management Review*, 15(4), pp. 29–34.

Rindova, V. P. and Petkova, A. P. (2007). When is a new thing a good thing? Technological change, product form design, and perceptions of value for product innovations. *Organization Science*, 18(2), pp. 217–232.

Rosenthal, S. R. and Capper, M. (2006). Ethnographies in the front end: designing for enhanced customer experiences. *Journal of Product Innovation Management*, 23(3), pp. 215–237.

Roy, R. (1994). Can the benefits of good design be quantified? *Design Management Journal*, 5(2), pp. 9–17.

Seybold, P. B. (2001). Get inside the lives of your customers. *Harvard Business Review*, 79(5), pp. 80–91.

Steensma, H. K., Marino, L., Weaver, K. M. and Dickson, P. H. (2000). The influence of national culture on the formation of technology alliances by entrepreneurial firms. *Academy of Management Journal*, 43(5), pp. 951–973.

Stein, E. and Iansiti, M. (1995). *Understanding User Needs* (Harvard Business School Publishing, Boston, MA).

Sutton, R. (2002). Weird ideas that spark innovation. *Sloan Management Review*, 43(2), pp. 83–87.

Sutton, R. I. (2001). The weird rules of creativity. *Harvard Business Review*, 79(8), pp. 95–103.

Swan, K. S., and Luchs, M. (2011). From the special issue editors: product design research and practice: past, present and future. *Journal of Product Innovation Management*, 28(3), pp. 321–326.

Swan, K. S. and Luchs, M. (2011). Product design research and practice: past, present and future. *Journal of Product Innovation Management*, 28(3), pp. 321–326.

Swan, K. S., Kotabe, M. and Allred, B. B. (2005). Exploring robust design capabilities: their role in creating global products, and their relationship to firm performance. *Journal of Product Innovation Management*, 22(2), pp. 144–164.

Swedish Industrial Design Foundation (2004). Swedish Agency for Economic and Regional Growth. Swedish Companies on Design: Attitudes, Profitability and Design Maturity, Stockholm.

Swink, M. (2000). Technological innovativeness as a moderator of new product design integration and top management support. *Journal of Product Innovation Management*, 17(3), pp. 208–220.

Talke, K., Salomo, S., Wieringa, J. E. and Lutz, A. (2009). What about design newness? Investigating the relevance of a neglected dimension of product innovativeness. *Journal of Product Innovation Management*, 26, pp. 601–615.

Thomke, S. and Von Hippel, E. (2002). Customers as innovators: a new way to create value. *Harvard Business Review*. Available at: https://hbr.org/2002/04/customers-as-innovators-a-new-way-to-create-value.

Trueman, M. and Jobber, D. (1998). Competing through design. *Long Range Planning*, 31(4), pp. 594–605.

Utterback, J., Vedin, B.-A., Alvarez, E., Ekman, S., Sanderson, S. W., Tether, B. and Verganti, R. (2006). *Design-Inspired Innovation* (Singapore: World Scientific Publishing).

Vanhaverbeke, W. (2017). *Managing Open Innovation in SMEs* (Cambridge, UK: Cambridge University Press).

Verganti, R. (2003). Design as brokering of languages: the role of designers in the innovation strategy of Italian firms. *Design Management Journal*, 3, pp. 34–42.

Verganti, R. (2006). Innovating through design. *Harvard Business Review*. Available at https://hbr.org/2006/12/innovating-through-design.

Verganti, R. (2009). *Design-Driven Innovation. Changing the Rules of Competition by Radically Innovating What Things Mean* (Boston, MA: Harvard Business Press).

Verganti, R. (2011). Designing breakthrough products. *Harvard Business Review*. Available at: https://hbr.org/2011/10/designing-breakthrough-products.

Verganti, R. and Dell'Era, C. (2013). Design-driven innovation: meaning as a source of innovation, in Dodgson, M. Gann D. and Philips, N. (Eds.), *The Oxford Handbook of Innovation Management*, ISBN 978-0-19-969494-5, (Oxford University Press, Oxford), pp. 139–162.

Verganti, R. and Öberg, Å. (2013). Interpreting and envisioning: a hermeneutic framework to look at radical innovation of meanings. *Industrial Marketing Management*, 42(1), pp. 86–95.

Veryzer, R. W. (2005). The roles of marketing and industrial design in discontinuous new product development. *Journal of Product Innovation Management*, 22(1), pp. 22–41.

Veryzer, R. W. and Hutchinson, J. W. (1998). The influence of unity and prototypicality on aesthetic responses to new product designs. *Journal of Consumer Research*, 24(4), pp. 374–394.

Wallace, R. (2001). Proving our value: measuring package design's return on investment. *Design Management Review*, 12(3), pp. 20–27.

Walsh, V. (1996), Design, innovation and the boundaries of the firm. *Research Policy*, 25(4), pp. 509–529.

Walsh, V., Roy, R. and Bruce, M. (1988). Competitive by design. *Journal of Marketing Management*, 4(2), pp. 201–216.

Whitney, P. and Kumar, V. (2003). Faster, cheaper, deeper user research. *Design Management Journal*, 14(2), pp. 50–57.

Zec, P. and Burkhard, J. (2010). Design Value: a strategy for Business Success. Red Dot.

Appendix A: The Role of Designers in the Furniture Industry

(Expected time: 10 minutes)

A.1. Design Management Approaches

A.1.1. Select the design management approach adopted by your company.

Design plays a crucial role in the definition of the aesthetics of products, style, shape, colors, and finishes. Moreover, it helps in consolidating the brand and distinguishing the company in the market. ☐

Design helps improve product functionality in a creative way. Moreover, it helps envision product problems and stimulate the generation of new solutions to solve them. ☐

Design increases the perceived value of the product introducing new forms of psychological satisfaction deriving from a deep sociocultural analysis. ☐

A.2. Investments in Design

A.2.1. Report the total number of employees in 2012. ☐

A.2.2. Report the total number of external designers with whom you collaborated in 2012. ☐

A.3. Collaborative Practices with Designers

A.3.1. Our collaboration with designers is aimed at obtaining different contributions. (1 = strongly disagree, 2 = disagree, 3 = neither disagree nor agree, 4 = agree, 5 = strongly agree)

	1	2	3	4	5
Designers propose new business strategies.	☐	☐	☐	☐	☐
Designers recognize and interpret new lifestyles and sociocultural trends.	☐	☐	☐	☐	☐
Designers develop new concepts.	☐	☐	☐	☐	☐
Designers identify new technologies useful to improving products.	☐	☐	☐	☐	☐
Designers contribute in the definition of the production processes.	☐	☐	☐	☐	☐
Designers define new customer experiences.	☐	☐	☐	☐	☐
Designers analyze emerging customer needs.	☐	☐	☐	☐	☐
Designers broker solutions from other industries and contexts of use.	☐	☐	☐	☐	☐
Designers take care of the product aesthetics.	☐	☐	☐	☐	☐

A.3.2. Report the average time spent by designers in collaborating with the different organizational units.

Board	%
Research & Development	%
Engineering	%
Marketing	%

A.3.3. We adopt different tools in the collaboration with designers. (1 = Never, 2 = Rarely, 3 = Sometimes, 4 = Often, 5 = Always)

	1	2	3	4	5
Customer needs analysis (i.e., focus group, ethnographical research)					
Cocreation (i.e., user toolkit, crowdsourcing platforms)					
Concept generation (i.e., user profiles, personas)					
Visualization (i.e., moodboard, storyboard)					

A.4. Competitive Performance

	2012	2013
A.4.1. Report the revenues achieved in 2012 and 2013 (million euro).	12) mln	13) mln

	2012	2013
A.4.2. Report the % of revenues achieved in 2012 and 2013 from new products.	%	%

Chapter 11

The Link Between Entrepreneurial Attributes and SME Ecosystem Orchestration: A Case from the Dutch HR Services Industry

Oana-Maria Pop, Nadine Roijakkers, Diana Rus and Marjolein Hins*

* *oanamaria.pop@uhasselt.be*

Abstract

This chapter provides a detailed account of the evolution and internal dynamics (network relationships) at Q-Search, a Dutch ecosystem of small and medium-sized enterprises (SMEs) active in the human resources services industry. More specifically, the chapter explores how the personality of the entrepreneurial orchestrator reflects upon the SME ecosystem forming around her. Using multiple data sources — i.e., interviews, archival search, and direct

* Corresponding author.

observation — we first explore the various phases of the ecosystem's 16-year evolution. Next, we examine the complex interplay between Q-Search's goals in each phase, the orchestrator's corresponding attributes/actions, SME partner roles, and the innovative value created. Our findings suggest that studying the psychology of entrepreneurial orchestrators can reveal a great deal about the networks and ecosystems they help shape. At Q-Search, the founding entrepreneur's vision, authenticity/honesty, passion for a cause, personal initiative, social skill, persistence/grit, ability to conduct self-reflection, and self-efficacy and willingness to change, have proven critical for the ecosystem's survival and performance.

11.1. Introduction

The importance of SMEs for economic growth (World Bank, 2016) as well as their propensity to form enterprise groups (Eurostat, 2016) or other types of collaborative arrangements, e.g., alliances (Lee *et al.*, 2012), networks (Gardet and Fraiha, 2012), and even meta-organizations/ecosystems (Gulati *et al.*, 2012) to sustain this growth continues to be emphasized. Despite this fact, relatively little is known about *how* SMEs build such important open innovation (OI) relations and constellations and, moreover, how they secure the strength of their collaborative arrangements in the long run — a notable exception is Vanhaverbeke (2017). To shed more light on this topic, this chapter explores how an entrepreneur orchestrates the ecosystem of SMEs formed around her. Specifically, we study the case of Q-Search, a Dutch human resources (HR) services ecosystem whose aim is to increase "workers' joy" (job satisfaction), and through this provide better societal outcomes.

In the literature, managing OI relations in SMEs has recently taken center stage. Brunswicker and van de Vrande (2014), in particular, underline that the OI relationships of SMEs (both formal and informal) are crucial for their survival, their ability to change their strategic direction, and their overall (innovative) performance (Brunswicker, 2011; Drechsler and Natter, 2012; Van de Vrande *et al.*, 2009). SMEs also create OI relations to enrich their knowledge capital (Laperche and Liu, 2013), that is, the information and knowledge produced, acquired, and used in the value creation process. In addition, being embedded in a network or ecosystem of OI relations enables SMEs to compensate for the lack of complementary resources and skills they typically need for innovation; hence, being part of a shared enterprise with an explicit customer focus (a typical characteristic of ecosystems) leads to a so-called "ecosystem advantage" (Williamson and De Meyer, 2012) in the face of changing customer demands and volatile markets. All in all, as Ceci and Iubatti (2012) note: "the presence of multiple types of relationships (in a network) modifies inter-firm dynamics, creating a space where traditional innovation activities take place in an unusual way."

While SMEs can potentially use their network/ecosystem partners' resources and skills to their advantage (e.g., become more effective at product/service innovation), realizing this potential requires the development of distinct capabilities (Vanhaverbeke, 2017). Specifically, SMEs need to build up the ability to *coordinate or orchestrate their OI network relations.* To effectively coordinate knowledge flows in a multi-actor environment, SMEs and the entrepreneurs that founded them must explore and learn about a variety of orchestration aspects including: degrees of formality, trust, division of benefits, guarantees (against potential loss), and conflict resolution modes, to name but a few (Gardet and Fraiha, 2012; Paquin and Howard-Grenville's, 2013). At the same time, entrepreneurs must understand orchestration processes such as managing innovation leverage, managing innovation coherence, and managing innovation appropriability (Nambisan and Sawhney, 2011). Finally, entrepreneurs must acknowledge that in SMEs embedded in ecosystems (Autio *et al.*, 2014), orchestration actions are "tailored" and reflect the personal attributes and goals of the orchestrating entrepreneur (Ucbasaran *et al.*, 2001). For example, entrepreneurial orchestrators, acting based on their personal attributes, can discretely influence the creation of the groundwork for value creation/capturing by all ecosystem partners (SMEs), other stakeholders, and society at large (Nambisan and Sawhney, 2011; Ritala *et al.*, 2013).

All in all, orchestrating entrepreneurs appear to create personal visions where each SME partner plays a distinctive role, markets the ecosystem — or other type of collaborative arrangement — to outside parties, sets up partner selection mechanisms reflecting their personal values, and creates a personalized resource network around the ecosystem from which partners can draw (see also the "player–structure duality" that Dhanaraj and Parkhe, 2006 note).

For the purpose of this chapter, we zoom in on *ecosystems* of SMEs. Hence, the notion requires defining. To date, a few notable publications have described ecosystems' purpose well, mainly by comparing and contrasting them to the more established notion of "alliance networks" (Autio *et al.*, 2014; Leroi-Werelds *et al.*, 2017). From these works, we learn that ecosystems seem to function on the basis of not one, but several central parties that activate and connect partners. Although there may be one party initiating the ecosystem, the organizational structure typically evolves over time to a state of shared leadership and a sustainable social community, functioning largely on the basis of shared norms and values (i.e., institutions), where several partners can take the lead on initiatives pursued by the ecosystem. While alliance networks and leadership positions herein are usually associated with superior performance and enhanced competitive positions of individual firms, several ecosystems that have come into existence, so far, seem to be more about benefiting the ecosystem at large and creating societal value (Ritala *et al.*, 2013). All in all, OI ecosystems create value at the level of the ecosystem, whereby partners integrate resources, i.e., serve each other, for the well-being of the ecosystem and all partners involved (including the customer and society). Specifically, partners contribute to the ecosystem according to their areas of expertise and then rely on others to provide access to complementary resources; as such, they depend on each other for long-term strategizing, survival, and mutual effectiveness.

In the light of the above, we define an ecosystem as a "relatively self-contained, self-adjusting system of resource-integrating actors connected by shared institutional arrangements and mutual value creation through service exchange" (Vargo and Lusch, 2016).

To illustrate the aforementioned aspects, i.e., OI ecosystem relations and capabilities to orchestrate these relations, this chapter examines the development of the Q-Search ecosystem in relation to the psychology of the orchestrating entrepreneur. Observing orchestrators' individual characteristics and how they drive OI in (networked) SMEs is also a topic tackled in Chapter 5 of this book as well as in an earlier work by Ahn *et al.* (2013). In the former case, the authors mention risk-taking propensity and endurance as well as knowledge and network as factors that influence the adoption of OI.

In our pursuit to better understand the dynamics of OI ecosystems formed by SMEs, we also answer prior calls for research by Gardet and Fraiha (2012), Brunswicker and van de Vrande (2014), and Vanhaverbeke (2017). These calls refer to the OI relations of SMEs and how these small firms can effectively coordinate their relations with (SME) partners. Finally, this chapter answers calls for research on different levels of analysis, i.e., the individual and the ecosystem level, by Chesbrough *et al.* (2014) and West and Bogers (2017), as well as on the human element in OI (Salampasis and Mention, 2017).

The remainder of this chapter is structured as follows. In Section 11.2, we introduce the research design by providing background to the case as well as notes on the research method. In Section 11.3 (analysis), we examine each evolutionary phase (initiation, growth, and maturity) at Q-Search and explicate the link between the orchestrator's attributes, the ecosystem's goals, orchestration actions, and ecosystem outcomes (generated value). In Section 11.4, we present the implications of our research.

11.2. Research Design

11.2.1. *Research context*

Founded in the year 2000 by Finance and HR expert, independent policy advisor, and entrepreneur, Marjolein Hins, Q-Search represents an SME ecosystem specialized in HR and development & transformational services consulting. A few examples of innovative services created by the ecosystem include the Development Vouchers and The Art of Letting Go training (see Section 11.3.3 for a more detailed explanation of these services).

When asked to describe the HR services industry in the Netherlands, Hins points to a conservative, divided market, undergoing slow and uneven change. The result is a gap between the corporate world and the real world, powered by a misalignment in values. For an SME like Q-Search trying to penetrate and generate positive societal impact in a reactive, old-fashioned market ("to compete with the McKinseys and the Ebigners") was therefore difficult, but not impossible. Q-Search's mission from the onset was not to meet market targets *per se*, although partners acknowledged the need to get their market act together in the mid and long run. In stark contrast to other HR service organizations, Q-Search set out to generate value for its partners and instill the Q-feeling, as defined in a set of written testimonials. All in all, at a time when collaboration/Information and Communication Technologies (ICT) platforms for recruiting and management were unknown in the Netherlands, Q-Search was pioneering ways to educate specialized SMEs and, through them, turn the workplace into an arena where people operate on the basis of trust, cooperation, and sharing.

Choosing to build an OI ecosystem beyond a typical organization or recruitment network hence set Hins apart. The aim was to create an ecosystem of professionals and SMEs that could, in time, function independently from her guidance and intervention. As such, Hins was laying the groundwork for an ecosystem with distinct OI relations. If built well, this vehicle could bring SME partners and expertise together, deliver better, innovative value propositions to customers, and generate social change.

Between 2000 and 2016, Q-Search has undergone a number of distinct evolutionary phases that resemble the development of single firms and even those of biological ecosystems. These phases are: initiation, growth, and maturity (see Figure 11.1).

The initiation phase takes place between 2000 and 2001 and is characterized by the orchestrator's drive to bring about change, fueled by her disappointing experience working as a manager for a large Dutch auto lease company. For one year, she searches for

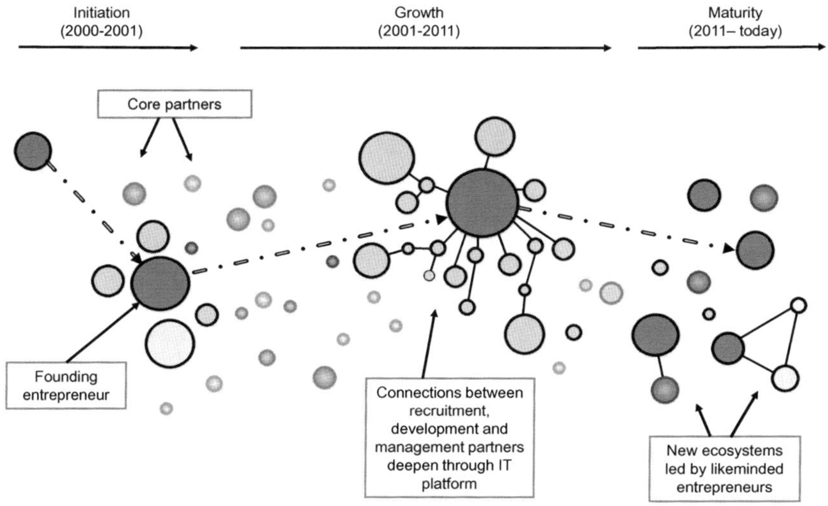

Figure 11.1. The evolution of Q-Search.

like-minded people and SMEs willing to join her ecosystem and ultimately starts off with a core group of around 10 SME partners. During this first phase, the ecosystem is not profit-driven.

An increase in the number of partners joining shifts the ecosystem into its second phase: growth. This phase spans the period 2001–2011 and is characterized by formalization and centralization. The orchestrator introduces a fee-based system and virtual platform to connect all members. Hins also provides member services, e.g., marketing collateral, newsletters, guest lectures and meetings, leads, in exchange for a yearly contribution. Furthermore, she commits values and ways of working to paper.

As partners increasingly expect the entrepreneur to take the lead, a fact that is reinforced by the dawn of the international financial crisis, she decides to abandon all formal administration and to collaborate on demand. This action enables the ecosystem to evolve to its current phase: maturity (2011–2016). This phase is characterized by several spin-offs, which maintain the spirit of Q-Search and work toward the orchestrator's initial purpose. One such spin-off is the "The Future of Work," whose philosophy is to lay a new foundation for the way people work, which is actively communicated to the Dutch government.

11.2.2. *Research method*

Q-Search offers a unique opportunity to study the link between the psychology of the entrepreneur and how her attributes reflect upon her ecosystem's evolution (2000–2016). Figure 11.2 captures this relationship.

Following Eisenhardt (1989) and Eisenhardt and Graebner (2007), we conduct a theory-building, case-based investigation, as this approach is particularly relevant in new topic areas leading to theories that are novel, testable, and empirically valid. Specifically, we use a single case design, which involves phenomena that are critical, unusual, and longitudinal (Yin, 2013) with the aim to expand existing theories on entrepreneurial firms (SMEs) embedded in ecosystems. Single case studies are especially valuable because they allow for a thorough analysis of complex, distinct phenomena. Bennett and Elman (2010), for example, point at the advantages of using single cases that are implicitly comparative, such as deviant or crucial cases.

To increase the generalizability of our findings, we employ a strategic selection of the case (Flyvbjerg, 2006). Such information-oriented selection (as opposed to random selection) is appropriate when having high expectations about the information content of the case, as well as when the goal is to maximize the utility of this content. In our specific case, information-oriented selection ensured that the greatest possible amount of information on the phenomena of interest was collected, facilitating a rich analysis.

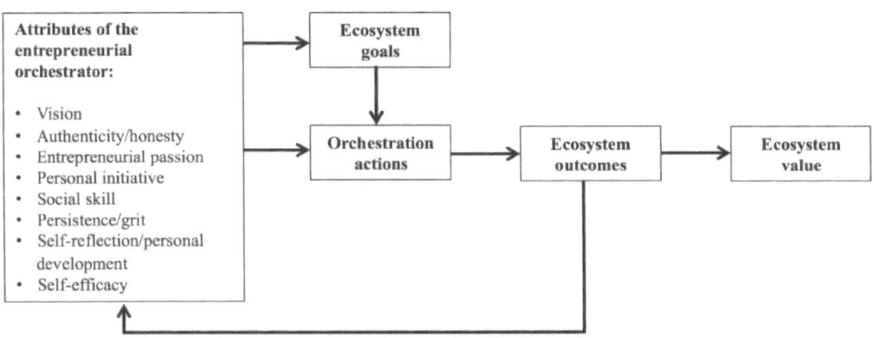

Figure 11.2. The relations under study.

We have ensured the quality of the research by taking several measures. First, we have tapped into multiple sources of evidence (see Table 11.1). These sources include: in-depth, semistructured interviews with the orchestrator and Q-Search partners active at various stages of the ecosystem's evolution; public and privately available information — for example, press releases about Q-Search and documents available on the back end of the platform (access into the archives was granted by the founder), brochures describing the Q-Search services; as well organizing a physical event (intervision group meeting) with Q-Search partners to receive feedback on the analysis and validate the milestones in the ecosystem's evolution. Second, we have conducted our analysis independently according to our own areas of expertise (strategy/innovation management and psychology), periodically comparing and contrasting results until consensus was reached. Third, we have used a case study protocol and have drawn up a database where all information was stored.

Following Gehman *et al.* (2017), we acknowledge that our data is only a static photograph of an otherwise dynamic phenomenon (ecosystem-building); hence, we attempt to explicate the evolution of Q-Search using process thinking (Langley *et al.*, 2013). As part of this approach, we have interviewed partners about specific factual events (initiation, growth, maturity, and even rebirth of the Q-Search ecosystem), rather than being interested in their interpretations and cognitions of these phases. The exception was the founding entrepreneur, who we also asked about her attitude toward the evolution in order to grasp her attributes and leadership style. Next, we have employed visual mapping and have examined, in-depth, each of the evolutionary phases. Following Van de Ven (1992), we focus on changes over time within the ecosystem as to identify underlying patterns. As we progress through the analysis, we explain what phenomena have unfolded at each phase of the ecosystem's evolution, and we set this against the specific attributes/orchestration actions of the entrepreneur. All data collected were stored and analyzed using NVivo10.

In Section 11.3, we analyze (phase by phase) the evolution of the Q-Search ecosystem closely by noting: OI ecosystem goals, key

Table 11.1. Sources of data.

Data type	Description
Interviews	In-depth, semistructured personal interview with Q-Search founder and orchestrator.
	In-depth, semistructured telephone interview with Q-Search partner in Stage II.
	In-depth, semistructured personal interview with member of The Future of Work project (a Q-Search spin-off) and Q-Search founding partner in Stage I.
Documents	Document outlining rules and attitudes prospective Q-Search partners adhered to prior to being granted membership, — e.g., "I am aware that working successfully in a networked organization means that I take other people's business as seriously as my own"; "I firmly believe that by working together within Q-Search I can serve more customers than I could on my own (I can also generate more revenue with the group than by working alone)."
	Document outlining values Q-Search members shared and/or developed while in the ecosystem. For example, Q-Searchers are "interested in others, informal, communicative, empathetic, respectful, have a sense of humor, helpful, and provide pleasant company"; professionally, Q-Searchers know what they are talking about, know their trade (via experience and skills), and are professional (you are in good hands with them). The document also includes partner testimonials.
	Document outlining benefits received in exchange for Q-Search membership fee in Stage II — e.g., network book, marketing collateral (Q-Search logos, banners, personalized signature in emails), OTYS user manual (the IT infrastructure), newsletters, direct weekly info, list of leads, access to shared meeting rooms, etc.
	Document outlining input for new tax plan in the Netherlands. There is reference to The Future of Work project in an end note.
	PowerPoint presentation for the Ministry of Internal Affairs and Transition Commission for the Social Domain in the Netherlands on achieving societal impact through new leadership and flexible forms of organization (networks).
	Document outlining The Future of Work project's philosophy.
Direct observation	Personal meeting with Q-Search founder and SME partners; in this meeting, the preliminary findings of the chapter were discussed. Feedback was incorporated into the final version of the chapter.

attributes of the entrepreneurial orchestrator, key orchestration actions, the value created within the SME ecosystem, and outcomes of the collaboration.

11.3. Analysis

11.3.1. *Initiation (2000–2001)*

The first phase of the ecosystem's development took place between 2000 and 2001 and was marked by the orchestrator's decision to abandon her position as manager at a large, hierarchical organization in favor of a more meaningful pursuit: to alter existing HR practices, increase job satisfaction, and boost the quality of work-life in the Netherlands. This period (Table 11.2) was synonymous with laying the groundwork for sustained value creation and included three goals: (1) establishing ecosystem legitimacy; (2) fostering a knowledge-sharing culture; and (3) creating space for continued development. The driving force across this phase is the orchestrator's vision (disseminated through careful communication), an image of the future that represents the shared values to which the ecosystem's SMEs should aspire (Shamir *et al.*, 1993).

To begin with, the orchestrator spends much time broadcasting her desire to build a strong HR services community. To do so, she takes part in individual and group conversations, holds speeches, and makes presentations, inspiring and attracting SMEs whose values are aligned with the vision. Within a short period of time, a core group of 10 partners forms around Hins, establishing a type of informal steering committee whose job would be to agree on the direction Q-Search should take.

Next, the orchestrator moves on to foster a knowledge-sharing culture (second key orchestration action) following her conviction that taking a humane approach to business will help create a community based on trust and collaboration, responsibility, openness, and experimentation. To realize this goal, she spends time deepening (as opposed to multiplying) the conversations around better HR practices and increasing job satisfaction, consolidates prospective

Table 11.2. The Q-Search ecosystem's development. Phase I: Initiation (2000–2001).

Core aspects		Evidence from the data
Ecosystem goals	Establish ecosystem legitimacy	Entrepreneur: So first we were a recruitment network but I was also talking about [...] the ultimate advisory, consultancy bureau, not in the old form, but in the new network form.
	Foster knowledge-sharing culture	Entrepreneur: Q-Search is nothing more than a little society for me.
	Create space for continued development	Partner: Doing business in a more humane way on the basis of equality, respect, warm-heartedness, and sincerity.
Key attributes of the entrepreneurial orchestrator	Vision	Entrepreneur: I was an entrepreneur and I started talking about the concept of... well, the consultancy firm in the network organization form. Partner: She said: I want to change [workplace] politics without going into politics. I want to increase [workers' joy].
	Authenticity/honesty	Partner: She is a reference point for a lot of people. And a reference point not solely in terms of business. Partner: A person in her authenticity, in her genuine being. Being herself. Doc1 Partner selection criteria.
	Entrepreneurial passion	Entrepreneur: I worked in companies and I saw that we were not cooperating easily or not even cooperating with each other (at all) although we were working on the same products or services. And I thought: This is strange, but you know, I am not going to change them.
	Personal initiative	Partner: She initiates a lot. Partner: She facilitates. She is always there. If [she] is not there, there is no group.

Key orchestration actions	Nonsystematized communication	Entrepreneur: The first model was: Everything for free, just being happy that people were even interested in me and came there.
	Member self-selection	Partner: She selected the persons she liked to join in the first stage. And she had, of course, her own reasons and her own assessment, for that. Entrepreneur: [The selections] I made them myself.
	Single-handedly supporting the ecosystem	Entrepreneur: At the beginning I paid everything myself but then I earned my money with recruitment.
	Coordination mechanisms creation	Partner: We had a starting meeting with, I think around 10–15 people.
Ecosystem value	Connecting like-minded HR service professionals	Partner: So yes, we were all, you can say in a major part in alignment with the values, ideas, philosophies that [the entrepreneur] exhibited.
Outcomes	Ecosystem takes shape with help from core partners in recruitment.	Partner: It all started up with I think, maybe, 10 people who were active in recruitment. Entrepreneur: Everyone was enthusiastic and I started with recruitment professionals. Entrepreneur: I realized that I have to write down all the rules we made up already with each other.

SME partners' points of view, and performs a careful selection of core partners based on a fit with her own personal values and vision. Partners are therefore selected based on their agreement with statements such as: "I/we believe that we can be more customer-oriented (service-minded) by working within Q-Search than we can by working separately" (Table 11.2). This pursuit of creating the right environment for sharing, collaborating, establishing mutual trust, and shared responsibility is a reflection of the orchestrator's personal belief system and behaviors.

The final goal relates to creating space for continued development or an environment that can help partners coevolve, withstand environmental contingencies, and safeguard exchanges. Taking a long-term perspective, the orchestrator supports the formation of Q-Search with her own resources, thereby communicating that she believes in the long-term viability of the collaborative arrangement. At the same time, Hins models unique OI principles and relations or the group-oriented behavior necessary for the ecosystem to remain viable.

By the end of 2001, the orchestrator had created a space where HR service professionals and SMEs enthusiastic about the same principles, ideas, and philosophies were connected. As the communication efforts (including word-of-mouth) regarding Q-Search's *raison d'être* began to attract many new partners, the orchestrator initiated a set of deliberate steps in order to facilitate growth and accommodate the deepening of connections among partners paving the way for innovative services creation.

11.3.2. *Growth (2001–2011)*

Growth at Q-Search takes place between 2001 and 2011 and is marked by the formalization of relationships between the orchestrator and her SME partners (Table 11.3). Having laid the groundwork for sustained value creation in initiation, the orchestrator's focus now shifts toward: (1) deepening the connection among partners; (2) changing and aligning their behavior; (3) creating a values community that transcends a singular focus on

Table 11.3. The Q-Search ecosystem's development. Phase II: Growth (2001–2011).

Core aspects		Evidence from data
Ecosystem goals	Deepening connection among partners (via IT platform)	Entrepreneur: What is also very interesting is that you can use ICT to change behavior. Partner: She was asking: What can I do for the whole group to create a better community? Entrepreneur: We were always pushing towards cooperation. Find each other without me.
	Changing/aligning partner behavior	Partner: She was more like: OK, if you go in that direction, what do you need to go in that direction? How can I help you with that? Entrepreneur: The formal issue was more necessary when the personal connection was not there.
	Creation of a values community	Partner: I think she was trying more to build, develop the values community more than the business community. And that's what a lot of partners didn't understand. Entrepreneur: All the selections, I made them myself. And after 3 years or so I also asked partners to look for new partners because I thought otherwise it's not a really a democratic process.
Key attributes of the entrepreneurial orchestrator	Vision	Entrepreneur: And 10 years ago, there was a partner in my network that said: You really have to change the website because nobody understands it. Well, I see that because I am talking about what will happen in the future, in 10–15 years from now.

(*Continued*)

Table 11.3. (*Continued*)

Core aspects		Evidence from data
	Social skill	Partner: She is a trust builder. Partner: She is a "bridger," she is a strong networker, and she knows quite a lot of people, and she is also a peacemaker, a world-changer, she is very strong and very quick and she is taking care. Entrepreneur: I am good at attracting new people, building a concept, testing new partners, writing a newsletter. But if someone else wants to take over some of these steps it's OK with me.
	Entrepreneurial passion (strong group-orientation)	Partner: She was asking: What can I do for the whole group to create a better community? Doc2 Q-Search values Entrepreneur: We have to be patient and take everyone with us.
	Persistence/grit	Partner: She is very strong at that [drive, persistence]. She really believes there is a better world. MIG meeting: Partners confirmed this trait.
	Self-reflection	Entrepreneur: I was constantly working on my inner self. I was so anxious to create this that I was constantly working on myself. It was a lot of hard work. MIG meeting: Partners confirmed this trait.
Key orchestration actions	Vision and mission document drafted	Entrepreneur: The contract was this intention, the vision, 1½-page, and then a 1½-page simple [outline of] rules; I delivered a few services, I organized meetings for the partners every month and I wrote a newspaper/newsletter every month in which I told them about each other because they were not always at the meetings. Doc2 Q-Search values + partner input

	IT platform commissioned and implemented	Entrepreneur: All the partners were connected to the IT system, which OTYS built with me for recruitment. Doc3 Services provided by orchestrator
	Full ecosystem facilitation for yearly fee (subscription model)	Entrepreneur: At the beginning I paid everything myself but then I earned my money with recruitment. After 3 years they paid 750 euros a year and therefore I did all the work but I quit recruitment.
Ecosystem value	Ongoing triage/self-organizing system	Entrepreneur: The network kind of organized itself also because the partners that did not share your vision, they leave. And the contract is not necessary. It's like a true biological ecosystem.
	Connectedness, resource diversity	Entrepreneur: I really let people connect their brains in a different way. Partner: The main value is the diversity and the sharing in that diversity.
Outcomes	Orchestration is formalized and partners polarize in the wake of a pending financial crisis	Partner: It was kind of a culture split between the recruitment partners and some of the consulting business, counseling business, and trainers. (...) The values that Marjolein embraced were more in the consulting, counseling side. The recruiters were the moneymakers and they were real goal getters. Partner: The crisis came and what you see is that everybody started acting human.
	Entrepreneur formally quits Q-Search	Entrepreneur: What people do a lot is to think that Q-Search is the ship [they] can go on, and they will help me survive. No, you have to survive yourself. Entrepreneur: So I quit after 11 years with the forms.

monetary benefits. These interconnected goals are reached through a series of synchronized actions and reveal new orchestrator attributes.

The first goal, deepening the connection among OI partners, is realized through two nearly simultaneous actions. First, the orchestrator and core recruitment partners draft and distribute a three-page vision and legal document (collaboration contract) in 2001 geared at reinforcing the vision, clarifying the engagement rules between the orchestrator and partners, outlining resources and facilities available to partners, and highlighting legal provisions. In 2002, the orchestrator commissions an external company (OTYS) to develop and implement an IT platform to facilitate knowledge creation and sharing, develop the basis for learning processes, and strengthen the ties between all categories of partners — recruitment, management, and development.

The decision to create a hybrid collaboration contract stems from the orchestrator's belief that successful collaboration can only happen if partners adhere to the underlying value principles and that the way to get partners on the same page is to rehearse the principles of collaboration. Originally confusing — that is, some partners felt vision and legal aspects cannot mix — the document eventually becomes the keystone of collaboration, enabling high-quality exchanges and transparency.

The creation of an IT platform to strengthen the ties between partners (management and development partners can now share vacancies with recruiters), blend routines (old and new), and accumulate a pool of complementary skills is a pioneering OI initiative for HR service professionals in the Netherlands, albeit one that is consistent with the orchestrator's initial vision of the future: a networked organization that can sustain long-term, innovative service creation generating value for society at large.

The second and third goals, changing and aligning partners' behavior and creating a values community that transcends a singular focus on monetary benefits, are also partly realized by the introduction of the IT platform. The important element is the orchestrator's decision to assume full ecosystem facilitation for a yearly fee by 2003. The fee is designed as an "all-inclusive" package for partners and covers elements such as group communications (OTYS user manual/guide to the IT platform, monthly newsletters about partners, coworking tips, guest lectures, facilitated meetings), marketing collateral (website, logos, banners, brochures, badges, personalized email signatures), access to facilities (shared meeting rooms), and personal coaching/feedback.

To iron out surfacing disparities between different types of partners, the orchestrator continues the group meetings, sometimes including only individual partner categories, and at other times bringing all types of partners together. During these meetings, her social skills, especially her ability to perceive others' interests and needs (Baron and Tang, 2008) accurately, her propensity to express her own reactions and feelings openly, and her skill in adapting her behaviors to the situation at hand, enable her to build trust ("she is a trust-builder"), to bridge various points of views ("she is a bridger"), and to make peace ("she is a peacemaker"). Her passion for making the OI ecosystem work, for sharing and helping others, and her persistence/grit and dedication to her SME partners, help create a climate of psychological safety, trust, and cooperation and a sense of shared purpose, mutual support, and voice — all prerequisites of shared leadership creation (Carson *et al.*, 2007).

The deliberate transition from loose to rigorous orchestration to create a values community occurs naturally and reveals the orchestrator's systematic approach to generating win–win situations. This attitude is consistent with the vision of a self-sustaining community (Vargo and Lusch, 2016) that can engineer behaviors (e.g., service attitudes and exchanges) and services for the greater good. The orchestrator also gravitates toward relinquishing ownership and control by letting more skilled partners take over tasks, democratizing partner selection criteria, and encouraging partners to collaborate without her direct involvement.

The outcomes are a formalized orchestration process and a diversified and visibly strengthened community, albeit one of intensified tensions between the idea-oriented (management and development) and business-oriented (recruiters) SME partners. Despite the orchestrator's effort to reconcile expectations and iron out disparities, the global financial crisis, whose effects are first felt in the Netherlands in 2008/2009, takes its toll and triggers the shift toward maturity. In the light of financial pressures, many partners had become dependent on the orchestrator's assistance, noncollaborative and unwilling to assume responsibility for their survival.

11.3.3. *Maturity (2011–2016)*

The final phase of the Q-Search ecosystem's development begins in late 2011 and continues at present (Table 11.4). The transition brings another round of structural changes, much like the previous transition had previously done through the establishment of the collaboration contract, the IT platform, and the subscription fee. The misalignment between the orchestrator's personal aspirations and her role within the OI ecosystem at the end of growth in conjunction with the growing divide between profit and inspiration-seeking SME partners, prompt her to dismantle the formal structure that had governed the ecosystem's functioning from 2003 throughout 2011. What remains is a borderless community whose partners represent seeds for new ecosystem initiation.

The main goal — pursuing the ecosystem's initial vision and continuing interpersonal support — is born from the orchestrator's personal development and renewed need to align her vision with actions. During this phase, the orchestrator's role is best described as an enabler and sustainer of transformation (Gastaldi *et al.*, 2015) for partners that have embraced the vision and are prepared to take it forward in their professional environments.

The ecosystem retains a subset of partners acquired throughout the first phases; those partners whose level of independence and responsibility matches that of the orchestrator and who are now free in the forms they adopt for service exchanges, innovative value creation, and coevolution. The closer collaboration with policymakers for innovative service generation benefiting society — for example, through the Art of Impact project — bears testimony to the general trend toward increased collaboration, flexibility, and a renewed focus on societally relevant issues.

Table 11.4. The Q-Search ecosystem's development. Phase III: Maturity (2011–2016).

Core aspects		Evidence from data
Ecosystem goals	Pursue vision and continue support	Entrepreneur: The only thing that we have to do is to interact on a deeper lever with each other and then we don't need all the rules.
Key attributes of the entrepreneurial orchestrator	Re-alignment between vision decisions	Partner: The development of Q-Search is nothing other than the mirror of her personal development. What she's doing now with and in Q-Search is very much more aligned with her deeper soul and attitudes. Partner: She was more and more interested in issues of transformation, transition, and culture change.
	Self-efficacy and persistence	Entrepreneur: My enthusiasm [remained constant] because I discovered that [what I had started was working] and I was more convinced that I have to continue with it. Entrepreneur: It takes time and effort and resilience to continue in doing [what you do]. Sometimes you take two steps forward and three back.
	Self-reflection/personal development	Entrepreneur: During those 15 years the hardest work I had to do was on myself. Entrepreneur: All the phases were necessary for my development and for my partners and for our surroundings. Entrepreneur: I have an ego myself. I've had to let a lot of it go and I still have it. [...] But it's a functional ego. Partner: Now she is very much more relaxed. She initiates and sees what happens.

(*Continued*)

Table 11.4. (*Continued*)

Core aspects		Evidence from data
Key orchestration actions	Quitting network as formal orchestrator	Partner: I think 70–80 per cent of her time she was busy, busy, busy with operational issues. So the [Phase] III was, in fact, her liberating act. Entrepreneur: What we are doing now with the intervision group is that we have a case, on which we share our wisdom so that [everybody] can go home with much more information and work on that. Doc5 Sample vision document (The Art of Impact) Doc6 The Future of Work project
	Collaboration with policymakers	Entrepreneur: My focus now is policies, the Hague, ministries, institutions. Doc4 Sample vision document (Input for new tax plan for self-employed)
Ecosystem value	Freedom in pursuit of vision, inspiration, and connectedness	Entrepreneur: We [serve more] as inspiration to each other than we [focus] on getting assignments with companies. MIG Meeting
Outcomes	Boundaries disappear	Entrepreneur: The idea of Q-Search being an organization, an organized organization is really gone. Entrepreneur: With me there are no more boundaries anymore. (…) I don't have official partners anymore. Partner: I left at the same time as the entrepreneur said: I will change the system of Q-Search. I thought OK, that is for me also a good time to change now.
	New ecosystems take shape	Entrepreneur: Q-Search now, it's the entrepreneur [myself], who is a good networker, but also the partners [and their new ecosystems]. So we are building a whole new ecosystem.

The new "Maatschappelijke Intervisie Groep" (MIG) (intervision group) meetings give rise to a condensed and effective means of interaction. As alternatives to the orchestration service package (meetings with facilitation, concept-building, newsletters, partner selection, personal coaching) delivered by the orchestrator during growth, MIG meetings take interaction to a new level by enabling participants, i.e., partners and experts keen on broadening their professional horizons, to reflect more deeply on their ideas, limitations, and concerns, and even perform role-play in order to solve conflicts. The fluid membership of these meetings reflects the orchestrator's and SME partners' belief that issues are best solved by bringing the right combination of people together, namely individuals with complementary skills that are passionate about the issue and open to new ideas. Trust is of fundamental importance in sustaining the new means of communication and, more broadly, the building of an effective knowledge-sharing network at the core of the ecosystem (Dyer and Nobeoka, 2000).

Examples of on-demand projects in collaboration with policymakers include: The Future of Work, a project whose aim is to lay the foundations for the workplace and work practices of the future and the Development Vouchers project. The Development Vouchers project represents a collaborative initiative involving a broad array of stakeholders, i.e., local policymakers, union representatives, career counselors and life management coaches (among them, Q-Search partners), and representatives of an industry-specific employer association for continuous learning), whose aim is to create a national platform for lifelong vocational development that will enable all members of society to fully participate in the workforce.

In addition, Q-Search partners are working closely with the Ministry of Social Affairs and Employment as well as the Ministry of Education, Culture and Science in the Netherlands to achieve impact in society through new leadership and flexible forms of organization (networks). The new projects — The Art of Impact and The Art of Letting Go training — are the result of the orchestrator's newly found self-efficacy and the fit between her vision and the preferred way of working with others, in an exploratory, collaborative, open manner. The alignment between the orchestrator's attributes and those of the ecosystem appears to be a strong predictor of success. This finding is in line with previous entrepreneurship research (Rauch and Frese, 2007).

The most important change occurs internally as the orchestrator succeeds in aligning her values and beliefs with those of the ecosystem. Maturity is a time of renewal for everyone involved in the OI ecosystem i.e., entrepreneurial orchestrator, SME partners, emergent stakeholders, and elicits the reintroduction of some of the successful practices undertaken during initiation. Informal conversations re-emerge in a new format ("What we are doing now with the intervision group is that we have a case, on which we share our wisdom so that [everybody] can go home with much more information and work on that"). Mutual sharing and responsibility for one's learning are again decentralized, as is the orchestrator's role in the facilitation of connection-making i.e., partners now connect independently, with or without her participation.

Maturity places the emphasis back on realizing the initial vision (changing existing HR practices and increasing job satisfaction and quality of work life in the Netherlands) within a community of dedicated SME partners, who have fully embraced the Q-Search vision and who choose to collaborate and support each other unencumbered by any formal agreements. These foundational elements currently serve as seeds for new ecosystem initiation and more fruitful OI relations.

11.4. Discussion and Conclusions

This chapter has examined the evolution of a Dutch OI ecosystem of SMEs active in the HR services industry. The relations under study (Figure 11.2) emphasized the complex interplay between the OI ecosystem's goals, the attributes of the entrepreneurial orchestrator, and her consequent actions, as well as ecosystem outcomes and innovative value creation. As such, we have answered to distinct calls, such as the one from Brunswicker and van de Vrande (2014) and Vanhaverbeke (2017), for more knowledge on how SMEs coordinate the network of OI relations forming around them. Furthermore, we have built on existing work by Gardet and Fraiha (2012) and Autio *et al.* (2014) on OI relations and coordination in a networked setting. Uniquely, the framework used for the analysis has combined different streams of literature, that is, entrepreneurship, strategy/innovation management, and psychology, to showcase the orchestrator's profound influence on her OI ecosystem of SMEs.

Drawing upon rich data, and taking a longitudinal view of Q-Search's evolution, we find that it takes a specific type of entrepreneur or orchestrator to set up SME ecosystems and OI relations that are ultimately self-sustaining and can thus create innovation and value. This entrepreneur has: (1) a strong vision of changing society for the better and is able to communicate this vision to her partners (Grant, 2012); (2) is willing to forego her own interests for the benefit of the greater good (passion) and is capable of cultivating this attitude in others through her authenticity; (3) leads by example, and (4) helps others become empowered, that is, makes making herself replaceable (Conger *et al.*, 2000; Shamir *et al.*, 1993). At the same time, this type of entrepreneur is: (5) passionate about her cause and, by extension, shows the persistence that is necessary to take her ecosystem through the learning phases of chaos, structure, and ultimately structured chaos (Baum and Locke, 2004); (6) willing to show her vulnerability; (7) able to use the ecosystem's feedback to her actions as input for self-reflection and personal development (Dhanaraj and Parkhe, 2006); and (8) willing to change her behavior to accommodate the ecosystem's needs.

In studying the psychological profile of the entrepreneurial orchestrator and how it affects ecosystem development and evolution, we build on the important work by Ahn *et al.* (2013) who also find that in an SME context the founder and his/her traits have a profound influence on the way OI relations unfold.

In addition, building on Salampasis and Mention (2017), we shed light on the human element in OI ecosystems where our findings show that personal relationships play a crucial role in supporting OI in each of the evolutionary phases (Ceci and Iubatti, 2012; Vanhaverbeke, 2017). During initiation, the regular meetings held by the entrepreneurial orchestrator with core partners help create familiarity and also attract new, complementary SMEs to Q-Search. This, in turn, leads to establishing ecosystem legitimacy, to knowledge-sharing, to continued development, and to innovative service creation. During growth, personal relationships contribute to the creation of the IT platform for closer activity coordination, better communication, and aligned expectations. Finally, during maturity, personal relationships (feedback from partners) enable the orchestrator to "free" herself from the formal orchestration role and let Q-Search become a truly self-sustaining (ecosystem) structure. Ultimately, Q-Search partners move on to create new OI relationships in HR consulting and value constellations of their own, drawing upon the experience from Q-Search.

While this chapter presents one of the first attempts to visualize and analyze ecosystem evolution over a prolonged period of time, its main weakness lies in that we draw up this view based on archival data and past accounts of Q-Search's development. Despite this limitation, we found consensus among interviewees as to when initiation, growth, and maturity occurred at Q-Search.

Our research provides several implications for academics, practitioners, and policymakers alike. First, by understanding SME ecosystems' evolutionary phases (also, the dynamics of each "transition"), governments and policymakers may design better instruments to stimulate their emergence and growth. Second, understanding the psychology of the SME orchestrators (entrepreneurs) can lead to more effective tools and training, and ultimately healthier OI relationships. Third, our research suggests several avenues for further investigation. We stimulate fellow researchers in different disciplines to conduct more in-depth, multi-disciplinary explorations of SME ecosystems, their internal dynamics, and the interactions between the psychology of the orchestrating entrepreneur and his/her partners. In particular, we encourage exploring the human element in OI.

As several other chapters in this book illustrate, understanding the human factor can provide answers to complex questions, including which organizational ingredients and "ways" contribute to the creation of the right corporate OI environment and culture, both within and across organizational boundaries (Salampasis and Mention,

2017). Chapter 5, for example, shows that understanding top executives' characteristics — e.g., risk-taking propensity, endurance, as well as their knowledge and network can reveal a great deal about how OI is promoted and adopted. Chapter 8 illustrates how misalignments between SMEs' business models and that of the ecosystem they are part of can lead to various challenges. Here, the entrepreneurs and SMEs must share business objectives in order to carry out successful OI.

References

Ahn, J. M., Mortara, L. and Minshall, T. (2013). The effects of open innovation on firm performance: a capacity approach. STI. *Policy Review*, 4, pp. 74–93.

Autio, E., Kenney, M., Mustar, P., Siegel, D. and Wright, M. (2014). Entrepreneurial innovation: the importance of context. *Research Policy*, 43, pp. 1097–1108.

Baron, R. A. and Tang, J. (2008). Entrepreneurs' social skills and new venture performance: mediating mechanisms and cultural generality. *Journal of Management*, 35, pp. 282–306.

Baum, Jr. and Locke, E. (2004). The relationship of entrepreneurial traits, skill, and motivation to subsequent venture growth. *Journal of Applied Psychology*, 89, pp. 587–598.

Bennett, A. and Elman, C. (2010). Case study methods. In: Reus-Smit, C. and Snidal, D. (eds.). *The Oxford Handbook of International Relations*. Oxford, UK: Oxford University Press.

Brunswicker, S. (2011). An empirical multivariate examination of the performance impact of open and collaborative innovation strategies. Unpublished dissertation, Universität Stuttgart.

Brunswicker, S. and van de Vrande, V. (2014). Exploring open innovation in small and medium-sized enterprises. In: Chesbrough, H., Vanhaverbeke, W. and West, J. (eds.). *New Frontiers in Open Innovation*. Oxford, UK: Oxford University Press.

Carson, J. B., Tesluk, P. E. and Marrone, J. A. (2007). Shared leadership in teams: an investigation of antecedent conditions and performance. *Academy of Management Journal*, 50, pp. 1217–1234.

Ceci, F. and Iubatti, D. (2012). Personal relationships and innovation diffusion in SME networks: a content analysis approach. *Research Policy*, 41, pp. 565–579.

Chesbrough, H., Vanhaverbeke, W. and West, J. (2014). *New Frontiers in Open Innovation*. Oxford: Oxford University Press.

Conger, J. A., Kanungo, R. N. and Menon, S. T. (2000). Charismatic leadership and follower effects. *Journal of Organizational Behavior*, 21, pp. 747–767.

Dhanaraj, C. and Parkhe, A. (2006). Orchestrating innovation networks. *Academy of Management Review*, 31, pp. 659–669.

Drechsler, W. and Natter, M. (2012). Understanding a firm's openness decisions in innovation. *Journal of Business Research*, 65, pp. 438–445.

Dyer, J. H. and Nobeoka, K. (2000). Creating and managing a high-performance knowledge-sharing network: the Toyota case. *Strategic Management Journal*, 21, pp. 345–367.

Eisenhardt, K. M. (1989). Building theories from case study research. *Academy of Management Review*, 14, pp. 532–550.

Eisenhardt, K. M. and Graebner, M. E. (2007). Theory building from cases: opportunities and challenges. *Academy of Management Journal*, 50, pp. 25–32.

Eurostat. (2017). Statistics on Small and Medium-Sized Enterprises Explained. Available at: http://ec.europa.eu/eurostat/statistics-explained/index.php/Statistics_on_small_and_medium-sized_enterprises, Accessed: 6 June, 2017.

Flyvbjerg, B. (2006). Five misunderstandings about case-study research. *Qualitative Inquiry*, 12, pp. 219–245.

Gardet, E. and Fraiha, S. (2012). Coordination modes established by the hub firm of an innovation network: the case of an SME bearer. *Journal of Small Business Management*, 50, pp. 216–238.

Gastaldi, L., Appio, F. P., Martini, A. and Corso, M. (2015). Academics as orchestrators of continuous innovation ecosystems: towards a fourth generation of CI initiatives. *International Journal of Technology Management*, 68, pp. 1–20.

Gehman, J., Glaser, L., Eisenhardt, M., Gioia, D., Langley, A. and Corley, K. (2017). Finding theory-method fit: a comparison of three qualitative approaches to theory building. *Journal of Management Inquiry*, Available at: https://doi.org/10.1177/1056492617706029, Accessed 5 November, 2017.

Grant, A. M. (2012). Leading with meaning: beneficiary contact, prosocial impact, and the performance effects of transformational leadership. *Academy of Management Journal*, 55, pp. 458–476.

Gulati, R., Puranam, P. and Tushman, M. (2012). Meta-organization design: rethinking design in interorganizational and community contexts. *Strategic Management Journal*, 33, pp. 571–586.

Langley, A., Smallman, C., Tsoukas, H. and Van de Ven, A. H. (2013). Process studies of change in organization and management: unveiling temporality, activity, and flow. *Academy of Management Journal*, 56, pp. 1–13.

Laperche, B. and Liu, Z. (2013). SMEs and knowledge-capital formation in innovation networks: a review of literature. *Journal of Innovation and Entrepreneurship*, 2, pp. 1–16.

Lee, H., Kelley, D., Lee, J. and Lee, S. (2012). SME survival: the impact of internationalization, technology resources, and alliances. *Journal of Small Business Management*, 50, pp. 1–19.

Leroi-Werelds, S., Pop, O. M. and Roijakkers, N. (2017). Understanding value creation in alliance ecosystems: enriching the strategic alliance literature with insights from marketing. In: Das, T. K. (ed.). *Managing Alliance Portfolios and Networks*. Charlotte: Information Age Publishing.

Nambisan, S. and Sawhney, M. (2011). Orchestration processes in network-centric innovation: evidence from the field. *The Academy of Management Perspectives*, 25, pp. 40–57.

Paquin, R. L. and Howard-Grenville, J. (2013). Blind dates and arranged marriages: longitudinal processes of network orchestration. *Organization Studies*, 34, pp. 1623–1653.

Rauch, A. and Frese, M. (2007). Let's put the person back into entrepreneurship research: a meta-analysis on the relationship between business owners' personality traits, business creation, and success. *European Journal of Work and Organizational Psychology*, 16, pp. 353–385.

Ritala, P., Agouridas, V., Assimakopoulos, D. and Gies, O. (2013). Value creation and capture mechanisms in innovation ecosystems: a comparative case study. *International Journal of Technology Management*, 63, pp. 244–267.

Salampasis, D. and Mention, A. L. (2017). *Open Innovation: Unveiling the Power of the Human Element*, Vol. 2. Singapore: World Scientific Publishing.

Shamir, B., House, R. J. and Arthur, M. B. (1993). The motivational effects of charismatic leadership: a self-concept based theory. *Organization Science*, 4, pp. 577–594.

Ucbasaran, D., Westhead, P. and Wright, M., (2001). The focus of entrepreneurial research: contextual and process issues. *Entrepreneurship Theory and Practice*, 25, pp. 57–80.

Van de Ven, A. H. (1992). Suggestions for studying strategy process: a research note. *Strategic Management Journal*, 13, pp. 169–191.

Van de Vrande, V., De Jong, J. P., Vanhaverbeke, W. and De Rochemont, M. (2009). Open innovation in SMEs: trends, motives and management challenges. *Technovation*, 29, pp. 423–437.

Vanhaverbeke, W. (2017). *Managing Open Innovation in SMEs*. Cambridge: Cambridge University Press.

Vargo, S. and Lusch, R. F. (2016). Institutions and axioms: a extension and update of service-dominant logic. *Journal of the Academy of Marketing Science*, 44, pp. 5–23.

West, J. and Bogers, M. (2017). Open innovation: current status and research opportunities. *Innovation: Organization Management*, 19, pp. 43–50.

Williamson, P. J. and De Meyer, A. (2012). Ecosystem advantage. *California Management Review*, 55, pp. 24–46.

World Bank. (2016). Entrepreneurs and Small Business Spur Economic Growth. Available at: http://www.worldbank.org/en/news/feature/2016/06/20/entrepreneurs-and-small-businesses-spur-economic-growth-and-create-jobs. Accessed: 6 June, 2017.

Yin, R. K. (2013). *Case Study Research: Design and Method*, 4th edition. London: Sage Publications.

Chapter 12

Crowdfunding of SMEs and Startups: When Open Investing Follows Open Innovation

*Giancarlo Giudici and Cristina Rossi-Lamastra**

School of Management, Politecnico di Milano, Italy
**cristina1.rossi@polimi.it.*

Abstract

Crowdfunding is currently booming, and it appears as a valuable alternative to finance startups and SMEs. In this chapter, we advise scholars to study the relationship between crowdfunding and open innovation. On the one hand, under this new way of financing, firms receive money as well as feedbacks and suggestions from crowdfunding investors who pledge money on the Internet. On the other hand, earlier open innovation activities and social networking allow SMEs to reduce information asymmetries and exploit the market of potential funders. Then, we take stock of contributions on the prominence of social capital in crowdfunding to discuss the linkages among crowdfunding, social capital, and open innovation. To this end, we define the concept of social capital and review the

* Corresponding author.

contributions, which study it in the crowdfunding realm. Finally, we describe three interesting cases of SMEs from the Crowdcube platform (goHenry, JustPark, Monzo), showing how these SMEs leveraged their social relations with extant customers and innovative users to raise equity capital.

12.1. Introduction

Expanding on the idea of crowdsourcing (Afuah and Tucci, 2012), crowdfunding implies mobilizing Internet users to finance entrepreneurial ideas, or personal initiatives, which their proponents post either on personal pages or on dedicated websites (i.e., crowdfunding platforms). Crowdfunding investors provide money through the Internet in exchange for some monetary claims on the project revenues, for some form of reward, or simply for donation (Belleflamme *et al.*, 2014; Giudici *et al.*, 2013). Scholars and practitioners agree that crowdfunding can change traditional funding practices in general, but — more interestingly — especially in case of startups and SMEs. Indeed, young and small firms usually suffer from liability of smallness and newness (Stinchcombe, 1965) and have severe difficulties in attracting external finance due to information asymmetries (see Hoenig and Henkel, 2015 for a recent discussion). These firms often do not have collaterals to give for guarantee, and, in case of startups, they lack a track record of past accomplishments. Accordingly, they often struggle to convince traditional investors (e.g., banks, venture capitalists, business angels, or equity providers) to finance their initiatives. In this context, crowdfunding can be a valuable financial alternative for equity capital (Giudici, 2015). Moreover, as we discuss in the following section, besides relaxing the financial constraints of startups and SMEs, crowdfunding allows them to interact online with investors and receive feedbacks and suggestions to improve their (innovative) products and services (Bogers *et al.*, 2017).

The crowdfunding phenomenon is rapidly growing. As of 2015, there were about 1,250 crowdfunding platforms in the world, which amounted to around $34.4 billion (Crowdsourcing LLC, 2015). Such an astonishing expansion is catching remarkable scholarly attention. Since 2010, 563 papers in the Scopus database include the term crowdfunding in their title. Similarly, the SSRN Electronic Library contains 519 papers on crowdfunding; the three most popular ones count more than 50,000 downloads.[1] The large majority of these scholarly contributions focus on the factors that determine the *success of crowdfunding projects* (see Butticè *et al.*, 2017a for a review of this literature). Specifically, crowdfunding platforms often adopt an "all or nothing" rule: when posting a project, a proponent has to display a *target capital*, which the project must collect by a certain deadline. If this happens, the project is successful and the proponent receives the collected amount. Conversely, the project is unsuccessful and the proponent receives nothing. Among the other factors, studies on success of crowdfunding projects acknowledge the importance of proponents' *social capital* (Adler and Kwon, 2002; Coleman, 1988), namely of the social relations, which proponents have *within* (Colombo *et al.*, 2015) and *outside* the crowdfunding platform (Mollick, 2014).

In the spirit of this book on open innovation, this chapter discusses the linkages among crowdfunding, social capital, and open innovation. Specifically, we intend to answer the following research questions: Do crowdfunding investors also provide suggestions for improvement of products/services *during* and *after* the campaign, thus activating open innovation processes? Alternatively, are extant customers and *innovative users* (i.e., those who provide feedbacks and suggestions by actively interacting with SMEs and startups in community of practices, focus groups, and online forums, and so on) the first crowdfunding investors?

In answering these research questions, we also take into account the role of *localized social capital* (Laursen *et al.*, 2012a). Indeed, this is an important form of social capital, which the literature on crowdfunding has largely disregarded (see Giudici *et al.*, 2017 for a notable exception). The term "localized social capital" indicates to the level of social capital in the area where a proponent resides. In particular,

---

[1] http://www.scopus.com. and http://www.ssrn.com. Both the databases have been accessed on November 4, 2017.

it refers to the social relations among residents (i.e., localized social relations) and to their compliance with social norms (i.e., localized social norms).

In shedding light onto the relations among crowdfunding, social capital, and open innovation, our work advances extant knowledge in two main directions. First, it originally contributes to the open innovation research proposing a novel way through which firms, in general, and SMEs and startups, in particular, can engage in *inbound* open innovation (Dahlander and Gann, 2010). In so doing, it attracts scholarly attention to another aspect of crowdfunding success. Current crowdfunding literature equates success with the collection of money. We make a step further and claim that a project is successful also if it collects many feedbacks from the crowd. Second, our work complements contributions about the impact of social capital on crowdfunding success by suggesting scholars to consider that social capital is a multiform construct and has a (potentially) many relevant dimension.

We base our work on a qualitative approach. Specifically, we report about three crowdfunding campaigns of three SMEs from the most important equity crowdfunding platform in Europe, Crowdcube, i.e., *goHenry*, *Monzo*, and *JustPark*. We focus on equity crowdfunding — and not on reward-based crowdfunding, as most of the scholars do — for two main reasons. First, this crowdfunding model specifically targets startups and (innovative) SMEs for which generating and exploiting innovation is a fundamental source of competitive advantage (Yli-Renko *et al.*, 2001). Second, in equity crowdfunding, investors obtain equity shares, thus becoming *owners* of the firms, and directly benefit from the potential creation of value for the long run. Accordingly, it is reasonable to expect that they are keen on giving suggestions besides providing financial support.

Our case studies offer interesting (and unexpected) insights on the relationship among crowdfunding, social capital, and open innovation. Given how we selected cases and collected empirical evidence, we do not make any claim of the generalizability for our qualitative evidence. However, we are confident that it can pave the way for more structured qualitative and quantitative analyses. First, we show that the three SMEs rely on suggestions of their early customers and innovative users (i.e., on their social capital) to improve the quality of their products and offer valuable services. This, in turn, seems to provide them credibility for collecting money. In other words, resorting to an open innovation approach appears to be a way for stimulating fundraising. Second, we (surprisingly) find no evidence that the community of crowdfunding investors brings new potential for innovation. None of the founders of the three SMEs did acknowledge original contributions from crowdfunding investors, who were not part of the community of early customers and innovative users, in terms of ideas or suggestions *after* the financing round. On the contrary, it seems that early customers and innovative users, who usually bring value-added for innovation, easily become crowdfunders. Third, localized social capital and distance seem to play no role in influencing the decision of crowdfunding investors. If an SME already has relationships with early customers and innovative users before the campaign, it can attract crowdfunding investors from any place.

The chapter proceeds as follows. In Section 12.2, we explain why crowdfunding scholars should look at crowdfunding as a novel way for engaging into inbound open innovation. In Section 12.3, we briefly illustrate the concept of social capital and review the crowdfunding literature on this theme. In Section 12.4, we present the three case studies to highlight the relationship among crowdfunding, social capital, and open innovation. Section 12.5 concludes the chapter by summarizing its contributions, acknowledging its limitations, and discussing directions for further research and practical implications.

12.2. Crowdfunding and Open Innovation

Firms in general and SMEs in particular can resort to diverse approaches and strategies for engaging in open innovation. More specifically, scholars distinguish between *inbound* and *outbound* open innovation (Dahlander and Gann, 2010). While the former consists in fostering the innovation process by insourcing relevant knowledge from external third parties, the latter prescribes to give away internally produced knowledge (for instance, through licensing; Bianchi *et al.*, 2014). The wide literature on outbound open innovation strategies (see West and Bogers, 2017, for a literature review) has recently noted that firms can do open innovation by mobilizing (online) *communities of innovative users*, such as communities of practices (e.g., Wolf *et al.*, 2011; Franke and Shah, 2003) or communities of Open Source software developers (Dahlander *et al.*, 2008). These communities are potentially highly beneficial for SMEs and startups as they allow these firms to access and share knowledge at low cost (see Colombo *et al.*, 2014 for a discussion of this issue).

Expanding on this debate, we argue that crowdfunding investors can also form online communities, providing feedbacks and suggestions to SMEs and startups that run crowdfunding campaigns, and we advise scholars to approach crowdfunding as the natural financial complement of open innovation. Our reasoning is based on three main arguments.

First, crowdfunding investors are interested not only in providing financial support to SMEs and startups during the campaigns, but also in what happens to these firms once the campaigns end. This holds particularly true in equity crowdfunding where crowdfunding investors receive equity shares and enter the firms' equity capital. Accordingly, we envisage that crowdfunding investors have incentives to interact with proponents for offering feedbacks and suggestions on how to improve their products and services and drive their firms toward success. Clearly, not all these feedbacks and suggestions prove to be useful, also because many crowdfunding investors are nonprofessionals with limited industry-specific and investment-specific experience (Piva and Rossi-Lamastra, 2017). However, if one assumes that the wisdom of the crowd (Surowiecki, 2004) is at work also in the crowdfunding realm, it is possible to conclude that this (large) amount of feedbacks and suggestions likely contains insights and ideas that the proponents can leverage to improve their offering. The growing crowdfunding literature is starting to document the role of crowdfunding investors' feedbacks and suggestions for project success (Brown *et al.*, 2017; Stanko and Henard, 2016).

Second, we contend that the crowdfunding setting favors the provision of feedbacks and suggestions. Crowdfunding platforms contain dedicated areas where crowdfunding investors can write their comments and interact with proponents. In turn, proponents can access these areas, see crowdfunding investors' comments, and react to them. In other words, we claim, as for instance, in case of Open Source software (Bonaccorsi and Rossi, 2003), that the technology enables the proponent–investor interactions. Third, we expect that the online medium stimulates crowdfunding investors to unleash their creativity. Literature on crowdsourcing platforms shows that interacting online leads individuals to think out of the box (Schemmann *et al.*, 2016), with less attention to conformity and less fear of social stigma. In principle, one can equate crowdfunding platforms to online crowdsourcing platforms where proponents collect not only money, but also novel ideas for innovating their products and services.

From the discussion above, the following key question emerges: *Is crowdfunding easier for firms adopting an open innovation strategy?* We conceive this chapter as the first attempt of providing an answer to this important question.

12.3. The Role of Social Capital in the Crowdfunding Realm

As aforementioned, in this chapter, we are interested to explore the relation between crowdfunding, social capital, and open innovation. Thus, this section briefly defines social capital and reviews studies on social capital in the crowdfunding realm. In so doing, we distinguish between social capital possessed by individuals (and organizations) and social capital existing in geographical areas, namely *localized social capital*.

A well-established tradition defines social capital as "the sum of the actual and potential resources embedded within, available through, and derived from the network of relationships possessed by an individual or a social unit. Social capital thus comprises both the network and the assets that network can mobilize" (Nahapiet and Ghoshal, 1998, p. 243). Scholars also relate social capital to other constructs like collective efficacy, social trust, reciprocity, voluntary participation in organizations, and social integration (Harpham *et al.*, 2002). Moreover, they apply the concept of social capital to a broad range of economic phenomena, including relations inside and outside the firm (Burt, 1992), organization–market interfaces (Baker, 1990), entry into entrepreneurship (Davidsson and Honig, 2003), and innovation (Tsai and Ghoshal, 1998). In this latter regard, the literature notes that social capital facilitates knowledge acquisition and exploitation by forging the conditions for the creation of value through the exchange and combination of existing intellectual resources.

In crowdfunding, proponents' social capital is a crucial driver of success. Mollick (2014) shows that personal contacts predict the success of reward-based crowdfunding projects. Colombo *et al.* (2015) elaborate on this and show that the social capital that proponents have accumulated on the platform by backing the projects of other proponents is fundamental for attracting initial backers. In a similar vein, Vismara (2016) shows that proponents of equity crowdfunding projects with larger social networks are more likely to succeed. Kang *et al.* (2017) study a Chinese crowdfunding platform and find that social capital, as reflected by the number of followers of a crowdfunding project's early investors, positively relates to the campaign success. They also find that the geographical distance enhances this positive relationship. Lehner (2014) proposes an analysis of 36 case studies related to crowdfunding of social projects and claims that social capital helps in recognizing opportunities, influences the forming of opportunities, and assists in overcoming normative, structural, and cultural barriers. Butticè *et al.* (2017b) document that serial crowdfunders benefit from the social contacts with those who backed their previous campaigns. A different result is documented by Ahlers *et al.* (2015). The authors find no evidence of a correlation between social capital and the probability of reaching the target in a crowdfunding project.

Moreover, scholars note that social capital exists not only at the individual level, but also at the *collective level*, namely within a group or a territory (Belliveau *et al.*, 1996). In this latter regard, social capital is an environmental feature which fosters cooperation and enhances economic activity and resource mobilization (Coleman, 1990). Accordingly, Laursen *et al.* (2012a) document that localized social capital has a positive impact on firms' capability to innovate and develop new products. In a companion study (Laursen *et al.*, 2012b), the authors find that localized social capital helps firms to expand in international markets. Interestingly, they hypothesize a U-shaped relation, so that too much social capital negatively affects internationalization propensity because of a "trapping" effect into local markets. Hauser *et al.* (2007) show that several measures of "localized" capital (in particular, the number of local associations) strongly correlate with SMEs' patenting activity. The concept of localized social capital translates in the crowdfunding literature. Agrawal *et al.* (2010) demonstrate that people residing in neighboring geographical areas are the first to finance crowdfunding projects, thus suggesting a positive impact of localized social capital on the success of crowdfunding projects. More recently, Giudici *et al.* (2017) document explicitly the importance of the *localized social capital* in determining the success of entrepreneurs proposing reward-based crowdfunding projects. Specifically, the authors show that social relations in the geographical area where proponents reside magnify the positive effect of the *local altruism* (i.e., the altruism of people residing in the area) on crowdfunding success.

12.4. Insights from Three Successful Cases

In the following, we offer qualitative insights on the relation among crowdfunding, social capital, and open innovation through three prominent cases of SMEs which engaged in equity crowdfunding campaigns on Crowdcube. Darren Westlake and Luke Lang established Crowdcube in 2011, and it is now one of the largest crowdfunding platforms in the world (Crowdsourcing LLC, 2015).

To date, Crowdcube has raised £358,293,889 and counts 453,254 registered investors.[2] Note that we did not select the projects based on traditional selection procedures in case study research (Eisenhardt, 1989), but we chose those projects which enjoyed a large success and visibility on the Internet. Indeed, as we mentioned in the Introduction, our aim is not to formulate generalizable propositions. Instead, we intend to gather vivid insights on the relations at the core of our investigation to pave the way to more structured qualitative and quantitative analyses. Specifically, *goHenry* is the top performer of the platform in terms of capital raised. *Monzo* was the quickest project in raising capital, while *JustPark* succeeded to attract the highest number of funders. We interviewed the proponents of these projects by email. The scheme of the questionnaire is available from the authors upon request. Moreover, we collected additional information from secondary sources.

12.4.1. *goHenry*

The idea of goHenry (www.gohenry.co.uk) originated in 2012 from a small group of parents who just wanted to help children and teenagers manage their money. The startup offers a digital banking solution, a prepaid debit card, and an app with unique parental controls to teach young people how to manage their money. With over 200,000 users and 10,000 joining every month, KPMG & H2 Ventures named goHenry one of the world's "Leading 100" Fintech innovators. Alex Zivoder is the CEO of the company and initiator of the campaign. He received his M.Sc. from Université Paris Dauphine and the Ph.D. in Economics from Université Nice Sophia Antipolis. He then worked at Tele2 Group, Expedia, viagogo, and lynda.com before joining goHenry in 2015. In May 2016, after a survey conducted among customers about their willingness to enter the equity capital of the company, goHenry went through a crowdfunding campaign to continue its growth both in the UK and globally. The target funding of £2 million (in exchange for 8.69% of the startup

[2] https://www.crowdcube.com/infographic, accessed on November 4th 2017.

equity capital) was widely overcome, resulting in £3.995 million raised from 2,209 crowdfunding investors (very close to the maximum allowed by EU regulation).

The CEO reports that there were no significant investments on public relations, marketing, or advertisement.[3] He leveraged on angel investors who had already funded the business and asked them to invest in the campaign. On day 1, goHenry launched the equity crowdfunding round with £857,000 already pledged from five existing investors and nine new funders. Earlier, the founders polled the existing customers asking if they would have been willing to invest in the business, and 80% responded positively. A 48-hour exclusive window was opened allowing them to invest, and 1,051 jumped at the opportunity investing £1.4 million.

According to Alex Zivoder,[4] to succeed in crowdfunding it is important to plan the actual details of the campaign before the start, having all the necessary documents ready and a well-produced video with good music, and thinking of every question that might be asked by prospective investors during the campaign. He set a deadline of answering any question posed on Crowdcube during the campaign within 6 hours, regardless of the time of day.

"Being largely unknown — Alex Zivoder says — the success of the crowdfunding campaign relied on how well our own customers thought of us and how many of these customers were ready to invest, according to their appetite and capacity to invest in a startup." This statement once again witnesses the role of social capital, in this case represented by existing customers, in determining the success of a crowdfunding campaign. In case of goHenry, prior to the campaign, customers have provided ideas to improve the product. The implementation of these ideas by the firm resulted in the creation of something valuable, which some customers were thus keen of financing. Referring to valuable feedbacks and suggestions collected during

---

[3] See www.thememo.com/2016/07/01/crowdfunding-5-tips-secrets-from-britains-most-successful-crowdfunder-gohenry. Quotations are taken from an interview with the founder.

[4] We thank Mr. Zivoder for having answered our questions in an interview.

the campaign, the CEO of goHenry said that during the campaign they "did not collect any new information to improve the product." However, they had understood the importance of the social relations with their customers for the product success. He claimed "we did collect one piece of information: we did not know we were playing such an important role in the life of our customers. This was quite a news to us." In sum, in the case of goHenry, we can say that social capital was a source of open innovation, which in turn was a source of funding.

Interestingly enough, according to Alex Zivoder, the company did not experience "any localized effect in the geographical distribution of investors." This seems to suggest that localized social capital has no role in the success of this specific campaign.

12.4.2. *JustPark*

JustPark (www.justpark.com) is a digital solution giving drivers instant access to a network of 150,000 parking spaces (public and private) across the UK and the US, taking the hassle out of parking and making it cheaper and easier. With venture backing from BMW in 2011 and Index Ventures in 2014 (through a convertible loan), JustPark has the mission to make parking stress-free. JustPark enables drivers finding, reserving, and paying for parking seamlessly, wherever and whenever they need it, both in advance and on-the-go. The founder is Anthony Eskinazi, a chess champion, who left a consulting company to set up the venture in 2006 after he experienced parking difficulties while attending a baseball game in San Francisco. He studied mathematics at the University of Leeds and visited the University of California at Berkeley for one year.

JustPark opened its crowdfunding campaign on Crowdcube in February 2015, offering 14.94% of its equity capital. The initial target capital was £1 million, and the startup succeeded in raising £3.5 million from 2,702 funders (record number on Crowdcube so far). The pledges started from £10. Before the campaign, JustPark polled at random the 10% of its current customers to find out whether they would actually be interested in investing in the company. The result was impressive. Over 800 of JustPark customers had invested already at the beginning of the campaign. "If people get the problem you're trying to solve — said Anthony Eskinazi[5] — then they will jump on it." Once again, this evidence supports the importance of customers' network, and in general of social capital, for the success of a crowdfunding campaign.

After reaching the initial target, Eskinazi organized parties where investors can meet the company team and released progress reports. In so doing, he aimed at keeping the crowdfunding investors' community involved, in an open innovation approach. He said "equity crowdfunding is the perfect way for JustPark to continue in the collaborative spirit that underpins the entire business." Again, we found no evidence of a particular role of geographical proximity and localized social capital, neither in the success of the campaign nor in community involvement.

12.4.3. *Monzo*

Monzo (www.monzo.com, formerly Mondo) is a Fintech startup, which offers banking services through mobile technology. One of Monzo's cofounders, the current CEO Tom Blomfield, studied Law at Oxford. He previously started GoCardless — an online payment processor, which currently processes more than $1 billion a year.[6] Other cofounders are Jonas Huckestein, Paul Rippon, Gary Dolman, and Jason Bates. Blomfield explains the mission of Monzo: "I think we're addressing a pain that everyone feels; we need a bank account to live our lives, but they (i.e., banks) provide a terrible customer experience."

In March 2016, Monzo asked £1 million on Crowdcube in exchange for 3.33% of the equity capital. The campaign closed in

---

[5] See www.telegraph.co.uk/technology/news/11408609/JustPark-invites-users-to-become-shareholders-in-1m-crowdfunding-campaign.html and www.londonlovesbusiness.com/entrepreneurs/famous-entrepreneurs/justpark-just-crowdfunded-1m-in-four-days-we-ask-its-founder-how-he-did-it/9786.article/.

[6] More information are available at http://tomblomfield.com/about.

just 96 seconds, tilting the platform for the number of bids. The funds, which Monzo raised through equity crowdfunding, are part of a £6 million funding round that saw the participation of prominent VCs like Passion Capital.[7] The successful fundraising of Monzo points to the real need for a change in the banking system and to the raising interest in the Fintech sector. The startup used the pledged capital to hire more people, build the IT systems, and aggressively push marketing and reputation.

Blomfield provided us interesting insights, which confirm the close relations among crowdfunding, social capital, and open innovation. First, in conducting its crowdfunding effort, Monzo explicitly targeted the network of its customers. Indeed, when launching the campaign, the startup could count on a mailing list of 50,000 people, which it created and updated mostly through its social relations and word-of-mouth, amplified on social media such as Facebook and Twitter. This social capital was fundamental for crowdfunding success. Second, it appears evident that this customers' community engages in open innovation, and crowdfunding was a way for boosting this process. Blomfield said: "We've had a very active community from the start. We amplified its effect by being very transparent with our users through a community forum on the web. People give feedbacks and suggestions, which make the product better. Crowdfunding was a way of deepening the engagement with those users." An (indirect) sign that Monzo intended to engage the crowdfunding investors (also) in open innovation processes is the fact that, along with owning shares in the Monzo, they received an exclusive prepaid debit card and access to Monzo iPhone app, skipping the waiting list of 30,000 people. Third, the startup together with its community created a valuable product, which many people were keen on supporting. Blomfield said: "We're trying to build the kind of bank that we'd be proud to call our own, and I think that really resonates with people." Finally, also in this case, localized social capital seems to play no role. Crowdfunder investors did not cluster at the local level, and Blomfield did not deem this aspect as important.

---

[7] http://www.passioncapital.com/.

12.5. Conclusions

Today, crowdfunding is a valuable opportunity to raise finance not only for charity and nonprofit projects, but also for startups and SMEs, to the extent that more and more countries (the US, Belgium, Italy, Lithuania, and Turkey are just some of the most recent cases) are issuing novel legislation to allow these firms to collect equity capital on the Internet. Tapping the crowd is also a means for proponents to receive feedback and suggestions from crowdfunding investors, so that, as discussed in Section 12.2, we advise scholars to think of crowdfunding as a further opportunity in open innovation activities. In this study, we take the first step in this direction by shedding light on the relationship between social capital, open innovation, and crowdfunding. The literature highlights that startups and SMEs can rely on their customers and innovative users to generate novel ideas and knowledge. From the three cases, which we report in this chapter, it appears evident that the social capital and the adoption of an open innovation approach can facilitate the collection of finance through the equity crowdfunding.

First, open innovation is a critical success factor in the three equity crowdfunding campaigns we studied. All the three SMEs brought to the market innovative products or services and used the Internet to solicit and receive feedback and suggestions and to improve their offering. goHenry manages a blog where parents and families can share best practices in children's education to money spending and saving. JustPark — like any player in the sharing economy — relies on its network of users to supply the service, since the company does not own a single parking spot. Monzo built a community of users and launched a merchant data crowdsourcing initiative allowing customers to send instant feedbacks on the credit card transaction data. In sum, open innovation practices seem to provide credibility for tapping the crowd's money.

Second, there is no evidence that crowdfunding investors brought new potential for innovation. None of the founders of the three successful SMEs acknowledged additional contributions from pledgers in terms of ideas or proposals. On the contrary, there is consensus on the idea that customers and innovative users, who bring value added for innovation, easily become crowdfunding investors when the SMEs launch a crowdfunding campaign. The practice of both predisclosing the terms of the financing round to the existing customers and allowing them to bid for equity stakes in exclusivity before other potential investors is coherent with this conjecture and appears as a reward for their trust. In sum, according to our qualitative evidence, an equity crowdfunding campaign does not engage crowdfunding investors in open innovation processes. Conversely, the engagement in open innovation by customers leads to crowdfunding success. This is an unexpected result, which emerges from our three cases and which we think calls for further investigation.

Last, as aforementioned, several studies show that space matters in attracting crowdfunding investors, as geographical proximity allows reducing information asymmetries in crowdfunding campaigns. Surprisingly none of the three cases supports a similar view. Instead, our qualitative evidence documents that this positive effect is much less strategic if a SME can count on the mass of customers and innovative users which it attracted before the campaign.

Our work has several limitations, which open up avenues for future research. In particular, we focus on three highly visible equity projects which three SMEs launched on a single successful platform. This allows us to uncover interesting insights on the phenomenon under investigation, but hampers the generalizability of our findings. In particular, on the basis of (exclusively) these successful cases, it might be problematic to fully comprehend the main pros and cons of crowdfunding and its linkages with social capital and open innovation for an average firm. Thus, we welcome further studies that collect more structured qualitative evidence by including also SMEs and startups, which enjoyed a mid-range or low level of crowdfunding visibility and even unsuccessful cases. Such studies should also consider projects with diverse crowdfunding models and posted in diverse crowdfunding platforms. Moreover, we think that these qualitative studies should pave the way to quantitative analyses, which can rigorously measure the relation among crowdfunding, social capital, and open innovation and account for its moderating factors.

In particular, it would be of interest to explore whether the unexpected findings, which emerge from our three cases (i.e., crowdfunding investors do not provide feedbacks during the campaign and irrelevance of localized social capital), still hold in studies which are based on cases selected through the standard maximum variance procedure or on large quantitative samples. Specifically, quantitative contributions may directly assess the relations between crowdfunding investors' comments on the project page and project success, both in fundraising and in subsequent steps. Such studies may also explore whether and how proponents' social capital moderates these effects. Finally, we focus on the relation among crowdfunding, social capital, and open innovation. However, other intervening factors (e.g., the social status of the founders of SMEs and startups, the possession of IPRs, the industry of operation) can influence the linkage between crowdfunding and open innovation; we invite scholars to explore them. Despite these limitations, this chapter offers interesting practical implications. Our key message to SMEs and startups, which intend to collect money through (equity) crowdfunding, is that they should be aware of the importance of social capital and open innovation practices and of their interplay in fundraising campaigns to success. Specifically, we think that having a large network of customers and innovative users is the *sine qua non* condition to succeed. This network engages in open innovation processes, which are fundamental for improving SMEs' offering. The salience of this social capital stems also from the fact that SMEs and startups have limited tangible and intangible resources to devote to innovation and thus benefit a lot from external knowledge sourcing. In turn, a better product/service attracts crowdfunding investors and leads to crowdfunding success. Likewise, managers of crowdfunding platforms can take stock from our findings to design mechanisms which favor the participation of extant customers and innovative users to crowdfunding campaigns.

References

Adler, P. S. and Kwon, S. (2002). Social capital: prospects for a new concept, *Academy of Management Journal*, 27(1), pp. 17–40.

Afuah, A. and Tucci, C. L. (2012). Crowdsourcing as a solution to distant search. *Academy of Management Review*, 37(3), pp. 355–375.

Agrawal, A., Catalini, C. and Goldfarb, A. (2010). The geography of crowdfunding, NET Institute Working Paper No. 10-08.

Ahlers, G., Cumming, D., Günther, C. and Schweizer, D. (2015). Signaling in equity crowdfunding, *Entrepreneurship Theory and Practice*, 39(4), pp. 955–980.

Baker, W. (1990). Market networks and corporate behavior. *American Journal of Sociology*, 96, pp. 589–625.

Belleflamme, P., Lambert, T. and Schwienbacher, A. (2014). Crowdfunding: tapping the right crowd. *Journal of Business Venturing*, 29(5), pp. 585–609.

Belliveau, M. A., O'Reilly, C. A. and Wade, J. B. (1996). Social capital at the top: effects of social similarity and status on CEO compensation. *Academy of Management Journal*, 39(6), pp. 1568–1593.

Bianchi, M., Frattini, F., Lejarraga, J. and Di Minin, A. (2014). Paths of technology exploitation: combining technological and complementary resources in new product development and licensing. *Journal of Product Innovation Management*, 31(S1), pp. 146–169.

Bogers, M., Zobel, A.-K., Afuah, A., Almirall, E., Brunswicker, S., Dahlander, L., Frederiksen, L., Gawer, A., Gruber, M., Haefliger, S., Hagedoorn, J., Hilgers, D., Laursen, K., Magnusson, M. G., Majchrzak, A., McCarthy, I. P., Moeslein, K. M., Nambisan, S., Piller, F. T., Radziwon, A., Rossi-Lamastra, C., Sims, J. and Ter Wal, A. L. J. (2017). The open innovation research landscape: established perspectives and emerging themes across different levels of analysis. *Industry and Innovation*, 24(1), pp. 8–40.

Bonaccorsi, A. and Rossi, C. (2003). Why Open Source software can succeed. *Research Policy*, 32(6), pp. 1243–1258.

Brown, T. E., Boon, E. and Pitt, L. F. (2017). Seeking funding in order to sell: crowdfunding as a marketing tool. *Business Horizons*, 60(2), pp. 189–195.

Burt, R. L. (1992). *Structural Holes: The Social Structure of Competition.* Cambridge, MA: Harvard University Press.

Butticè V., Franzoni C., Rossi-Lamastra C., Rovelli P. (2017a). The Road to Crowdfunding Success: A Review of Extant Literature, in Tucci, C., Afuah, A., Viscusi, G. (Eds.), Creating and Capturing value through Crowdsourcing, Oxford University Press, Oxford (UK), (in press).

Butticè, V., Colombo, M. G. and Wright, M. (2017b). Serial crowdfunding, social capital and crowdfunding success, *Entrepreneurship Theory & Practice*, 41(2), pp. 183–207.

Coleman, J. S. (1988). Social capital in the creation of human capital. *The American Journal of Sociology*, 94, pp. S95–S120.

Coleman, J. S. (1990). *Foundations of Social Theory*. Cambridge and London: The Belknap Press of Harvard University Press.

Colombo, M. G., Piva, E., and Rossi-Lamastra, C. (2014). Open innovation and within-industry diversification in small and medium enterprises: the case of open source software firms. *Research Policy*, 43(5), pp. 891–902.

Colombo, M. G., Franzoni, C. and Rossi Lamastra, C. (2015). Internal social capital and the attraction of early contributions in crowdfunding. *Entrepreneurship Theory and Practice*, 39(1), pp. 75–100.

Crowdsourcing, LLC. (2015). *2015 Crowdfunding Industry Report*.

Dahlander, L. and Gann, D. M. (2010). How open is innovation? *Research Policy*, 39, pp. 699–709.

Dahlander, L., Frederiksen, L. and Rullani, F. (2008). Online communities and open innovation. *Industry and Innovation*, 15(2), pp. 115–123.

Davidsson, P. and Honig, B. (2003). The role of social and human capital among nascent entrepreneurs. *Journal of Business Venturing*, 18(3), pp. 301–331.

Eisenhardt, K. M. (1989). Building theories from case study research. *Academy of Management Review*, 14(4), pp. 532–550.

Franke, N. and Shah, S. (2003). How communities support innovative activities: an exploration of assistance and sharing among end-users, *Research Policy*, 32(6), pp. 157–178.

Giudici, G., Guerini, M. and Rossi Lamastra, C. (2013). Crowdfunding in Italy: state of the art and future prospects, *Economia e Politica Industriale — Journal of Industrial and Business Economics*, 40(4), pp. 173–188.

Giudici, G. (2015). Equity crowdfunding of an entrepreneurial activity, In: Audretsch, D. *et al.* (eds.). *University Evolution, Entrepreneurial Activity and Regional Competitiveness*, Springer Switzerland, pp. 415–425.

Giudici, G., Guerini, M. and Rossi Lamastra, C. (2017). Reward-based crowdfunding of entrepreneurial projects: the effect of local altruism and localized social capital on proponents' success. *Small Business Economics Journal*, forthcoming.

Harpham, T., Grant, E. and Thomas, E. (2002). Measuring social capital within health surveys: key issues. *Health Policy Plan*, 17(1), pp. 106–111.

Hauser, C., Tappeiner, G. and Walde, J. (2007). The learning region: the impact of social capital and weak ties on innovation. *Regional Studies*, 41(1), pp. 75–88.

Hoenig, D. and Henkel, J. (2015). Quality signals? The role of patents, alliances, and team experience in venture capital financing. *Research Policy*, 44(5), pp. 1049–1064.

Kang, L., Jiang, Q. and Tan, C. (2017). Remarkable advocates: an investigation of geographic distance and social capital for crowdfunding. *Information & Management*, 54(3), pp. 336–348.

Laursen, K., Masciarelli, F. and Prencipe, A. (2012a). Regions matters: how localized social capital affects innovation and external knowledge acquisition. *Organization Science*, 23(1), pp. 177–193.

Laursen, K., Masciarelli, F. and Prencipe, A. (2012b). Trapped or spurred by the home region? The effects of potential social capital on involvement in foreign markets for goods and technology. *Journal of International Business Studies*, 43(9), pp. 783–807.

Lehner, O. M. (2014). The formation and interplay of social capital in crowd-funded social ventures. *Entrepreneurship & Regional Development*, 26(5–6), pp. 478–499.

Mollick, E. R. (2014). The dynamics of crowdfunding: determinants of success and failure. *Journal of Business Venturing*, 29(1), pp. 1–16.

Nahapiet, J. and Ghoshal, S. (1998). Social capital, intellectual capital, and the organizational advantage. *Academy of Management Review*, 23(2), pp. 242–266.

Piva, E. and Rossi-Lamastra, C. (2017). Human capital signals and entrepreneurs' success in equity crowdfunding. *Small Business Economics Journal*, forthcoming.

Schemmann, B., Herrmann, A. M., Chappin, M. M. H. and Heimeriks, G. J. (2016). Crowdsourcing ideas: involving ordinary users in the ideation phase of new product development. *Research Policy*, 45(6), pp. 1145–1154.

Stanko, M. A. and Henard, D. H. (2016). How crowdfunding influences innovation. *MIT Sloan Management Review*, 57(3), pp. 15–17.

Stinchcombe, A. L. (1965). Social structure and organizations, In: March, J. G. (ed.). *Handbook of Organizations* 7, Chicago (Rand McNally), pp. 142–193.

Surowiecki, J. (2004). *The Wisdom of Crowds: Why the Many are Smarter than the Few and How Collective, Wisdom Shapes Business, Economies, Societies and Nations*, Doubleday, Anchor. https://en.wikipedia.org/wiki/The_Wisdom_of_Crowds.

Tsai, W. and Ghoshal, S. (1998). Social capital and value creation: the role of intrafirm networks. *Academy of Management Journal*, 41(4), pp. 464–476.

Vismara, S. (2016). Equity retention and social network theory in equity crowdfunding. *Small Business Economics Journal*, 46(4), pp. 579–590.

West, J. and Bogers, M. (2017). Open innovation: current status and research opportunities. *Innovation: Organization & Management*, 19(1), pp. 43–50.

Wojciechowski, A. (2009). Models of charity donations and project funding in social networks. *Lecture Notes in Computer Science*, 5872, pp. 454–463.

Wolf, P., Späth, S. and Haefliger, S. (2011). Participation in intra-firm communities of practice: a case study from the automotive industry. *Journal of Knowledge Management*, 15(1), pp. 22–39.

Yli-Renko, H., Autio, E. and Sapienza, H. J. (2001). Social capital, knowledge acquisition, and knowledge exploitation in young technology-based firms. *Strategic Management Journal*, 22(6–7), pp. 587–613.

Part V

Measuring, Evaluating and Stimulating Open Innovation in SMEs

Chapter 13

Measuring Open Innovation in SMEs

Dylan Jones-Evans, Aineias Gkikas, Martin Rhisiart and Niall G. MacKenzie*

**dylan.jones-evans@southwales.ac.uk*

Abstract

Research on open innovation in firms has grown significantly over the last decade. While scholars have started to address the knowledge gaps around open innovation in small and medium-sized enterprises (SMEs), there is still much to be done to develop more systematic evidence on the open innovation practices of SMEs. Given this, and in order to develop a greater understanding of open innovation practices at the SME level, this chapter develops a methodology for capturing, measuring, and benchmarking open innovation practices of SMEs. The theoretical framework proposes and develops six composite indicators to measure the different elements of open innovation within SMEs. These six pillars of open innovation practices are knowledge and technology sourcing activities; innovation expenditure; sources of knowledge; human capital; innovation networks; and intellectual property protection. They allow the examination of this phenomenon across spatial,

* Corresponding author.

temporal, and firm-level dimensions. The theoretical framework proposed herein has the potential to provide useful insights on the open innovation activity of SMEs across different size, sectoral, and spatial measures. This has value for policymakers, practitioners, and academic scholars seeking to better understand the differences in open innovation activity at these levels. This can be achieved by utilizing data from Community Innovation Surveys as the main data source for measuring innovation in Europe. The harmonized nature of these surveys provides a unique opportunity for benchmarking and comparative analysis on open innovation activities of different sectors and regions utilizing the proposed framework of open innovation practices. A further impact of the methodology that we propose in this chapter is its potential integration with the annual European Innovation Scoreboard and Regional Innovation Scoreboards. Currently, such an approach is lacking from both these EU-level scoreboards.

13.1. Introduction[1]

The ability to exploit external knowledge is a critical component of innovative performance (Cohen and Levinthal, 1990), and there is now a well-established argument that rather than depending only on self-generated research and development activities, we have entered a new area of "open innovation" where organizations can and should utilize external and internal ideas to advance their technology (Chesbrough, 2003). As a result, it has been proposed that firms that are more open to external sources of knowledge in their innovative activities are more likely to have a higher level of innovative performance (Laursen and Salter, 2006; Huizingh, 2011; Alvarez and Iske, 2015).

[1]This chapter is dedicated to Professor Martin Rhisiart, our friend, colleague, and coauthor who unexpectedly passed away during the production of this chapter. Martin was a warm, engaging, and collegiate scholar who will be very sadly missed by all who knew him. We are grateful for his intellectual contribution to the chapter, but more than that we are forever grateful for his friendship. Diolch, Yr Athro Martin Rhisiart.

While there has been substantial attention given by academics and practitioners to the study of open innovation, there is little systematic evidence on open innovation practices within small to medium-sized enterprises (SMEs) despite the important role that SMEs are playing in knowledge generation and innovation diffusion within the economy (Love and Roper, 2015).

This chapter examines the relationship between open innovation and SMEs. In doing so, it proposes a theoretical framework that captures six pillars of open innovation practices. These include the following: knowledge and technology sourcing activities; innovation expenditure; sources of knowledge; human capital; innovation networks; and intellectual property protection. Such a framework enables the development of an index of open innovation that is substantively more complete than earlier work and allows the examination of this phenomenon across spatial, temporal, and firm-level dimensions.[2]

13.2. Open Innovation: Concepts, Practices, and the SME Level

Creating and capturing value from innovation is increasingly viewed as a strategic process that crosses the boundary of the firm. Open innovation emphasizes the importance of sourcing resources from outside the firm (e.g., R&D, technology, and specialized knowledge) to complement internal capabilities, engendering value-creating activities that would be less feasible on a stand-alone basis. Establishing the appropriate complementarity of internal and external resources is critical in unlocking innovation and improving the firm's "absorptive capacity" (Cohen and Levinthal, 1990). According to Chesbrough (2012, p. 21), "open innovation has become a new paradigm for organizing innovation. It assumes that firms can and should use external ideas as well as internal ideas, and internal and external paths to market, as they look to advance their innovations.

[2]Future research could utilize Community Innovation Survey Data in order to test the theoretical framework across different economies, such as local, regional, and national ones.

Open innovation processes combine internal and external ideas together into platforms, architectures, and systems. Open innovation processes utilize business models to define the requirements for these architectures and systems. These business models access both external and internal ideas to create value while defining internal mechanisms to claim some portion of that value." Consequently, access to market sources such as customers, suppliers, universities, consultants, government laboratories, and other organizations can enable firms to develop and commercialize innovative ideas (Laursen and Salter, 2006; Abreu *et al.*, 2010; Iammarino *et al.*, 2012; Cook *et al.*, 2013; MacKenzie and Zhang, 2014). This can range from simple access to knowledge to the development of more complex innovation networks that lead to collaborative research and innovation activity. This may be especially pertinent to SMEs that often lack resources internally and require these external linkages to access knowledge (Gronum *et al.*, 2012).

While there is increasing evidence that SMEs are playing an important part in knowledge generation and innovation, especially within emerging high-technology sectors, there has been relatively limited research on open innovation in SMEs (West *et al.*, 2014). This is despite the fact that the nature and size of SMEs provide a stimulus for seeking external resources for innovation with their agile, entrepreneurial behavior, reinforcing both the tendency toward openness and their capacity to exploit opportunities that emerge. To date, much of the research on open innovation adoption has focused to a significant degree on large firms (Chesbrough, 2003; Keupp and Gassmann, 2009; Lichtenthaler and Ernst, 2009; Mortara and Minshall, 2011) despite suggestions that the degree of openness is influenced by firm size (Drechsler and Natter, 2012). Even though it has been to an extent the poorer relation, there has been some work undertaken to examine open innovation in SMEs (Van de Vrande *et al.*, 2009; Lee *et al.*, 2010; Parida *et al.*, 2012; Spithoven *et al.*, 2013; Brunswicker and van de Vrande, 2014), especially in relation to inbound, outbound, and coupled activities (Dahlander and Gann, 2010; Bogers, 2011). For example, the organizational and competitive structure of SMEs in some sectors lends itself to inbound activities such as scouting, in-licensing intellectual property (IP), crowdsourcing, competitions, and spin-ins (Chesbrough and Bogers, 2014). The SME sector can also be direct beneficiaries of various outbound activities, especially the out-licensing of their own IP and technology as part of a proactive and nuanced strategy to exploit knowledge generated and developed within the business (Chesbrough *et al.*, 2006). In key sectors, such as life sciences, there is increasing evidence of coupled activities, that is, collaboration between two or more organizations including SMEs in the innovation ecosystem that combine inbound and outbound activities (Gassmann and Enkel, 2004).

In addition, SMEs may have to rely more on external sources of innovation than many large firms, given their limited resources although such openness has its challenges. For example, firms face high coordination and transaction costs in seeking external knowledge and maintaining relationships with other actors (Grant, 1996; Keupp and Gassman, 2009), and this may be compounded for SMEs. Government support may be needed to develop and maintain these relationships. Given this, another important factor to consider may be the relative expenditure by SMEs on internal and external sources of innovation as this may provide an indicator of the importance of open innovation as part of the firm's activities. They may also find it challenging to evaluate external ideas (Menon and Pfeffer, 2003) as they may not have the internal expertise and experience to do so, and the internal human capital to manage both external and internal ideas could be crucial.

In terms of improving the economic competitiveness of firms, prior research has found a relationship between open innovation, product innovativeness (Laursen and Salter, 2006), and profitability (Lichtenthaler and Ernst, 2009). Another recent study concluded that open innovation leads to improved innovation performance for new service success and financial performance (Cheng and Huizingh, 2014). Teece (2007) also suggests that open innovation is an important dimension of the ability of firms to reconfigure resources to adapt to anticipated changes in the external environment and thus become more competitive. Therefore, if open innovation has a positive influence on competitiveness and productivity, given the increasing evidence of the impact of SMEs on employment, wealth creation, and innovation (de Kok *et al*, 2011; de Wit and de Kok, 2014), there is a case for exploring how open innovation activities within SMEs can contribute to overall economic development in order to better understand the relationship. To measure this phenomenon, however, it is important to establish a common theoretical framework that researchers, policymakers, and practitioners can use in order to examine such a relationship.

13.3. Theoretical Framework for Measuring Open Innovation

Research on firms' openness has grown significantly over the last decade with an increasing focus on the study of open innovation practices. Researchers have approached the field from different perspectives, recognizing that the phenomenon of open innovation is complex and multidimensional (Chesbrough, 2003; Chesbrough, 2006; Grimaldi *et al.*, 2013; Felin and Zenger, 2014; Gambardella and Panico, 2014; Laursen and Salter, 2014; Mina *et al.*, 2014; West *et al.*, 2014). However, there are few metrics available to assess open innovation performance, which are consistent across national boundaries, with most studies utilizing bespoke data collection techniques for discrete studies. Given this, and in order to develop a greater understanding of open innovation practices at the SME level, this chapter develops a framework for the creation of an SME Open Innovation Index. Here we draw on items and data from the harmonized European Community Innovation Survey (CIS),[3] which offers the opportunity to compare open innovation performances across sectoral, spatial, and temporal levels. Consistent within this approach

[3] The questionnaire for the Community Innovation Survey 2010 can be accessed from http://ec.europa.eu/eurostat/documents/203647/203701/CIS_Survey_form_2010.pdf While this is a harmonized one, there is a small variation across different waves. One reason for this is the various updates that the survey has gone through over the different iterations. The framework proposed takes this into account as well, and the UK Innovation Survey 2010 can be accessed from http://webarchive.nationalarchives.gov.uk/20121212135622/http://www.bis.gov.uk/assets/biscore/science/docs/q/questionnaire-uk-innovation-survey-cis7.pdf.

is that the key pillars of open innovation identified in the index can be used by other researchers for their own surveys to assess open innovation practices, for example, by size of firm,[4] industry type, time period, and regional and national economies. In this sense, the proposed framework offers a significant opportunity to researchers seeking to offer standardized measures of open innovation performance that can be compared and utilized accordingly without the need to collect more bespoke, discrete data.

In the last few years, various researchers have used the CIS as an instrument for examining firm-level innovation within the UK (D'Este *et al.*, 2012; Frenz and Ietto-Gillies, 2007, 2009; Haskel *et al.*, 2014; Iammarino *et al.*, 2012; Laursen and Salter, 2004, 2006, 2014) and in other countries (De Marchi, 2012; Evangelista and Vezzani, 2010; Schmiedeberg, 2008; Smit *et al.*, 2015). In this respect, it is perceived to be one of the most comprehensive time series datasets available, and this chapter will add to the body of work utilizing its data. While there have been other studies which have developed their own methodologies and survey instruments to measure open innovation in SMEs (Brunswicker and Vanhaverbeke, 2015; Lee *et al.*, 2010; Popa *et al.*, 2017), these are highly specific to that research and cannot be utilized, like the CIS, for studying open innovation across regions and nations longitudinally. As with previous studies, we have selected a range of variables to create a set of composite indicators that reflect open innovation practices. Such indicators are already recognized as a "useful tool in policy analysis and public communication" (OECD, 2008, p. 13) and are widely used to compare regional and national performance (Huggins, 2003; Mahroum *et al.*, 2008; Hollanders *et al.*, 2012). Therefore, utilizing findings from the academic and policy literature, six composite indicators are developed in order to measure the different elements of open innovation within enterprises. Figure 13.1 presents the frame of analysis and corresponding indicators, which summarizes the SME Open Innovation Index with pertinent sources and measurements. Section 13.4 provides a methodical note on this.

---

[4] We focus specifically on SMEs here, but the survey collects data and reports on all firm sizes.

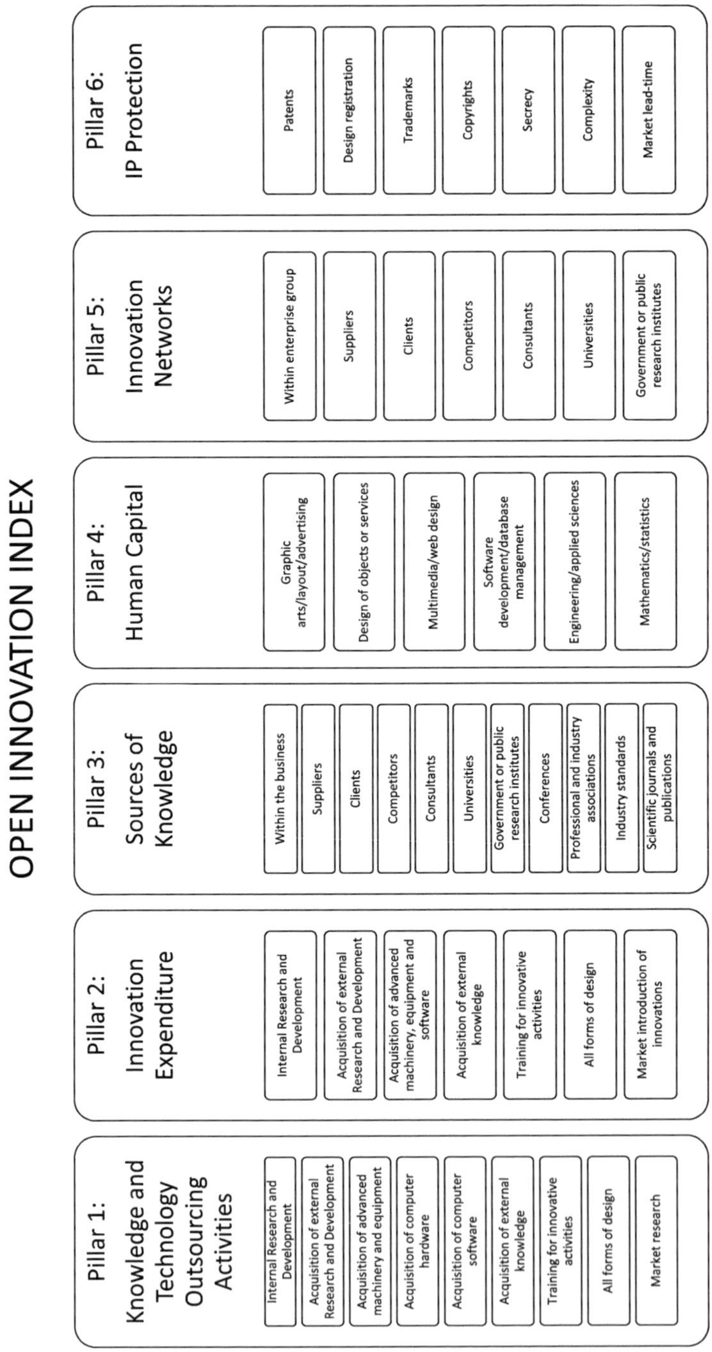

Figure 13.1. Frame of analysis for open innovation indices and indicators.

Similar to the work of other scholars (Ebersberger *et al.*, 2012), the framework uses means and relative frequencies for the descriptive analysis. However, the contribution is the creation of composite indicators for each of the six pillars with an overall index of open innovation practices. That is, the framework allows for a benchmarking analysis, allowing like-for-like comparison on open innovation practices in terms of the main pillars, as well as the breakdown indices. Next, we describe how this is achieved. In order to bring the six elements together, relevant composite indicators can be created.

13.4. Methodology

In this chapter, composite indices for the Open Innovation Index (OII) are based on a mean average of the standardized values for indicators of each index. For each indicator, we calculated the mean and standard deviation. Then for each indicator, the mean is subtracted and divided by the standard deviation. An average is then taken for each indicator to create the overall composite score (e.g., knowledge and technology sourcing activities index). Similar methodology has been applied by other innovation scholars (e.g., Mahroum *et al.*, 2008) who used composite indices in order to measure and map absorptive capacity of firms in the UK nations and regions. Following a methodological approach similar to Laursen and Salter (2006, 2014), a reliability test (Cronbach's alpha) was performed for each of the pillars in order to establish that all items within each pillar measure the same construct. For each pillar, a composite indicator was then created, which is based on a mean average of the standardized values for indicators of each index that could be calculated using, for example, MS Excel.

13.4.1. *Pillar 1: Knowledge and technology sourcing activities*

SMEs invest in a number of internal and external activities for the development of innovation. This can include internal and external

R&D that increases knowledge for developing new and improved products, services, and/or processes. On other occasions, R&D activities performed by external actors such as other businesses or public/private research institutes can be purchased by the firm. In addition to the R&D that is crucial for the development of innovation (Frenz and Ietto-Gillies, 2009), it has been argued that activities such as training; acquisition of machinery, equipment, and software engagement in design activities; and market research are likely to influence innovation and increase the firm's ability to develop and apply new knowledge. These indicators "consist of activities that may have an influence on the ability to learn and assimilate new knowledge" (Doloreux, 2015, p. 105).

As a means of measuring this component of the index, we focus on a number of indicators. This then allows us to examine the differences in knowledge and technology sourcing activities by firm size, region, sector, and industry. To obtain insight into the role of knowledge and technology sourcing activities within SMEs in the UK, we draw on the responses in the UKIS survey related to acquisition and performing certain activities related to this pillar. The indicators we use to measure *knowledge and technology sourcing activities* are as follows:

- Internal research and development (R&D) — this indicates creative work undertaken within the business that increases knowledge for developing new and improved goods or services and process. Measurement: firms' engagement in internal R&D as a percentage of all firms. Data on this can be obtained from the following source: *question 4.a. UK Innovation Survey* (*CIS7*).
- Acquisition of external research and development (R&D) — this indicates purchase of R&D performed by other businesses or by public or private research organizations. Measurement: firms acquiring external R&D as a percentage of all firms. Data on this can be obtained from the following source: *question 4.b. UK Innovation Survey* (*CIS7*).
- Acquisition of advanced machinery and equipment — this indicates purchase of advanced machinery and equipment for the

source of innovation. Measurement: firms acquiring advanced machinery as a percentage of all firms. Data on this can be obtained from the following source: *question 4.c. UK Innovation Survey* (*CIS7*).

- Acquisition of computer hardware — this indicates purchase of computer hardware as an essential ingredient of business innovation. Measurement: firms acquiring computer hardware as a percentage of all firms. Data on this can be obtained from the following source: *question 4.c. UK Innovation Survey* (*CIS7*).
- Acquisition of computer software — this indicates purchase of computer software as an essential ingredient of business innovation. Measurement: firms acquiring computer software as a percentage of all firms. Data on this can be obtained from the following source: *question 4.c. UK Innovation Survey* (*CIS7*).
- External knowledge — this indicates purchase or licensing of patents and nonpatented inventions, know-how, and other types of knowledge from other businesses or organizations. Measurement: firms acquiring external knowledge as a percentage of all firms. Data on this can be obtained from the following source: *question 4.d. UK Innovation Survey* (*CIS7*).
- Training for innovative activities — this indicates internal or external staff training specifically for the development and/or introduction of innovations. Measurement: firms engaging in training as a percentage of all firms. Data on this can be obtained from the following source: *question 4.e. UK Innovation Survey* (*CIS7*).
- All forms of design — this indicates engagement in all design activities, including strategic, for the development or implementation of new or improved goods, services, and processes. Design activities in the R&D phase are not included in this indicator. Data on this can be obtained from the following source: *question 4.f. UK Innovation Survey* (*CIS7*).
- Market research — this indicates engagement in market research for knowledge sourcing. Measurement: firms engaging in market research as a percentage of all firms. Data on this can be obtained from the following source: *question 4.g. UK Innovation Survey* (*CIS7*).

13.4.2. *Pillar 2: Innovation expenditure*

The amount that SMEs invest on open innovation practices is an important element that determines their sustained competitive advantage (Bayona-Sáez *et al.*, 2013). With research suggesting that higher levels of R&D investment lead to greater innovation (Frenz and Ietto-Gillies, 2009; Abreu *et al.*, 2010), this will examine the amount of expenditure on both internal and external innovation activities (such as the purchase or licence of patents and nonpatented related inventions, know-how, and other types of knowledge from other businesses and organizations). Given the fact that high levels of innovation activity are not always based on R&D expenditure, especially within the service sector (Abreu *et al.*, 2010), innovation related expenditures can also include spending on the acquisition of external knowledge, training directly related to innovation, design-related activities, and expenditures related to the introduction of innovations into the market.

As a means of measuring this component of the index, we focus on a number of indicators. This then allows us to examine the differences in innovation expenditure-related activities by firm size, region, sector, and industry. The indicators we use to measure *Innovation Expenditure Activities* are as follows:

- R&D — this indicates business expenditure on internal research and development. Measurement: this is expressed in monetary terms as the proportion of total expenditure by all firms. Data on this can be obtained from the following source: *question 5.a. UK Innovation Survey* (*CIS7*).
- Acquisition of external R&D — this indicates business expenditure on external research and development. Measurement: this is expressed in monetary terms as the proportion of total expenditure by all firms. Data on this can be obtained from the following source: *question 5.b. UK Innovation Survey* (*CIS7*).
- Acquisition of advanced machinery, equipment, and software — this indicates business expenditure on advanced machinery, equipment, and software to be used for new or significantly

improved products or processes. Measurement: this is expressed in monetary terms as the proportion of total expenditure by all firms. Data on this can be obtained from the following source: *question 5.c. UK Innovation Survey* (*CIS7*).

- Acquisition of external knowledge — this indicates business expenditure on external knowledge. Measurement: this is expressed in monetary terms as the proportion of total expenditure by all firms. Data on this can be obtained from the following source: *question 5.d. UK Innovation Survey* (*CIS7*).
- Training for innovative activities — this indicates business expenditure on Training for innovative activities. Measurement: this is expressed in monetary terms as the proportion of total expenditure by all firms. Data on this can be obtained from the following source: *question 5.e. UK Innovation Survey* (*CIS7*).
- All forms of design — this indicates business expenditure on all forms of design. Measurement: this is expressed in monetary terms as the proportion of total expenditure by all firms. Data on this can be obtained from the following source: *question 5.f. UK Innovation Survey* (*CIS7*).
- Market introduction of innovations — this indicates business expenditure on market introduction of innovations. Measurement: this is expressed in monetary terms as the proportion of total expenditure by all firms. Data on this can be obtained from the following source: *question 5.g. UK Innovation Survey* (*CIS7*).

13.4.3. *Pillar 3: Sources of knowledge*

Building on the work of Laursen and Salter (2006, 2014), the pillar on sources of knowledge captures the number of internal and external sources of knowledge used by SMEs in their innovation activities and are normally grouped as internal sources, market sources, institutional sources, and specialized sources (Doloreux, 2015). The Community Innovation Survey asks a number of possible external sources of information and knowledge with firms asked to indicate the level of importance. The UK version of this survey, for example, the seventh iteration of the UK innovation survey, has this on a scale of 0–3, for each one of these sources. Previous research has demonstrated that firms rely on a number of sources when it comes to accessing knowledge to develop internally (Chesbrough *et al.*, 2006; Lundvall, 2007). However, some of these have been perceived to be of greater importance including those internal to the firm and market sources such as clients and suppliers (Laursen and Salter, 2006, 2014).

As a means of measuring this component of the index, we focus on a number of indicators. This then allows us to examine the differences in sources of information for innovation-related activities by firm size, region, sector, and industry. The indicators we use to measure *sources of information for innovation* are as follows:

- Within the businesses — this indicates the importance of information obtained from within the business or enterprise group. Measurement: firms rating this source of information as "High," as a percentage of all firms. Data on this can be obtained from the following source: *question 15.a. UK Innovation Survey* (*CIS7*).
- Suppliers — this indicates the importance of information obtained from suppliers of equipment, materials, services. Measurement: firms rating this source of information as "High," as a percentage of all firms. Data on this can be obtained from the following source: *question 15.b. UK Innovation Survey* (*CIS7*).
- Clients — this indicates the importance of information obtained from clients, customers, or end-users. Measurement: firms rating this source of information as "High," as a percentage of all firms. Data on this can be obtained from the following source: *question 15.c. UK Innovation Survey* (*CIS7*).
- Competitors — this indicates the importance of information obtained from competitors or other businesses. Measurement: firms rating this source of information as "High," as a percentage of all firms. Data on this can be obtained from the following source: *question 15.d. UK Innovation Survey* (*CIS7*).
- Consultants — this indicates the importance of information obtained from consultants, commercial labs, or private R&D

institutes. Measurement: firms rating this source of information as "High," as a percentage of all firms. Data on this can be obtained from the following source: *question 15.e. UK Innovation Survey* (*CIS7*).

- Universities — this indicates importance of information obtained from universities or other higher education institutes. Measurement: firms rating this source of information as "High," as a percentage of all firms. Data on this can be obtained from the following source: *question 15.f. UK Innovation Survey* (*CIS7*).
- Government — this indicates the importance of information obtained from government or public research institutes. Measurement: firms rating this source of information as "High," as a percentage of all firms. Data on this can be obtained from the following source: *question 15.g. UK Innovation Survey* (*CIS7*).
- Conferences — this indicates the importance of information obtained from conferences, trade fairs, and exhibitions. Measurement: firms rating this source of information as "High," as a percentage of all firms. Data on this can be obtained from the following source: *question 15.h. UK Innovation Survey* (*CIS7*).
- Professional and industry associations — this indicates the importance of information obtained from professional and industry associations. Measurement: firms rating this source of information as "High," as a percentage of all firms. Data on this can be obtained from the following source: *question 15.i. UK Innovation Survey* (*CIS7*).
- Industry standards — this indicates importance of information obtained from technical, industry, or service standards. Measurement: firms rating this source of information as "High," as a percentage of all firms. Data on this can be obtained from the following source: *question 15.j. UK Innovation Survey* (*CIS7*).
- Scientific journals and publications — this indicates importance of information obtained from scientific journals and trade/technical publications. Measurement: firms rating this source of information as "High," as a percentage of all firms. Data on this can be obtained from the following source: *question 15.k. UK Innovation Survey* (*CIS7*).

13.4.4. *Pillar 4: Human capital*

In the innovation literature, skills and abilities (e.g., employees' skills) have been used as a proxy of the firm's human capital (Bartoloni and Baussola, 2001; Iammarino *et al.*, 2012; Cook *et al.*, 2013). Human capital has also been found as an important driver of innovation in services in order to build capabilities and competencies. Indeed, it has been found that "non-technological aspects of novelty creation, organisational change and the role of human capital trained in 'soft' disciplines such as arts and humanities represent and distinguish innovation in services" (Abreu *et al.*, 2010, p. 116). In addition, "human capital is seen as being complementary to innovation, constituting *per se* a competence of the firm and generating a super-additive effect in terms of both innovative and economic performance" (Iammarino *et al.*, 2012, p. 1289). It is via this human capital that firms can be aware of the new technology and be able to use it (Autant-Bernard *et al.*, 2010), and, in turn, it is "the introduction of this new technology and human capital and improvements in the organisation of production is what enables the firm to be more efficient and produce at lower costs than its rivals" (Hashi and Stojčić, 2013, p. 354). In this chapter, employees' skills such as engineering and applied sciences, mathematical and statistical skills, graphic art, design, and software development have been used as a proxy of firm's human capital.

Firms will employ individuals in-house or, alternatively, can obtain those skills from external sources to improve their knowledge base. It is this knowledge accumulation via a diverse skill set, such as graphic arts, design of objects and services, multimedia and web design, software development and database management, as well as more traditional skills such as engineering, applied sciences, mathematics, and statistics, that will allow the firm not only to develop its own research, but also to absorb external knowledge, assimilate it, and apply it (Belenzon and Schankerman, 2015; Lazzarotti *et al.*, 2015).

As a means of measuring this component of the index, we focus on a number of indicators. This then allows us to examine the differences in human capital related activities by firm size, region, sector, and industry. The indicators we use to measure *human capital* are as follows:

- Graphic arts/layout/advertising — this indicates businesses employing individuals in-house with these skills at any level, or obtained those from external sources. Measurement: share (%) of individuals employed in-house in all firms with this skill or obtained it from external sources. Data on this can be obtained from the following source: *question 25.a. UK Innovation Survey* (*CIS7*).
- Design of objects or services — this indicates businesses employing individuals in-house with these skills at any level, or obtained those from external sources. Measurement: share (%) of individuals employed in-house in all firms with this skill or obtained it from external sources. Data on this can be obtained from the following source: *question 25.b. UK Innovation Survey* (*CIS7*).
- Multimedia/web design — this indicates businesses employing individuals in-house with these skills at any level, or obtained those from external sources. Measurement: share (%) of individuals employed in-house in all firms with this skill or obtained it from external sources. Data on this can be obtained from the following source: *question 25.c. UK Innovation Survey* (*CIS7*).
- Software development/database management — this indicates businesses employing individuals in-house with these skills at any level, or obtained those from external sources. Measurement: share (%) of individuals employed in-house in all firms with this skill or obtained it from external sources. Data on this can be obtained from the following source: *question 25.d. UK Innovation Survey* (*CIS7*).
- Engineering/applied sciences — this indicates businesses employing individuals in-house with these skills at any level, or obtained those from external sources. Measurement: share (%) of individuals employed in-house in all firms with this skill or obtained it from external sources. Data on this can be obtained from the following source: *question 25.e. UK Innovation Survey* (*CIS7*).

- Mathematics/statistics — this indicates businesses employing individuals in-house with these skills at any level, or obtained those from external sources. Measurement: share (%) of individuals employed in-house in all firms with this skill or obtained it from external sources. Data on this can be obtained from the following source: *question 25.f. UK Innovation Survey* (*CIS7*).

13.4.5. *Pillar 5: Innovation networks*

In addition to sourcing knowledge, some firms will collaborate and work more closely with a range of actors to develop new products, services, or processes. This reflects many studies on the increasing links between firms and their market partners (Iammarino *et al.*, 2012; Love *et al.*, 2010; Tether, 2002). Such external cooperative arrangements include cooperation with clients, suppliers, competitors, universities, and government or public research institutes (Brunswicker and Vanhaverbeke 2015; Masiello *et al.*, 2015; Zhang *et al.*, 2016). As a means of measuring this component of the index, we focus on a number of indicators. This then allows the differences in innovation networks by firm size, region, sector, and industry to be examined. The indicators we use to measure *innovation networks* are as follows:

- Other businesses within enterprise group — this indicates business cooperation on innovation with other businesses within enterprise group. Measurement: firms that collaborated with this actor on innovation as a percentage of all firms. Data on this can be obtained from the following source: *question 16.a. UK Innovation Survey* (*CIS7*).
- Suppliers — this indicates business cooperation on innovation with suppliers of equipment, materials, services, or software. Measurement: firms that collaborated with this actor on innovation as a percentage of all firms. Data on this can be obtained from the following source: *question 16.b. UK Innovation Survey* (*CIS7*).
- Clients — this indicates business cooperation on innovation with clients, customers, or end-users. Measurement: firms that

collaborated with this actor on innovation as a percentage of all firms. Data on this can be obtained from the following source: *question 16.c. UK Innovation Survey* (*CIS7*).

- Competitors — this indicates business cooperation on innovation with competitors or other businesses in the same industry. Measurement: firms that collaborated with these actors on innovation as a percentage of all firms. Data on this can be obtained from the following source: *question 16.d. UK Innovation Survey* (*CIS7*).
- Consultants — this indicates business cooperation on innovation with consultants, commercial labs, or private R&D institutes. Measurement: firms that collaborated with these actors on innovation as a percentage of all firms. Data on this can be obtained from the following source: *question 16.e. UK Innovation Survey* (*CIS7*).
- Universities — this indicates business cooperation on innovation with universities or other higher education institutions. Measurement: firms that collaborated with these actors on innovation as a percentage of all firms. Data on this can be obtained from the following source: *question 16.f. UK Innovation Survey* (*CIS7*).
- Government — this indicates business cooperation on innovation with government or public research institutes. Measurement: firms that collaborated with this actor on innovation as a percentage of all firms. Data on this can be obtained from the following source: *question 16.g. UK Innovation Survey* (*CIS7*).

13.4.6. *Pillar 6: Intellectual property protection*

While firms undertake a number of measures to protect the value of innovation, it can be argued that the process of open innovation would encourage greater sharing of knowledge among actors without restrictions. For example, no longer should a company lock up its intellectual property but instead find ways to profit from others' use of that technology through licensing agreements, joint ventures, and other arrangements. However, it is important for SMEs to be able to protect the value they derive from internal innovation so that they can participate fully in the open innovation process (Enkel *et al.*, 2009; Laursen and Salter, 2014; van de Vrande *et al.*, 2009). This can be done in a variety of ways including applying for a patent; registering an industrial design; registering a trademark; producing materials eligible for copyright; using secrecy; using complexity of design; and getting lead-time advantage on competitors (OECD, 2005).

Prior research has suggested that small business owners place most emphasis on informal methods to protect intellectual property and consider formal registrable rights such as patents less important (Kitching and Blackburn, 1998). Traditionally, the use of patents has been given as one of the dominant methods for measuring innovation at the firm level (Greenhalgh and Rogers, 2006). Based on this, future research could explore how this varies across firm size, regions, and industries within the wider context of intellectual property protection mechanisms, similarly exploring how the various IP methods work as a tool for keeping them competitive.

As a means of measuring this component of the index, we focus on a number of indicators. This then allows for examining the differences in innovation networks by firm size, region, sector, and industry. The indicators we use to measure IP protection are as follows:

- Patents — indicates protection of innovations via the application of patents. Measurement: firms that are applying for a patent as a percentage of all firms. Data on this can be obtained from the following source: *question 19.a. UK Innovation Survey* (*CIS7*).
- Design — indicates innovation protection via the registration of industrial design. Measurement: firms that are registering an industrial design as a percentage of all firms. Data on this can be obtained from the following source: *question 19.b. UK Innovation Survey* (*CIS7*).
- Trademark — indicates innovation protection via the registration of trademarks. Measurement: firms that are registering a trademark as a percentage of all firms.

- Copyrights — indicates innovation protection via the production of material eligible for copyright. Measurement: firms that are producing material eligible for copyrights as a percentage of all firms. Data on this can be obtained from the following source: *question 19.c. UK Innovation Survey* (*CIS7*).
- Secrecy — indicates innovation protection via the use of secrecy and nondisclosure agreements. Measurement: firms using this IP method as a percentage of all firms. Data on this can be obtained from the following source: *question 19.d. UK Innovation Survey* (*CIS7*).
- Complexity — indicates innovation protection via the use of complexity of design. Measurement: firms using this IP method as a percentage of all firms. Data on this can be obtained from the following source: *question 19.e. UK Innovation Survey* (*CIS7*).
- Market-lead time — indicates innovation protection via the use of lead-advantage time on competitors. Measurement: firms using this IP method as a percentage of all firms. Data on this can be obtained from the following source: *question 19.f. UK Innovation Survey* (*CIS7*).

13.5. Future Research Avenues

The construction of the Open Innovation Index is based upon the questions and data gathered as part of the harmonized European Commission's Community Innovation Survey. Consequently, the index offers the potential for a number of future research avenues based upon the data collection methods and constructs associated with the survey. The survey is collected by member nations of the European Union (and associated territories) and is completed by individual companies whereby a number of characteristics are recorded (over and above the innovation activities outlined above) by each company including their sector (activity), size (number of employees), location (by postcode), and the name of the respondent company (this could therefore potentially allow for cross-referencing with other survey responses to check for longitudinal changes in the company innovation performance). Consequently, there are a number of future research avenues utilizing the index we have developed including: spatial differences at NUTS (Nomenclature of Territorial Units for Statistics) 1, 2, and 3 levels[5] looking at local, regional, and national performance; sectoral differences across Europe; open innovation activities according to company size differences across Europe; and individual-level company development toward (or away from) open innovation practices. As noted, the index can be used for large firms as well as SMEs. The pillars and the indicators used appear appropriate for both large firms and SMEs — acknowledging the diversity of both large firms and SMEs (size, industry characteristics, etc.). It seemed entirely appropriate — given the historical neglect of open innovation research in SMEs — to focus initial attention on SMEs for the index. The framework and index are powerful tools for integrating and measuring performance levels for open innovation, as well as opening up the opportunity to explore the relationships between the different aspects outlined above. The framework provided herein should help in examining the business innovation activity of companies across different regions, time periods, and industries through the lens of open innovation practices (Chesbrough, 2003; Laursen and Salter, 2006, 2014) and using a benchmarking methodology approach (Huggins, 2003; OECD, 2008). This would contribute markedly to our understanding of open innovation, both as a concept for academic research and as a potential policy focus for governments. For managers, the capacity to benchmark practices with SMEs within their own industry — and across different industries — is also a potentially powerful application of the index. This index has been produced using UK Innovation Survey data. With further development, the availability of CIS data in other EU member states enables wider benchmarking. National Innovation Surveys — following the model of CIS — have been adopted in other countries (e.g., South Korea; see Kang and

[5]NUTS refers to the EU's geocode standard for referencing administrative divisions of countries for statistical purposes.

Kang, 2014). This provides rich potential for developing the index on a broader international basis.

Existing work on open innovation to date has tended toward individual, discrete analyses, which are subject to specific contextual factors including periodization, spatial considerations, and a lack of easy comparability due to different techniques, parameters, and data. For example, it would be worthwhile to explore whether the open innovation practices performance observed during a particular reference period are broadly similar or change in different reference periods (with different conditions for innovation), given that the time frame examined could cover prerecessionary, recessionary, postrecessionary periods, or a combination of those. In creating our framework, we have sought to address the lack of easy comparability present in existing studies and offer a route forward for scholars and policymakers alike. Open innovation has gained increasing credence within both academic and policy circles — in order for the concept to be operationalized within the latter, metrics become very important. Without benchmarking, policymakers are unable to address measurement within their own domains confidently. Our framework and methodology will allow for this issue to be addressed and progressed accordingly.

13.6. Concluding Remarks

This chapter has sought to contribute to the literature on open innovation via the development of a framework for measuring open innovation activities at the enterprise level. The identified pillars of open innovation and corresponding indicators are used widely in the literature but not, until now, in an integrated fashion. Constructing these pillars allows for the development of a wider account of open innovation practices at the enterprise level. In addition to the operational definition of each pillar and indicators, we have presented factual materials from the UK innovation survey, which could be used by other researchers to compare other national and regional economies within the European Union (given the harmonized nature of the Community Innovation Survey). The framework has the potential to provide accessible information that is useful to a range of stakeholders, policy analysts, researchers, and businesses. It can be further used to explore linkages of open innovation practices and firm performance, especially as the literature investigating this phenomenon is now gaining momentum (Ebersberger *et al.*, 2012; Lasagni, 2012; Mazzola *et al.*, 2012). Such work may unpack some of these relationships and provide additional insight. In creating the SME Open Innovation Index, a purposefully integrative approach has been adopted to combine different dimensions of SMEs' innovation-related activities.

Acknowledgments

The framework developed in this chapter is based on one of the authors' (Aineias Gkikas) doctoral thesis undertaken at Cardiff University. All authors acknowledge useful comments obtained at international conferences (ISBE 2015, Glasgow, and ISPIM 2016, Porto), and in particular from Professor Gillian Bristow and Professor Robert Huggins on earlier iterations of the work. The usual disclaimers apply.

References

Abreu, M., Grinevich, V., Kitson, M. and Savona, M. (2010). Policies to enhance the "hidden innovation" in services: Evidence and lessons from the UK. *Service Industries Journal*, 30, pp. 99–118.

Alvarez, H. and Iske, P. (2015). Internal capabilities and external knowledge sourcing for product innovation in LMT SMEs. *Journal of Innovation Management*, 3, pp. 55–70.

Asheim, B. T. and Gertler, M. S. (2005). The geography of innovation: regional innovation systems. In: Fagerberg, J., Mowery, D. C., and Nelson, R. R. (eds.). *The Oxford Handbook of Innovation* (Oxford University Press, Oxford), pp. 291–317.

Autant-Bernard, C., Chalaye, S., Manca, F., Moreno, R. and Suriñach, J. (2010). Measuring the adoption of innovation. A typology of EU countries based on the Innovation Survey. *Innovation: The European Journal of Social Science Research*, 23, pp. 199–222.

Bartoloni, E. and Baussola, M. (2001). The determinants of technology adoption in Italian manufacturing industries. *Review of Industrial Organization*, 19, pp. 305–328.

Bayona-Sáez, C., García-Marco, T. and Sanchez-García, M. (2013). The impact of open innovation on innovation performance: The case of Spanish agri-food firms. In: Martinez, M. G. (ed.). *Open Innovation in the Food and Beverage Industry*, Woodhead Publishing Series in Food Science, Technology and Nutrition, No. 243 (Elsevier, Cambridge, UK) pp. 74–94.

Belenzon, S. and Schankerman, M. (2015). Motivation and sorting of human capital in open innovation. *Strategic Management Journal*, 36, pp. 795–820.

Bogers, M. (2011). The open innovation paradox: Knowledge sharing and protection in R&D collaborations. *European Journal of Innovation Management*, 14, pp. 93–117.

Brunswicker, S. and van de Vrande, V. (2014). Exploring open innovation in small- and medium-sized enterprises. In: Chesbrough, H., Vanhaverbeke, W. and West, J. (eds.). *New Frontiers in Open Innovation* (Oxford University Press, Oxford).

Brunswicker, S. and Vanhaverbeke, W. (2015). Open innovation in small and medium-sized enterprises (SMEs): External knowledge sourcing strategies and internal organizational facilitators. *Journal of Small Business Management*, 53(4), pp. 1241–1263.

Cheng, C. C. and Huizingh, E. K. (2014). When is open innovation beneficial? The role of strategic orientation. *Journal of Product Innovation Management*, 31(6), pp. 1235–1253.

Chesbrough, H. (2003). *Open Innovation: The New Imperative for Creating and Profiting from Technology* (Harvard Business School Press, Boston, MA).

Chesbrough, H. (2006). Open innovation: a new paradigm for understanding industrial innovation. In: Chesbrough, H., Vanhaverbeke, W. and West, J. (eds.). *Open Innovation: Researching a New Paradigm* (Oxford University Press, Oxford).

Chesbrough, H. Vanhaverbeke, W. and West, J. (Eds.) (2006). *Open Innovation: Researching a New Paradigm* (Oxford University Press, Oxford).

Chesbrough, H. (2012). Open innovation: where we've been and where we're going. *Research-Technology Management*, 55(4), pp. 20–27.

Chesbrough, H. and Bogers, M. (2014). Explicating open innovation clarifying an emerging paradigm for understanding innovation. In: Chesbrough, H.,Vanhaverbeke, W. and West, J. (eds.). *New Frontiers in Open Innovation.* (Oxford University Press, Oxford).

Cohen, W. and Levinthal, D. (1990). Absorptive capacity: a new perspective on learning and innovation. *Administrative Science Quarterly*, 35(1), pp. 128–152.

Cook, G. A., Pandit, N. R., Lööf, H. and Johansson, B. (2013). Clustering, MNEs, and innovation: who benefits and how? *International Journal of the Economics of Business*, 20(2), pp. 203–227.

Dahlander, L. and Gann, D. M. (2010). How open is innovation? *Research Policy*, 39(6), pp. 699–709.

de Kok, J., Vroonhof, P., Verhoeven, W., Timmermans, N., Kwaak, T., Snijders, J. and Westhof, F. (2011). Do SMEs create more and better jobs. Report prepared by EIM for the European Commission DG Enterprise and industry, Brussels, European Commission.

De Marchi, V. (2012). Environmental innovation and R&D cooperation: empirical evidence from Spanish manufacturing firms. *Research Policy*, 41(3), pp. 614–623.

D'Este, P., Iammarino, S., Savona, M. and von Tunzelmann, N. (2012). What hampers innovation? Revealed barriers versus deterring barriers. *Research Policy*, 41(2), pp. 482– 488.

de Wit, G. and de Kok, J. (2014). Do small businesses create more jobs? New evidence for Europe. *Small Business Economics*, 42(2), pp. 283–295.

Doloreux, D. (2015). Use of internal and external sources of knowledge and innovation in the Canadian wine industry. *Canadian Journal of Administrative Sciences/ Revue Canadienne des Sciences de l'Administration*, 32(2), pp. 102–112.

Drechsler, W. and Natter, M. (2012). Understanding a firm's openness decisions in innovation. *Journal of Business Research*, 65(3), pp. 438–445.

Ebersberger, B., Bloch, C., Herstad, S. J. and Van De Velde, E. (2012). Open innovation practices and their effect on innovation performance. *International Journal of Innovation and Technology Management*, 9(6), pp. 1–22.

Enkel, E., Gassmann, O. and Chesbrough, H. (2009). Open R&D and open innovation: exploring the phenomenon. *R&D Management*, 39(4), pp. 311–316.

European Commission. (2015). *Regional Innovation Scoreboard* (Brussels: European Commission).

Evangelista, R. and Vezzani, A. (2010). The economic impact of technological and organizational innovations. A firm-level analysis. *Research Policy*, 39(10), pp. 1253–1263.

Felin, T. and Zenger, T. R. (2014). Closed or open innovation? Problem solving and the governance choice. *Research Policy*, 43(5), pp. 914–925.

Frenz, M. and Ietto-Gillies, G. (2007). Does multinationality affect the propensity to innovate? An analysis of the third UK Community Innovation Survey. *International Review of Applied Economics*, 21(1), pp. 99–117.

Frenz, M. and Ietto-Gillies, G. (2009). The impact on innovation performance of different sources of knowledge: evidence from the UK Community Innovation Survey. *Research Policy*, 38(7), pp. 1125–1135.

Gambardella, A. and Panico, C. (2014). On the management of open innovation. *Research Policy*, 43(5), pp. 903–913.

Gassmann, O. and Enkel, E. (2004). Towards a theory of open innovation: three core process archetypes. Paper presented at the *R&D Management Conference*. Lisbon.

Grant, R. M. (1996). Toward a knowledge-based theory of the firm. *Strategic Management Journal*, 17(S2), pp. 109–122.

Greenhalgh, C. and Rogers, M. (2006). The value of innovation: the interaction of competition, R&D and IP. *Research Policy*, 35(4), pp. 562–580.

Grimaldi, M., Quinto, I. and Rippa, P. (2013). Enabling open innovation in small and medium enterprises: a dynamic capabilities approach. *Knowledge and Process Management*, 20(4), pp. 199–210.

Gronum, S., Verreynne, M. L. and Kastelle, T. (2012). The role of networks in small and medium-sized enterprise innovation and firm performance. *Journal of Small Business Management*, 50(2), pp. 257–282.

Hashi, I. and Stojčić, N. (2013). The impact of innovation activities on firm performance using a multi-stage model: evidence from the Community Innovation Survey 4. *Research Policy*, 42(2), pp. 353–366.

Haskel, J., Hughes, A. and Bascavusoglu-Moreau, E. (2014). *The Economic Significance of the UK Science Base* (UK Innovation Research Centre, Cambridge, UK).

Hollanders, H., Derbyshire, J., Lewney, R., Tijssen, R, Tarantola, S. and Rivera Leon, L. (2012). Regional Innovation Scoreboard 2012 — Methodology report. INNO Metrics 2011–2012 report, Brussels: European Commission, DG Enterprise.

Huggins, R. (2003). Creating a UK competitiveness index: regional and local benchmarking. *Regional Studies*, 37(1), pp. 89–96.

Huizingh, E. K. (2011). Open innovation: state of the art and future perspectives. *Technovation*, 31(1), pp. 2–9.

Iammarino, S., Piva, M., Vivarelli, M. and Von Tunzelmann, N. (2012). Technological capabilities and patterns of innovative cooperation of firms in the UK regions. *Regional Studies*, 46(10), pp. 1283–1301.

Kang, K. H. and Kang, J. (2014). Do external knowledge sourcing modes matter for service innovation? Empirical evidence from South Korean service firms. *Journal of Product Innovation Management*, 31(1), pp. 176–191.

Keupp, M. M. and Gassmann, O. (2009). Determinants and archetype users of open innovation. *R&D Management*, 39(4), pp. 331–341.

Kitching, J. and Blackburn, R. (1998). Intellectual property management in the small and medium enterprise (SME). *Journal of Small Business and Enterprise Development*, 5(4), pp. 327–335.

Lasagni, A. (2012). How can external relationships enhance innovation in SMEs? New evidence for Europe. *Journal of Small Business Management*, 50(2), pp. 310–339.

Laursen, K. and Salter, A. (2004). Searching high and low: what types of firms use universities as a source of innovation? *Research Policy*, 33(8), pp. 1201–1215.

Laursen, K. and Salter, A. (2006). Open for innovation: the role of openness in explaining innovation performance among U.K. manufacturing firms. *Strategic Management Journal*, 27(2), pp. 131–150.

Laursen, K. and Salter, A. J. (2014). The paradox of openness: appropriability, external search and collaboration. *Research Policy*, 43(5), pp. 867–878.

Lazzarotti, V., Manzini, R. and Pellegrini, L. (2015). Is your open-innovation successful? The mediating role of a firm's organizational and social context. *The International Journal of Human Resource Management*, 26(19), pp. 2453–2485.

Lee, S., Park, G., Yoon, B. and Park, J. (2010). Open innovation in SMEs — An intermediated network model. *Research Policy*, 39(2), pp. 290–300.

Lichtenthaler, U. and Ernst, H. (2009). Opening up the innovation process: the role of technology aggressiveness. *R&D Management*, 39(1), pp. 38–54.

Love, J. H. and Roper, S. (2015). SME innovation, exporting and growth: a review of existing evidence. *International Small Business Journal*, 33(1), pp. 28–48.

Love, J. H., Roper, S. and Hewitt-Dundas, N. (2010). Service innovation, embeddedness and business performance: evidence from Northern Ireland. *Regional Studies*, 44(8), pp. 983–1004.

Lundvall, B. Å. (2007). National innovation systems — analytical concept and development tool. *Industry and Innovation*, 14(1), pp. 95–119.

MacKenzie, N. G. and Zhang, Q. (2014). A regional perspective on the entrepreneurial university: practices and policies. In: Fayolle, A. and Redford, D. (eds). *Handbook on the Entrepreneurial University* (Edward Elgar, Cheltenham), pp. 188–206.

Mahroum, S., Huggins, R., Clayton, N., Pain, K. and Taylor, P. (2008). *Measuring and Mapping Absorptive Capacity in UK Nations and Regions* (NESTA, London).

Masiello, B., Izzo, F. and Canoro, C. (2015). The structural, relational and cognitive configuration of innovation networks between SMEs and public research organisations. *International Small Business Journal*, 33(2), pp. 169–193.

Mazzola, E., Bruccoleri, M. and Perrone, G. (2012). The effect of inbound, outbound and coupled innovation on performance. *International Journal of Innovation Management*, 16(06), 1240008.

Menon, T. and Pfeffer, J. (2003). Valuing internal vs. external knowledge: explaining the preference for outsiders. *Management Science*, 49(4), pp. 497–513.

Mina, A., Bascavusoglu-Moreau, E. and Hughes, A. (2014). Open service innovation and the firm's search for external knowledge. *Research Policy*, 43(5), pp. 853–866.

Mortara, L. and Minshall, T. (2011). How do large multinational companies implement open innovation? *Technovation*, 31(10), pp. 586–597.

OECD (2005). Intellectual property as an economic asset: key issues in valuation and exploitation. EPO-OECD-BMWA Conference Summary Report, OECD, Paris.

OECD Joint Research Centre-European Commission. (2008). *Handbook on Constructing Composite Indicators: Methodology and User Guide* (OECD Publishing).

Parida, V., Westerberg, M. and Frishammar, J. (2012). Inbound open innovation activities in high-tech SMEs: the impact on innovation performance. *Journal of Small Business Management*, 50(2), pp. 283–309.

Popa, S., Soto-Acosta, P. and Martinez-Conesa, I. (2017). Antecedents, moderators, and outcomes of innovation climate and open innovation: an empirical study in SMEs. *Technological Forecasting and Social Change,* 118, pp. 134–142.

Schmiedeberg, C. (2008). Complementarities of innovation activities: an empirical analysis of the German manufacturing sector. *Research Policy*, 37(9), pp. 1492–1503.

Smit, M. J., Abreu, M. A. and Groot, H. L. (2015). Micro-evidence on the determinants of innovation in the Netherlands: the relative importance of absorptive capacity and agglomeration externalities. *Papers in Regional Science*, 94(2), pp. 249–272.

Spithoven, A., Vanhaverbeke, W. and Roijakkers, N. (2013). Open innovation practices in SMEs and large enterprises. *Small Business Economics*, 41(3), pp. 537–562.

Teece, D. J. (2007). Explicating dynamic capabilities: the nature and microfoundations of (sustainable) enterprise performance. *Strategic Management Journal*, 28(13), pp. 1319–1350.

Tether, B. S. (2002). Who co-operates for innovation, and why: an empirical analysis. *Research Policy*, 31(6), pp. 947–967.

Van de Vrande, V., De Jong, J. P., Vanhaverbeke, W. and De Rochemont, M. (2009). Open innovation in SMEs: trends, motives and management challenges. *Technovation*, 29(6), pp. 423–437.

West, J., Salter, A., Vanhaverbeke, W. and Chesbrough, H. (2014). Open innovation: the next decade. *Research Policy*, 43(5), pp. 805–811.

Zhang, Q., MacKenzie, N. G., Jones-Evans, D. and Huggins, R. (2016). Leveraging knowledge as a competitive asset? The intensity, performance and structure of universities' entrepreneurial knowledge exchange activities at a regional level. *Small Business Economics*, 47(3), pp. 657–675.

Chapter 14

The *Open Innovation Project Canvas* for SMEs

Marc Steen and Wim Vanhaverbeke*

**marc.steen@tno.nl*

Abstract

Open innovation involves multiple parties to succeed in collaborative innovation. We propose to create a temporary form for collaboration during the innovation process, for example, a project consortium, a (public–private) partnership, a network, or any other (informal) type of collaboration. This collaboration then becomes the vehicle for the first steps of the innovation process, typically from idea to the development of a working prototype and a first implementation for a launching customer. The next stages of innovation will involve, for example, the creation of a viable business model, further development and implementation, scaling-up, and commercialization. Sometimes the form of the collaboration needs to change in these phases, for example, by creating a new organizational entity that will be the vehicle for scaling-up and commercialization.

In order to facilitate the creation of a consortium, partnership or network that can conduct an innovation project that can develop an idea into a prototype, we developed the *Open Innovation Project Canvas*, which can be used together with, for example, the *Business*

* Corresponding author.

Model Generation (BMG) or the *Value Proposition Design* (VPD) Canvases, both created by Osterwalder and Pigneur.

The *Open Innovation Project Canvas* is distinct and complementary in the following ways: (1) it focuses on collaboration, and it has the collaboration as focal entity — rather than having one (key) actor as focal entity, as is typical in BMG; (2) it focuses on the first phases of innovation, for example, from idea to prototype — not on operations (after the innovation project), as BMG typically does; (3) it focuses on project results, for example, a prototype — in addition to VPD, which typically focuses on a finished product and its production and delivery; and (4) it focuses on the project results' value and identifying a party that can use these — in addition to VPD, which typically focuses on the finished product's value.

14.1. Introduction

Open innovation (OI) offers many opportunities to improve innovation processes. This is also — or maybe especially — the case for small and medium enterprises (SMEs). If organized well, OI can help them to collaborate with, for example, potential customers, knowledge institutes, suppliers, and public partners, and thereby improve their competitive strength (Chesbrough *et al.*, 2016; Ullrich and Vladova, 2016). SMEs are often not large enough and do not have the required resources and competencies to develop (radical) innovation internally. Therefore, they have strong incentives to collaborate with innovation partners, for example, with a university, for state-of-the-art knowledge; with a manufacturer, for the necessary up scaling; or with a sales organization for marketing purposes. OI has proven to be an interesting way for SMEs to overcome their liability of smallness when it comes to developing new product, services, processes, or business models (Van de Vrande, 2009; Vanhaverbeke, 2017). However, most SMEs in Europe did not embark yet on an OI journey. Most SME managers are afraid to deal with risks inherent in collaborative innovation projects and they have the right mind-set to develop innovative projects with different partners creating joint value for customers in ways that cannot be achieved by a single firm.

As OI still has to be promoted among SMEs, the European Commission is currently providing "€2.8 billion to SMEs for funding research and innovation businesses and accelerating the exploitation of innovations" (European Commission, 2016). The European Commission has recently adopted OI as a crucial instrument in its innovation policy. Despite this recent interest of European policymakers in OI in SMEs, relatively little is known about the ways to organize such projects effectively and efficiently. There is an urgent need to develop practical tools to support managers in their ambition to develop OI projects that also involve SMEs.

With this chapter, we aim to contribute to a small but growing body of knowledge on managing and organizing OI projects involving SMEs. More specifically, we will introduce a practical guideline for SME managers — we call it the *Open Innovation Project Canvas* — who have the ambition to collaborate with partners to conceptualize, develop, and introduce new products or services into the market. The *Open Innovation Project Canvas* is a management tool that SMEs' managers can use during their meetings with different partners. It functions as a visual guideline that shows a stepwise process for the partners to follow in order to develop value-creating and lasting relationships. We have used the Canvas for a period of two years within a FP7 project about new applications of solid-state lighting. We expect that this management tool can help managers to organize joint innovation projects successfully. The tool can also help policymakers, project evaluators, and managers to select promising projects and to monitor and evaluate these projects.

14.2. OI in SMEs

The literature on OI has been dominated by the firm-level approach (Chesbrough *et al.*, 2006, 2014). This is a consequence of the position OI takes versus the so-called "closed innovation." In closed innovation, a firm is relying mainly on its own internal capabilities to generate new products and businesses, and a firm-level perspective is appropriate to understand the benefits and functioning of closed innovation — see left figure in Figure 14.1. OI was originally seen as the antipode of closed innovation and was therefore also analyzed at the firm-level perspective: OI was explained within the context of a (large) company, and it was thought of as a process of managing inbound or outbound flows of ideas and knowledge, for example, in the form of R&D partnerships (Roijakkers and Vanhaverbeke, 2014); see middle figure in Figure 14.1. Relatively little, however, is known on organizing OI in "inter-organizational value networks" (West *et al.*, 2006, p. 26; Vanhaverbeke, 2006; Vanhaverbeke and Cloodt, 2006), for example, in strategic alliance networks, joint ventures with multiple partners, project consortia, innovation ecosystems, or digital platforms. Such forms of OI will typically involve knowledge flows between multiple partners: flows of ideas and knowledge between multiple organizations — see right figure in Figure 14.1. One can imagine collaboration between multiple companies with ideas and knowledge flowing in-between them. Imagine a simple case of multiple partnerships: company A created a product, based on knowledge from a university (inbound); company B has a proprietary technology, which it also sells to others (outbound); and company C (right) that has access to the market, to a launching customer, e.g., a school or hospital. The complexity of this ecosystem-centric, cross-organizational innovation

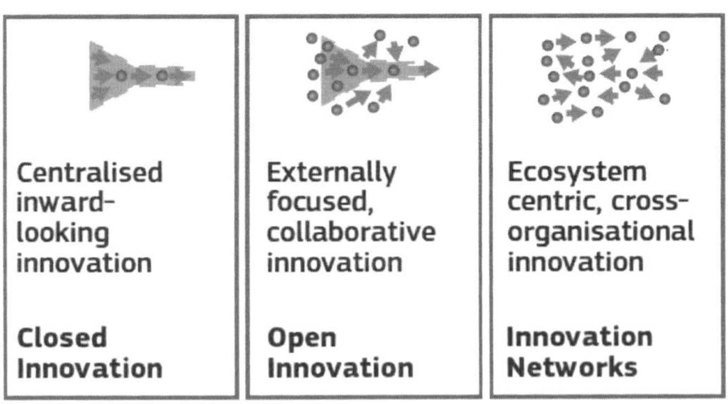

Figure 14.1. Evolution of innovation.

Source: Open Innovation 2.0: Open innovation 2013, p. 10 — after Chesbrough, Forrester, and Von Hippel.

illustrates the complexity of OI in an "inter-organizational value network" involving SMEs.

There are a number of major differences between the traditional "OI" (figure in the middle) and the innovation network approach (figure on the right). Innovation networks do not focus any longer on a firm-level approach. The focus is on the network for value creation, as all partners jointly create value for a specified customer. How to create value can only be understood if one understands how different partners have to collaborate to create and deliver joint value to the customer. That does not mean that the firm level is not important: each firm has to be able to benefit from being embedded in the network. Organizations have to capture part of the value that is jointly created, and this is a process that needs coordination. Therefore, in innovation networks, new roles — such as a network orchestrator — tend to emerge (Nambisan and Sawhney, 2011), and network dynamics are different than dynamics within an alliance between two innovation partners (Autio and Llewellyn, 2014; Brusoni and Prencipe, 2013; Gomes-Casseres, 1996).

Fortunately, there is a small but rapidly growing number of studies of OI in SMEs. One found that SMEs pursue OI primarily for market-related motives, for example, to meet customers' demands or to keep up with competitors, and that they face organizational and cultural challenges (Van de Vrande *et al.*, 2009). Another study found that strong relationships with innovative suppliers, users, customers, and research institutes could lead to better innovation performance (Lasagni, 2012). One study suggested that strong, heterogeneous ties can improve innovation in SMEs, provided that these ties are directly related to innovation (Gronum *et al.*, 2012). And another found links between technology sourcing and radical innovation, and between technology scouting and incremental innovation (Parida *et al.*, 2012).

Indeed, it has been noted that more research is needed on OI in SMEs (Brunswicker and Van de Vrande, 2014; Brunswicker and Vanhaverbeke 2015). Brunswicker and Van de Vrande proposed four themes for future research into OI in SMEs (2014, pp. 135–156); one theme relates to "network relationships and coordination." They claimed that "It is not fully understood how SMEs can make use of and manage different kinds of relationships and networks, also personal ones, when engaging in open innovation" (Brunswicker and Van de Vrande, 2014, p. 144), and they concluded that "SMEs require capabilities for managing these network relationships, which they regularly don't have and which are different from those of large firms" (p. 155).

14.3. Examples of OI Projects Involving SMEs

We (the authors) were involved in the *SSL-erate* project, which was funded by the European Commission (under call FP7-ICT-2013-11) and which aimed "to accelerate the uptake of high-quality SSL technology in Europe by means of OI with and by bringing validated information to all relevant stakeholders" (http://ssl-erate.eu). The project partners included five Lighting Clusters: the Danish Lighting Innovation Network (DLIN) in Denmark, Luce in Veneto (LiV) in Italy, Cluster d'Il-Luminació de Catalunya (CICAT) in Spain, Groen Licht Vlaanderen (GLV) in Belgium, and Cluster Lumière (CL) in France. The Cluster coordinators organized more than 20 OI projects in the course of 2015–2016. They organized workshops with companies and facilitated the process of opportunity finding, idea generation, partner selection, and consortium building. The projects typically involved two or three SMEs, a larger company, a launching customer, and a university department. The projects were typically managed by the SMEs' directors and delivered prototypes or first implementations within 12 or 18 months.

Our role was to support the Cluster coordinators regarding the organization of OI. We were thus able to observe the projects from the sideline, with little or no direct involvement. For our current purpose — discussing the challenges and lessons learnt — we selected ten successful projects; two of each Lighting Cluster. We propose that this biased selection (it is not a random selection as we were excluding unsuccessful projects) is justified since we are interested in ways to organize OI projects *successfully*. We intend to identify critical success factors by studying successful projects. Moreover, we found that it was nearly impossible to find participants in projects who want to talk about unsuccessful projects. Our study is based on the analysis of documents prepared by the Cluster coordinators and on discussions with them (two weekly conference calls) and several interactive workshops and conference calls with some of the organizations involved. Our approach was similar to the proposal by Brunswicker and Van de Vrande (2014, p. 156) to study "network relationships and coordination," and to do that via a longitudinal study, which "allow a 'deep dive' into the specific nature of open innovation in SMEs." Some basic information on the 10 projects that we studied is presented in Table 14.1. Projects 1–4 are described in some detail in a publicly available report (Steen, 2015). We have similar descriptions of the other projects, but these are not publicly available, since the companies involved prefer to not disclose their names and projects.

Pictures of some results of some of these projects are illustrated in Figure 14.2.

14.4. Success Factors for Organizing OI Involving SMEs

Through reflecting on the processes in these OI projects together with the Cluster coordinators and the companies, we were able to identify several recurring success factors. OI projects need to be carefully organized and managed in order to be successful.

The success factors can be categorized as follows: Some are related to *network structure* (a, b, c), others to *network coordination* (d, e, f) (loosely based on Tjemkes *et al.*, 2012). In line with evidence from the literature, R&D projects involving multiple partners need to be managed well in order to be successful, loosely managed projects tend to lead to inferior performance (Du *et al.*, 2014).

14.4.1. *Network structure*

(a) **There is a need for clear and shared short-term project goals** (e.g., to deliver a prototype for a customer in six months). The goals need to be in line with individual partners' goals, and partners need

Table 14.1. The ten projects of the SSL-erate program.

Projects	Project partners	Project goal
Lighting Surface (LiV)	• Metalco, outdoor furniture (production and marketing) • Hikari (SME), lighting systems • Heliv (SME), optics (patent) • Municipality of Bassano del Grappa, customer	To develop a transparent varnish that can be printed and that emits light when activated by a LED light source ("LIT")
H-Lamp (LiV)	• Fairwind (SME), lighting systems • Arte Light (SME), prototyping • University of Bologna, testing • Industrial food producer and a healthcare provider, customers	To develop a product that produces better quality light and better quality air, using UV LEDs, to prevent airborne diseases
The Window (DLIN)	• Dovista, windows manufacturer • Simplight (SME), technology • GXN (SME, architect), customer • Nordic Power Converters (SME) • Aalborg University, knowledge • DTU, knowledge	To develop a window system that combines natural daylight and artificial light to support people's circadian rhythm
Better Light, Better Control (DLIN)	• Simplight (SME), technology • Dansk Farveinstitut, design • Gladsaxe Municipality, customer • Aalborg University, knowledge • DTU, knowledge	To develop an easy-to-control human-centric lighting system that supports circadian rhythms in people over 65 years old, to promote health and well-being

Street Lighting (CICAT)	• Ignia Light, manufacturer • Municipality of Santa Pau, customer	To replace luminaires in the village of Santa Pau with LED lights of amber color, to save energy and to preserve wildlife in the Natural Parc of Garrotxa
LED in the office (CICAT)	• SSIE (SME), manufacturer • Plai Ingenieros (SME), engineer • CBRE (SME), engineer • Mercedes Benz España offices in Azuqueca de Henares, costumer	To install LED lights and automated on/off switching in an office, to provide better light for staff and to reduce energy
Lighting for Elderly (GLV)	• Angoma (SME), manufacturer • Helbig (SME), manufacturer • Prizma, installer • Sint-Jozefskliniek Izegem (hospital), customer • KU Leuven, knowledge	To implement a human-centric lighting system for dementia care in hospitals
Industrial and office lighting (GLV)	• Multiple manufacturers (SMEs) • Lotus Bakeries, customer • KU Leuven lighting lab, knowledge	Make LED more interesting for, e.g., human resources and prevention, by highlighting positive effects on employees' well-being and performance
Healthy Airport Lighting (CL)	• Lighting Engineering Firm (SME) • Airport Authority, customer • Neuroscience Laboratory, knowledge	Offer lighting in lounges which reduce fatigue of travellers during their stopover in Paris airport and to install a prototype
Ceiling Luminaires (CL)	• Lighting company (SME) • Lighting Engineering firm (SME) • Paris Metro Authority, customer	To create flat SSL ceiling luminaires with high flux and low maintenance, and to install 3,000 units in the Paris Metro

Figure 14.2. From top left to bottom right: Lighting surface; the window; street lighting; and H-Lamp.

to have a clear plan how to get to these goals. It has been argued that the alignment of strategic goals is especially needed when diverse partners are involved in OI and that OI then turns into a joint strategizing process (Vanhaverbeke *et al.*, 2016). For example, in the *H-Lamp* project, the partners had short-term goals to conduct a market analysis, to develop a prototype, and to evaluate this prototype in collaboration with a university. In addition, they discussed long-term goals for creating a legal entity to share commercial benefits.

(b) **Partners need to have complementary (not competing) competences and interests**, and they have to be able to effectively combine them (e.g., a developer and a manufacturer can collaborate, provided that they are not direct competitors). For example, in the *Lighting Surface* project, one SME had an innovative, patented technology, another SME had expertise in LED, and another company had expertise in outdoor furniture and international sales; together, they were able to develop and market an innovative product.

(c) **A consortium has to be relatively small, and it has to involve a customer.** Observing the different projects, we come to the conclusion that an innovation project of three to five partners is the optimal size to reach success. Larger consortiums have the tendency to disintegrate or stall during the process. Successful consortiums typically involve two or three SMEs for the development and manufacturing, a launching customer, and a knowledge institute. Involvement of a customer — or a partner that can reach customers — is critical. For example, in *The Window* project, an architectural agency served as a market entry. A launching customer helps to steer the project toward delivering a practical and valuable solution, and offers a first foothold in the market.

14.4.2. *Network coordination*

(d) **Involvement of management and effective leadership are needed.** During the development of the projects, we also observed that specific leadership styles are required to facilitate collaborative innovation (see also (e) and (f)). Leadership and decision-making are relatively easy within SMEs because of their small size; directors can access key personnel relatively easily. In addition, personal contacts can help; for example, in *The Window* project, the project manager used to work at a company associated to one of the main project partners. This gave her access to several personal contacts in this company, which enabled smooth communication collaboration.

(e) **Collaboration between organizations and between disciplines.** The projects were all examples of collaboration between organizations from different industries combining prospective customers, users, knowledge institutes, and suppliers. This combination of a broad variety of partners with different backgrounds and objectives can only be managed effectively if sufficient attention is paid to communication between and commitment of the partners. This is obviously one of the reasons why it is easier to obtain results in a relatively small consortium (see (c)). In most projects, we observed that collaboration with prospective customers and with knowledge institutes is critical to success. For example, in the *H-Lamp* project, the consortium collaborated with an industrial site for food/beverage processing as a potential client, and with the University of Bologna in testing of the H-Lamp's sanitizing properties. Moreover, we found examples of successful triple-helix innovation, involving industry, local government (e.g., municipalities in the *Lighting Surface* and *Better Light, Better Control* projects), and academia; this combination can deliver solutions that matter for industry, academy, and society.

(f) **Organizing iterations, experimentation, and learning**. Successful projects were characterized by organizing early prototyping and early user-testing. Rapid prototyping and early user involvement accelerate the learning process and allow the project members to develop improvements in an iterative way (which is easier if there are clear and shared goals — see (a)). The development of an idea into a prototype can take a long time and requires iterations of experimentation and learning: it is therefore mandatory that the project partners understand how they can speed up these cycles: for example, *The Window* project was preceded by a process of collaborative creativity, which in turn required mutual trust; when the project ended, the participants could not trace back who first came up with the idea as it was developed during many cycles of experimentation and learning where each partner contributed to the improvements.

Furthermore, we found that the more successful projects managed to find additional funding, for example, via national innovation programs (note that the participating companies received no funding from the *SSL-erate* project). Such funding can be used, for instance, for activities that are important to the project, but which cannot be executed effectively by the project partners: for example, a patent search, a market analysis, and a scientific experiment in the *H-Lamp* project. Moreover, we found that the more successful projects were first kick-started and then monitored and supported by the cluster coordinators. Coordinators were able to bridge gaps between partners with different interests, and they could act as a neutral party to build consensus and create trust.

14.5. The *Open Innovation Project Canvas*

As briefly mentioned in the introduction, there is currently a lack of practical tools to support SME managers in organizing and managing OI projects. We (the authors) also observed the problems SME managers faced when they started the *SSL-erate* projects. None of them had prior experience in managing collaborative innovation projects, and problems popped up very quickly. The *SSL-erate* project stipulated that an OI guideline would be developed for the participants. We developed a little booklet with a detailed description how partners could and should work together. The problem with this approach was twofold: first, participants did not take the time to read the booklet as SME managers were very busy and they were only interested in making progress with the project; second, coordination between project partners was *ex ante* not considered as an issue that should be taken care of. Most managers were focusing on the realization of the new product, not on potential issues when collaborating with partners.

In an attempt to correct this situation, we changed our strategy and opted for a two-page visualization of the guidelines that could be used as a guideline for SME managers during the project meetings with their partners. We got inspired by existing tools like the Business Model Canvas (Osterwalder and Pigneur, 2009), which supports the process of creating a viable and feasible business model, and the Value Proposition Canvas (Osterwalder *et al.*, 2014), which supports the process of creating a value proposition that matches the needs of customers and which can be inserted as a mechanism to develop the value proposition into the Business Model Canvas (see Figures 14.3 and 14.4).

The Business Model Canvas and the Value Proposition Canvas have been around for several years and have been adopted widely by practitioners because of their simplicity and ease of use. However, we realized that these tools are only partly applicable as tools to organize OI projects involving a network of SMEs, and we realized that substantial changes had to be introduced to transform it into a management instrument that would be valuable for SME mangers in the context of collaborative projects as we envisioned in *SSL-erate*. There are several reasons.

First, the Business Model Canvas was designed to be used by a single company — *not* by a network of companies or organizations that want to collaborate. Managers of OI projects involving multiple SMEs could of course attempt to use the Business Model Canvas; they would then have to write down, for example, under the header of "Key Activities," the activities of Company A, the activities of Company B, and the activities of Company C (and similarly for the other headers), and describe their partnerships under the header of "Key Partnerships." Such an approach may work, but it would not be easy in this way to keep track of who does what.

In addition, the Business Model Canvas is aimed primarily at generating business models for *operations* — that is, for creating viable and feasible business models — *not* for defining a *project* in the early phases of an innovation project, progressing from an idea to a prototype. Managers of OI projects involving multiple SMEs can of course use the Business Model Canvas, but instead of specifying their *operations* they would have to understand to manage their joint *innovation project*. It is not about their operational activities and resources, or current customer relationships and channels, but they need to figure out the activities, resources, current customer relationships, and channels that they have to develop during the project together with their partners.

Furthermore, the Value Proposition Canvas helps to identify customer segments, the pains these customers would like to avoid,

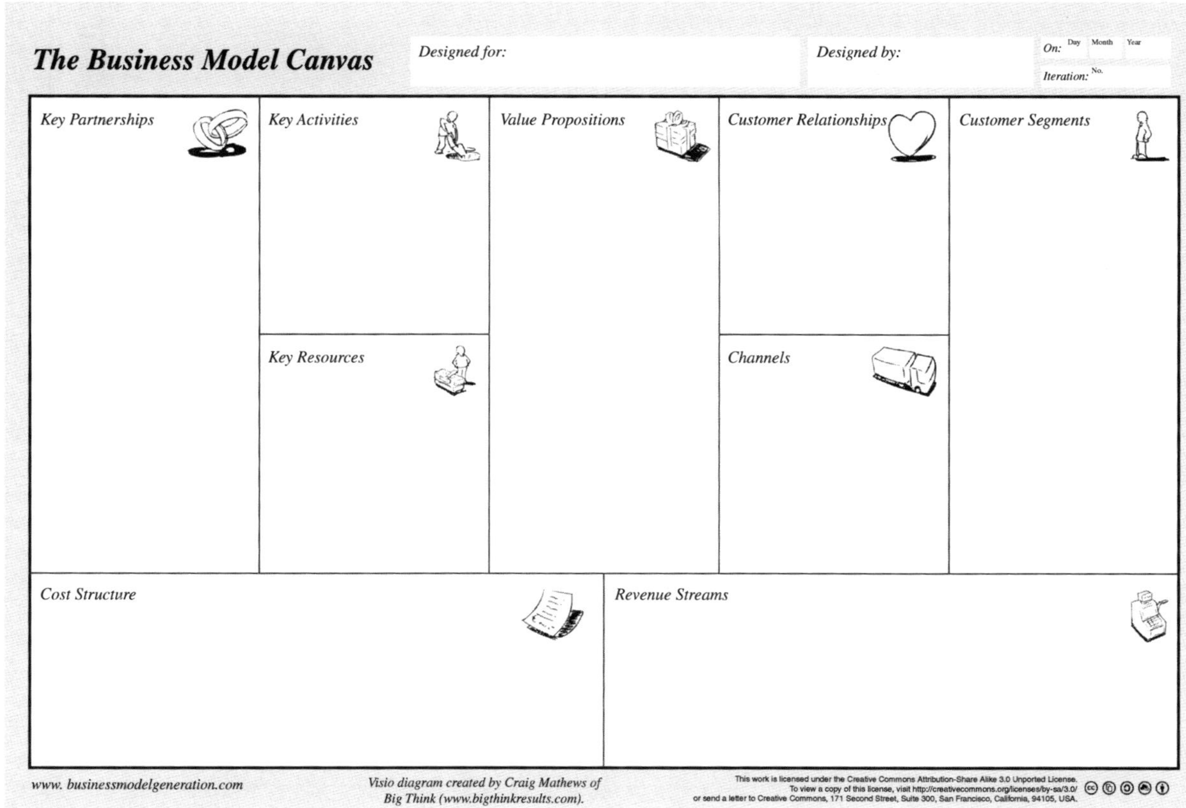

Figure 14.3. The business model canvas (Osterwalder and Pigneur, 2009).

Source: http://diytoolkit.org/tools/business-model-canvas/ (accessed May 30, 2017).

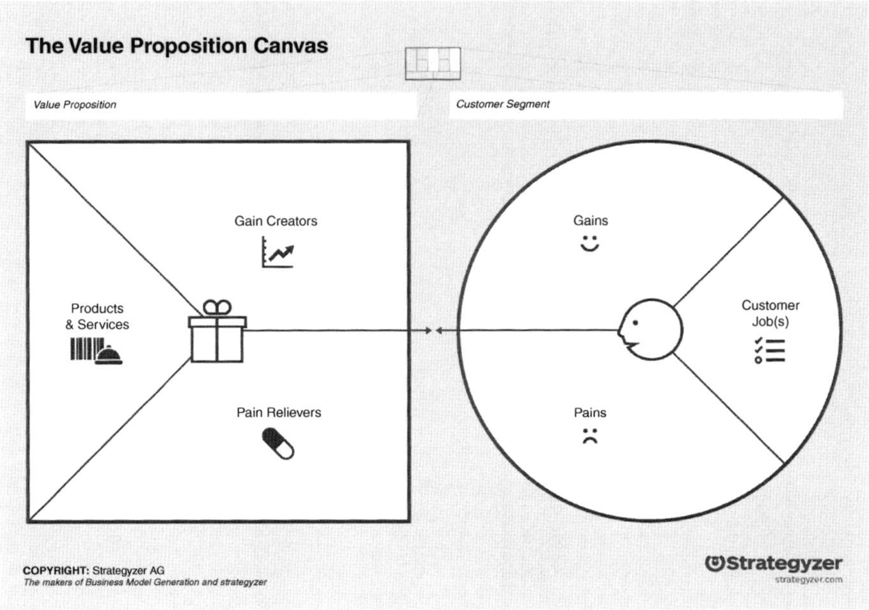

Figure 14.4. The value proposition canvas (Osterwalder *et al.*, 2014).
Source: http://businessmodelalchemist.com/blog/2012/08/achieve-product-market-fit-with-our-brand-new-value-proposition-designer.html (accessed May 30, 2017).

the gains they would like to realize, and how to generate a product or service that meets these needs. This tool, however, is only partly applicable to the organizing of an OI project among SMEs because it focuses on creating a product or a service, whereas the managers of an OI project need a tool that can help them in creating an attractive project outcome. This is a small but important difference. The product or service generated with the Value Proposition Canvas will typically be part of the ongoing business and will, as such, be offered to the company's customers, whereas an OI project outcome will often be a prototype and therefore not (yet) ready to be introduced at full commercial scale into the market. The managers involved in an OI project to turn an idea into a prototype and later on into a product. This can typically not be done with current customers, but it will be done in collaboration with a lead user, which is usually new to the innovating firms.

In order to support the SME managers during OI projects, we developed the *Open Innovation Project Canvas* (see Figure 14.5). This tool is to some extent based on the Business Model Canvas and the Value Proposition Canvas. The *Open Innovation Project Canvas* is, however, distinct from the above-mentioned canvases in several ways:

- It focuses on *collaboration* and on organizing the conditions and processes of collaboration — rather than putting one (key) actor at the center (and addressing collaboration on the side, i.e., the top left corner), as is typically done in the Business Model Canvas.
- It focuses on the project, and typically on the *first phases of the joint innovation process*, for example, the activities and resources needed to move from idea to prototype — rather than creating business model for feasible and viable operations, as is typical for the Business Model Canvas.
- It focuses on creating value in the form of *project results*, for example, a prototype, and finding a party to turn these into a product or service — rather than creating a finished product or service for customers, as the Value Proposition Canvas typically does.
- Finally, we mirrored the Business Model and Value Proposition Canvases, so that we can work from left to right (compatible with our habit of reading from left to right) and start with the customers and their needs (left), then move to defining a value proposition and project results (middle), and then to the formation of a consortium (right) based on the competencies and resources that are needed to be successful with the project.

The *Open Innovation Project Canvas* is meant to be used to manage and organize *collaboration* in the *first phases* of innovation, for example, to create a project consortium and to move from idea to a prototype or first implementation. We used the *Open Innovation Project Canvas* during the meetings with the *SSL-erate* project groups. We guided/coached every group when they used the Canvas

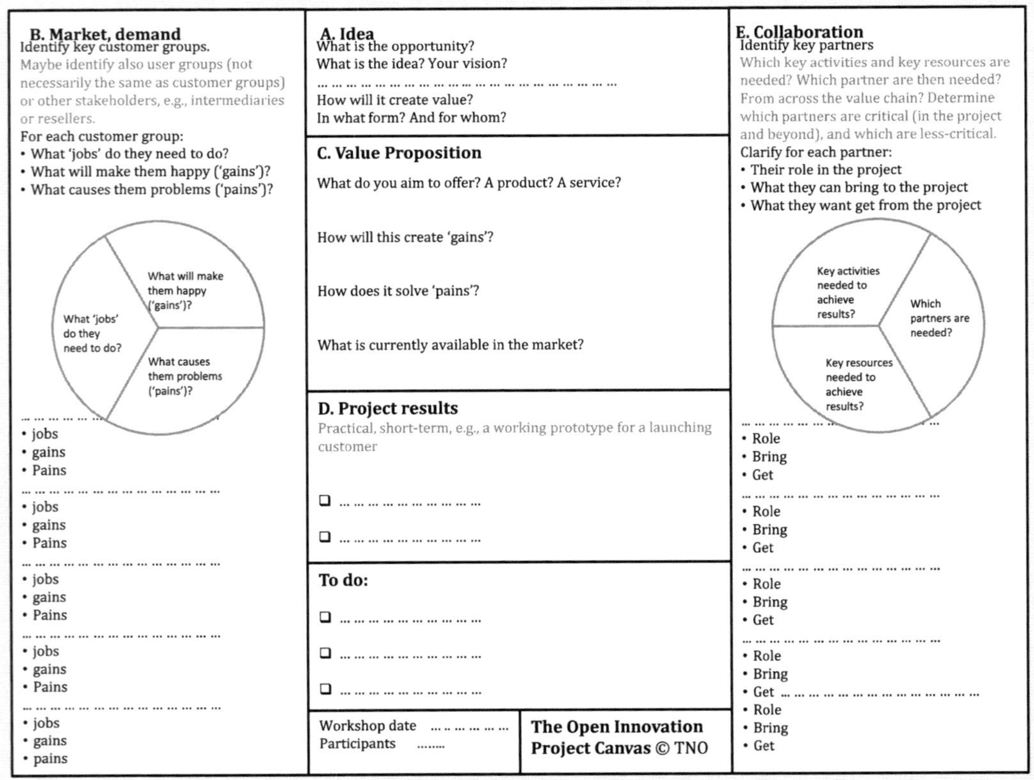

Figure 14.5. Two representations of the *Open Innovation Project Canvas*.

for the first time. This coaching was necessary as most SME managers and other participants only understand and appreciate the value of the *Open Innovation Project Canvas* during its use in a project meeting. We also observed that some groups tried to speed up the meetings by skipping one or more parts of the Canvas because participants thought they already understood, for instance, the needs of the customers or the contribution of a specific partner. Yet, in almost all cases, these "shortcuts" tend to have disastrous consequences for the project as overlooked problems tended to backfire later in the projects.

The main advantage of the *Open Innovation Project Canvas* is that it combines the major benefits of the Value Proposition Canvas — understanding the needs of the customer and aligning the offering to these needs — and the Business Model Canvas — how to configure the business model for that specific project — with a detailed understanding of which competencies and resources have to be brought together into the project and who should be contacted and integrated as partners into the OI project. The *Open Innovation Project Canvas* also has its limitations: it does not, for instance, mention anything about who has to lead the project or how the coordination should be done. It is a management tool that should be used for the purposes it is designed for. It can be quite useful, but it is no silver bullet to solve all the problems in establishing an OI project.

The *Open Innovation Project Canvas* is typically used in the early stages of an OI project up to the development of a prototype. The next phases typically involve the further development of the technology, scaling-up, and commercialization. Note that sometimes the collaboration has to change during these phases, for example, by creating a new entity that can be the vehicle for scaling-up and commercialization.

The *Open Innovation Project Canvas* is meant to be used in a similar way as the Business Model Canvas and the Value Proposition Canvas: in workshop settings, with the relevant people around the table. For OI projects involving multiple SMEs, delegates of all the participating companies and organizations have to be present.

The *Open innovation Project Canvas* is presented in Figure 14.5. It has to be used in a stepwise process: participants start with A, then work on B, C, D, and E. We explain shortly the different steps in Box 14.1.

Box 14.1. Different steps to be followed when using the Open Innovation project Canvas

A. **Idea:** What is the opportunity? What is the idea? What is your vision? How will it create value? Creating value in which way and for whom?

B. **Market, demand:** Identify key customer groups

Maybe identify also user groups (not necessarily the same as customer groups) or other stakeholders, e.g., intermediaries or resellers.

For each customer group (these questions are from the Value Proposition Canvas):

- What "jobs" have to be done?
- What will make them happy ("gains")?
- What causes them problems ("pains")?

C. **Value proposition:**

What do you aim to offer? A product? A service? A new process? A new business model?
How will this create "gains"? How does it solve "pains"?
What is currently available in the market? How is it better than that?
Benefits for "people," e.g., health or well-being?
Benefits for "planet," e.g., energy or materials?

D. **Project results:** Practical, short-term, e.g., a working prototype for a launching customer

E. **Collaboration:** Identify key partners

Which key activities and key resources are needed? Which partners are then needed? Partners from within and across the value chain?

Determine which partners are critical (in the project and beyond), and which are less critical.

Clarify for each partner:

- their role in the project
- what they can bring to the project
- what they want/can get from the project

The Canvas has room for five key customer groups and for five strategic partners. Obviously, one can easily add additional customer groups or additional partners, if needed. Finally, there is room on the Canvas to write down actions that came out of the workshop; this will help to take the results from the workshop forward into the project.

14.6. Applying the *Open Innovation Project Canvas*

We applied this Canvas in several of the projects listed above (in projects of DLIN in March 2015 and in projects of LiV in June 2015 and May 2016) and were able to improve it based on the results. This iterative approach allowed us to test and improve it "on-the-go."

The success factors (a)–(f) for organizing OI projects involving SMEs (see Section 1.4) relate to specific parts of the *Open Innovation Project Canvas* (A)–(F), both on a content level and on a process level.

On a content level, the parties involved can express and discuss their ideas and visions (Canvas: A) and focus on customers' needs (Canvas: B), for example, they can share their insights, or articulate questions which they do not yet have answers for. Based on these ideas and insights, they can articulate a value proposition (Canvas: C) and practical results they aim to deliver in the project (Canvas: D), which relate to the need for clear and shared short-term project goals (success factor (a)). Based on their ambitions to create this value proposition and to deliver specific project results, they can clarify which activities and resources are critically needed, and discuss which partners can bring these to the project (Canvas: E). They can discuss how the different partners can collaborate effectively, and in which roles (success factors (b) and (e)). They can negotiate for each partner, what they need to bring to the project, and what they can get from the project (Canvas F). These discussions will also help to determine which partners are critical, and which are less critical, and to make decisions regarding the size of the consortium (success factor (c)).

On a process level, the usage of the Canvas, typically in a work shop setting, will need to involve the relevant people, for example, to involve management (success factor (d)), and people from different disciplines (success factor (e)). Moreover, the exercise of discussing the different topics in the Canvas will need to be done several times, in an iterative process (success factor (f)), for example, during project definition and scoping, and every couple of months, for example, for each milestone, also to update the project deliverables (Canvas: D) vis-à-vis the overall idea and vision (Canvas A), in order to revisit the project's ambitions and to synchronize plans and goals (success factor (a)).

We used the *Open Innovation Project Canvas* several times during the *SSL-erate* projects, and it worked well. We have to keep in mind that using the *Open Innovation Project Canvas* is more difficult than using the Business Model Canvas. There are several reasons:

- A Business Model Canvas is usually applied in a single company. Therefore, the people around the table are from the same firm. In contrast, The *Open Innovation Project Canvas* is working with different organizations, each with their own objectives and agendas. The participants in the *Open Innovation Project Canvas* are likely to have different backgrounds and different organizational processes and discourses. This multi-disciplinarily is necessary, but it is also a major challenge to ensure that partners understand each other and develop a common language to minimize misunderstandings during the project.
- More partners with different objectives potentially lead to trust issues. First, trust has to be built, and that may be a major challenge between organizations that never have seen each other before. Once trust is established, it is crucial to maintain it and to avoid conflicts that may break down trust.
- The use of the Canvas is complex because an innovation project should not only be managed content wise but also process wise. A project develops over time, and one cannot leave the process out of the equation.

14.7. Discussion and Conclusions

More and more SMEs are applying OI to face competition and develop new products and services. The benefits as well as the challenges of OI for SMEs have been discussed in the literature (Huang and Rice, 2009; Bianchi *et al.*, 2010; Lee *et al.*, 2010; Parida *et al.*, 2011; Van de Vrande *et al.*, 2009). OI in SMEs also has become a prominent policy topic in the European Union, and several policy instruments are tried out to enhance SME innovativeness while relying on external technology and value chain partners.

Despite the growing attention among academia, practitioners, and policymakers for OI in SMEs, we witnessed the difficulties of SME managers to organize and manage OI networks with *SSL-erate* — an FP7 project about new application areas of LED lighting. SME managers can benefit a lot from guidance and coaching during which it comes to partnering with technology and value chain partners. We developed for *SSL-erate* members the *Open Innovation Project Canvas*, a practical guideline for meetings and discussions among the members of a specific project within *SSL-erate*. The Canvas has been explained in this chapter, and we shared our experience in working with this management tool. In our view, such a guideline is necessary, as most SME managers have no experience in managing inter organizational innovation networks. We learned over time that a guideline based on visualization of the process in combination with process coaching could significantly accelerate the progress of innovation projects.

The *Open Innovation Project Canvas* is of course not a silver bullet. It is a management tool that helps SME managers to align the objectives of different innovation partners, but it does not tell them how to organize and manage innovation networks. In addition, we noticed that people with a different background, for example, practitioners, academics, and public officers, tend to choose to work following their own methods, or they prefer to work intuitively or informally not using this Canvas. We are interested to make the Canvas more attractive to practitioners. Furthermore, we would be interested in further usage, evaluation, and further improvement of the Canvas. Therefore, we would like to invite anyone interested in OI in SMEs to use the Canvas and contribute to its further development and adoption.

References

Autio, E. and Llewellyn, T. (2014). Innovation ecosystems: implications for innovation management. In: Dodgson, M., Gann D. and Phillips, N. (eds.). *The Oxford Handbook of Innovation Management*. Oxford University Press, Oxford: U.K., pp. 204–228.

Bianchi, M., Campodall'Orto, S., Frattini, F. and Verseci, P. (2010). Enabling open innovation in small- and medium-sized enterprises: how to find alternative applications for your technologies. *R&D Management*, 40(4), pp. 414–431.

Brunswicker, S. and Van de Vrande, V. (2014). Exploring open innovation in small and medium-sized enterprises. In: Chesbrough, H., Vanhaverbeke, W. and West, J. (eds.). *New Frontiers in Open Innovation*. Oxford: Oxford University Press, pp. 135–156.

Brunswicker, S. and Vanhaverbeke, W. (2015). Open innovation in small and medium-sized enterprises (SMEs): external knowledge sourcing strategies and internal organizational facilitators. *Journal of Small Business Management*, 53(4), pp. 1241–1263.

Brusoni, S. and Precipe, A. (2013). The organization of innovation in ecosystems: problem framing, problem solving, and patterns of coupling. In: Adner, R., Oxley, J. E. and Silverman, B. S. (eds.). *Collaboration and Competition in Business Ecosystems (Advances in Strategic Management, Volume 30)*. Emerald Group Publishing Limited, Howard House, Wagon Lane,Bingley: U.K., pp. 167–194.

Chesbrough, H., Almirall, E. and Vanhaverbeke, W. (2016). *Why Does Open Innovation Work?* Available online: https://ec.europa.eu/digital-single-market/en/news/why-does-open-innovation-work. European Commission.

Chesbrough, H., Vanhaverbeke, W. and West, J. (2006) (eds.). *Open Innovation: Researching a New Paradigm*. Oxford: Oxford University Press.

Chesbrough, H., Vanhaverbeke, W. and West, J. (2014) (eds.). *New Frontiers in Open Innovation*. Oxford: Oxford University Press.

Du, J., Leten, B. and Vanhaverbeke, W. (2014). Managing open innovation projects with science-based and market-based partners. *Research Policy*, (43), pp. 828–840.

European Commission (2013). *Open Innovation 2.0: Open Innovation 2013*. Doi:10.2759/87245. Luxembourg: Publications Office of the European Union.

European Commission (2016). *SME Instrument*. Available online: https://ec.europa.eu/digital-single-market/en/sme-instrument-0

Gomes-Casseres, B. (1996). *The Alliance Revolution: The New Shape of Business Rivalry*. Cambridge, MA: Harvard University Press.

Gronum, S., Verreynne, M.-L. and Kastelle, T. (2012). The role of networks in small and medium-sized enterprise innovation and firm performance. *Journal of Small Business Management*, 50(2), pp. 257–282.

Huang, F. and Rice, J. (2009). The role of absorptive capacity in facilitating "open innovation" outcomes: a study of Australian SMEs in the manufacturing industry. *International Journal of Innovation Management*, 13(2), pp. 201–220.

Lasagni, A. (2012). How can external relationships enhance innovation in SMEs? New evidence for Europe. *Journal of Small Business Management*, 50(2), pp. 310–339.

Lee, S., Park, G., Yoon, B. and Park, J. (2010). Open innovation in SMEs — An intermediated network model. *Research Policy*, 39(2), pp. 290–300.

Nambisan, S. and Sawhney, M. (2011). Orchestration processes in network centric innovation: evidence from the field. *Academy of Management Perspectives*, 25(3), pp. 40–57.

Osterwalder, A. and Pigneur, Y. (2009). *Business Model Generation: A Handbook For Visionaries, Game Changers, and Challengers*. Amsterdam: Modderman Drukwerk.

Osterwalder, A., Pigneur, Y., Bernarda, G. and Smith, A. (2014). *Value Proposition Design*. Hoboken, NJ: John Wiley and Sons.

Parida, V., Westerberg, M. and Frishammar, J. (2012). Inbound open innovation activities in high-tech SMEs: the impact on innovation performance. *Journal of Small Business Management*, 50(2), pp. 383–309.

Roijakkers, N. and Vanhaverbeke, W. (2014). Open innovation through R&D partnerships: implementation challenges and routes to success. In: Chesbrough, H., Vanhaverbeke, W. and West, J. (eds.). *New Frontiers in Open Innovation*. Oxford: Oxford University Press, pp. 115–134.

Steen, M. (2015). SSL-erate Open Innovation Toolkit. SSL-erate Consortium. Available at: http://lightingforpeople.eu/2016/wp-content/uploads/2016/04/SSLerate-open-innovation-toolkit_v4.pdf.

Tiemkes, B., Vos, P. and Burgers, K. (2012). *Strategic Alliance Management*. Tayor and Francis. Abingdon, Oxon:UK.

Ullrich, A. and Vladova, G. (2016). Weighing the pros and cons of engaging in open innovation. *Technology Innovation Management Review*, 6(4), pp. 34–40.

Van de Vrande, V., De Jong, J. P. J., Vanhaverbeke, W. and De Rochemont, M. (2009). Open innovation in SMEs: trends, motives and management challenges. *Technovation*, 29, pp. 423–437.

Vanhaverbeke, W. (2006). The inter organizational context of open innovation. In: Chesbrough, H., Vanhaverbeke, W. and West, J. (eds.). *Open Innovation: Researching a New Paradigm*. Oxford: Oxford University Press, pp. 205–219.

Vanhaverbeke, W. and Cloodt, M. (2006). Open innovation in value networks. In: Chesbrough, H., Vanhaverbeke, W. and West, J. (eds.). *Open Innovation: Researching a New Paradigm*. Oxford: Oxford University Press, pp. 258–281.

Vanhaverbeke, W., Jingshu, D., Leten, B. and Aalders, F. (2014). Exploring open innovation at the level of R&D projects. In: Chesbrough, H., Vanhaverbeke, W. and West, J. (eds.). *New Frontiers in Open Innovation*. Oxford: Oxford University Press, pp. 115–134.

Vanhaverbeke, W., Roijakkers, N., Lorenz, A. and Chesbrough, H. (2016). The importance of conncecting open innovation to strategy. In: Pfeffermann, N. and Gould, J. (eds.). *Strategy and Communication for Innovation* (3rd edn.). Springer. Spinger–Verlag, Berlin Heidelberg: Germany.

West, J., Vanhaverbeke, W. and Chesbrough, H. (2006). Open innovation: a research agenda. In: Chesbrough, H., Vanhaverbeke, W. and West, J. (eds.). *Open Innovation: Researching a New Paradigm*. Oxford: Oxford University Press, pp. 285–307.

Chapter 15

Open Innovation and Public Policy in the EU with Implications for SMEs*

Henry W. Chesbrough[†] *and Wim Vanhaverbeke*[‡,§]

[†] *Haas School of Business, UC Berkeley*
[‡] *Hasselt University, ESADE Business School, and National University of Singapore, Singapore*
[§] *wim.vanhaverbeke@uhasselt.be*

* This chapter is an extension of a report commissioned by ESADE Business School in Barcelona and the Science|Business Innovation Board AISBL. The Board is a Belgian not-for-profit scientific association that commissions original policy research, engages with policymakers and the press, and works generally to improve the climate for innovation in Europe. Its three cofounders are Science|Business, ESADE and INSEAD, with participation and support from Microsoft, BP, SKF and Imperial College London. In addition, Pfizer contributed support for this particular study. Further information is at www.sciencebusiness.net. The authors thank ESADE Business School and the Science|Business Innovation Board for granting the permission to reprint the report as a chapter in this book. The authors are also grateful to Tuba Bakici and Henry Lopez-Vega for the interviews and support when they were writing the first draft of this report.
[§] Corresponding author.

Abstract

Industrial innovation processes are becoming more open. The large, vertically integrated Research and Development laboratory systems of the 20th century are giving way to more vertically disintegrated networks of innovation that connect numerous companies into ecosystems. Since innovation policy ultimately rests on the activities and initiatives of the private sector, it is vital that policy follows this evolution.

Previous innovation policies relied on large companies to act as the engines of innovation in the EU. While large companies remain quite relevant to innovation within the EU, they themselves report that their processes involve many more small and medium-sized enterprises (SMEs) and other contributors outside their own walls. Therefore, innovation policy must also move outside the walls of these large companies and consider the roles of human capital, competition policy, financing, intellectual property (IP), and public data in promoting an environment of open innovation that is equally beneficial to SMEs.

In this chapter, we combine new research and analysis on open innovation with focused interviews of major participants in the European innovation system. The result is a series of recommendations for public policies that could, if implemented, improve the climate for open innovation to take place in the European Union — and thereby improve the competitiveness of the European economy overall. Taken together, these recommendations comprise an informal "charter" for EU open innovation policy.

15.1. Introduction

Open innovation relies heavily upon the availability of external knowledge that companies assimilate and integrate into their businesses. Yet the stock of available knowledge and its availability to firms of different sizes cannot be taken for granted. This knowledge is the result of numerous, and often unconnected, public policies regarding science, technology, intellectual property (IP), and education within society. In this report, we will bring these background elements to the fore, and ask how governments can craft policies that support innovation in a world of widely dispersed knowledge, mobile workers, and venture capital (VC).

Many current public policy measures have their roots in the closed innovation era. They stem from a logic focused on developing large national or regional markets, protecting local companies, restricting foreign workers and students, and subsidizing large local firms to keep them innovating. These prescriptions assume economic autarky, where national economies operate largely independently of one another.

Yet, science and technology are nowadays widely diffused across the world. Most technologies are, nowadays, developed through a global network of both large and small technology partners. The number of technologies (even those that are thought to be crucial for national security) that can be developed and exploited within national borders is decreasing rapidly. Currently, no national or European government can reasonably hope to exclude a hostile government or interest group from having access to these technologies.

A similar reasoning applies to national procurement in EU member states for military and other technologies. Most national procurement regulations — especially those with military or national security applications — were born in a mindset of closed innovation. The increasing globalization and rapid proliferation of open innovation imply that governmental agencies cannot effectively exclude others from accessing widely available technologies. The same erosion factors that have caused private firms to move away from the closed innovation mindset are also forcing innovation policies to change. In the United States, for instance, experiments along these lines came from the CIA when it contributed financial capital to start a venture firm, In-Q-Tel. This VC firm is chartered with finding innovative startups to commercialize important software and communication technologies. Importantly, In-Q-Tel does not need to follow any federal procurement regulation guidelines and provides the CIA access to technologies that were previously difficult to access. In the UK, Qinetiq represented during its first years a similar initiative to set up commercial applications for military technologies. These initiatives make far better use of today's knowledge environment than policies based on a closed innovation logic.

Chesbrough (2003) examined several erosion factors that led to the decline of closed innovation. They included the following:

- increasing mobility of trained engineers and scientists;
- increasing importance of VC;
- greater dissemination of knowledge throughout the world;
- increased quality of university research;
- increased rivalry between companies in their product markets.

These factors help enable a new division of labor in the funding, conduct, and focus of Research and Development (R&D) in innovation systems. This new division has caused businesses to shift the focus of their internal efforts from more basic research discoveries toward more external sources of knowledge and has caused businesses to seek new uses for their knowledge more aggressively than in the recent past.

However, one important difference between the perspective of a firm and the perspective of a society is that a firm benefits from a single clear and coherent business model, while knowledge-intensive societies benefit from a multiplicity of business models competing to create value out of ideas. VC has become an integral part of the innovation system in leading OECD countries, and combined with increased labor mobility, the result has been a larger role for small and medium sized enterprises (SMEs) in the industrial innovation systems of these countries. These SMEs offer society a variety of possible business models vying to create value out of knowledge.

Starting up new companies and growing them into global businesses are crucial for the economic growth of an economy. The US economy has spawned new global players in industries that were embryonic or nonexistent 20 or 30 years ago; examples include Microsoft, Dell Computer, Cisco Systems, America Online, Genentech, Amgen, Millennium, eBay, Google, and Facebook.

Both the American and European economies have lost market share in manufacturing to the more efficient and responsive manufacturing systems of Japan and some other emerging Asian economies. The difference is that the European innovation system has been unable to copy the dynamism of the American innovation system over the last 20 years. Much of the American resurgence came from the ability of new firms to discover new industries and of society's ability to redirect human, financial, and technological resources to these new firms and away from distressed industries. Moreover, this change went hand in hand with a more fundamental change in how innovation systems functioned. Internal R&D within large businesses became less important and gave way to external sourcing of technology, as SMEs and universities became strong technology players.

If Europe wants to keep or improve its competitive position in the globalizing knowledge economy in the next decade, then public policy has to develop some basic guidelines that are in line with the imperative of open innovation. We will develop some suggestions for these policy guidelines in the following sections. First, we focus on education and human capital development and diffusion. We then analyze how the transition from closed to open innovation requires new funding systems. Third, we tackle policy issues related to IP. Fourth, with a particular focus on SMEs, we look at how open innovation encourages policymakers to look at networks rather than individual firms — and to promote competition and rivalry in product markets. Fifth, we look at some topics related to open government. Finally, we examine how public policy can stimulate open innovation initiatives in low- and medium-tech SMEs. We conclude this report by drawing some conclusions that can be considered a charter for open innovation policies in Europe.

15.2. Education, Development, and the Diffusion of Human Capital

15.2.1. *Human capital creation*

Top-level research and technology development hinge on the availability of excellent scientists and researchers. Universities play a key role in educating new generations of researchers and scientists and in generating new knowledge through research. Yet, a quick look at the worldwide ranking of EU universities compared with American universities in terms of publications and citation indices, Nobel prizes, valuable patents, and university spin-offs shows that the Americans do better in academic research. The relative position of Europe is also worsening as several non-Western countries rapidly upgrade their educational and knowledge infrastructures and quickly climb in the international rankings.

The main reason for the relative decline of the European educational system is its lack of transparency. European educational system. It is not easy to compare universities in the same country, and international comparisons within Europe are much harder. It is crucial that European policymakers set up a ranking system for universities in Europe against which all institutions can be benchmarked (as the European Commission is currently considering). Any metric is simplistic. But better rankings would offer students information about how much value they can expect for their money. As a result, good students would look for good universities and so offer Europe much better researchers. When rigorous research assessments were introduced in the UK, university administrators began to think about their strengths and weaknesses. As a result, they either addressed their weaknesses or started differentiating their offerings from other universities by building on their strengths.

As well as educating new students and researchers as a key resource, universities and related research institutes also play an important role in advancing basic research. Only two decades ago, large industrial companies had enormous corporate R&D centers where research was oriented toward the mission of the company, and each center had greater scientific and technological capabilities than most universities. The majority of these central labs were dismantled — especially during the 1990s — because large companies were forced by shareholders to focus on short-term profits or just plain survival. At the same time, the governments (especially in the US) were investing in research systems, national labs, and major universities. In this way, the incentives weakened for large companies to tackle (basic) research themselves, rather than working with major universities and, more generally, the innovation ecosystem existing in different countries. In consequence, as companies focused on applied sciences and the development and commercialization of technologies, universities became the major (and maybe only) institutions driving basic science research. As a result, governments have to make investments in fundamental science — which, if managed appropriately, is a major source of new technological developments.

The success of the Defense Advanced Research Projects Agency in funding basic research in the US in information technologies is a demonstration of how government funding, directed to decentralized research institutions, can yield cumulatively important research outcomes.

During our interviews with leading R&D managers in major industrial companies in Western Europe, there was a surprising unanimity that research in Europe is not "in good shape" because of institutional inhibitors. While there is great research in Europe, getting more of it hinges on top researchers working in top institutes. Large manufacturing companies are interested in accessing the fundamental research capabilities of top-performing universities and research labs, but not second-tier universities. Hence, what counts is the presence of world-leading research labs. Top researchers will work in universities and research institutes that can offer leading-edge knowledge infrastructures, interesting connections, or collaboration opportunities with other top researchers, and large, long-term projects (5–10 years depending on the technological field). The latter is necessary as it enables researchers to build a faculty that is large enough to cope with important scientific problems, and there is enough time to move the scientific frontier through scientific publications.

Europe faces problems in generating sufficient top-level research that can compete with universities and research institutes on a global scale. Unlike agricultural funding, R&D budgets are still mostly a national matter; 93% to 95% of all public-sector research spending in Europe is funded nationally. Of course, the European Commission has launched a number of central initiatives such as the European Research Council (ERC), but budgets are limited in comparison with those of the US National Science Foundation, the National Institutes of Health, and a number of private American foundations. As a result, there is no pan-European competition between universities as in the US. What provides the drive at American universities to have the best researchers and labs? Every lab must be funded every four to five years through national competition. Permanent competition is the best way to match budgets to the best technology. To this end, the European Commission should convince member states to transfer more of their R&D budgets to the ERC, provided that the basis for resource allocation is meritocratic and not political.

The current system used in the rest of the EU's Seventh Framework Programme (FP-7) projects is not really a contribution to pan-European competition between universities/research labs. The requirement in many FP-7 projects that research partners collaborate with many different universities and many different companies adds cost and slows the pace of work. Participants lose their competitive edge or seek funding elsewhere where administrative procedures are quicker and grants are usually larger. In sum, research programs should be made competitive on a pan-European scale, and universities should collaborate only if it actually improves the proposition.

15.2.2. *Knowledge diffusion*

Diffusion of knowledge is as important as creation to spur innovation within society. Yet many European countries have long-standing policies that constrain the diffusion of knowledge from universities to industry. For example, university lecturers in many European countries are civil servants, prohibited from working with and for private companies while drawing a public salary. Consequently, universities cannot learn from management practices in industry. Graduate students in many of these same countries are effectively indentured servants of the lecturers they work for, and so cannot seek out the best places to apply their cutting-edge knowledge. Lack of mobility has other unintended side effects. When faculty members select their next research initiative, they do so in ignorance of the burning issues that need to be addressed in other areas, including industry. This ignorance multiplies when university staff reviews the research proposals of their peers to allocate funding, or oversee the training of their students. Research by Van Looy *et al.* (2004) demonstrates that researchers who work closely with companies doing applied research achieve higher quality rankings for their fundamental research than peers who do not collaborate with industry. Therefore, contrary to the traditional thinking, academics do not face a trade-off between collaborating with industry and doing fundamental research. Both activities are highly complementary.

Diffusion of knowledge between universities and business would be dramatically improved if academics could temporarily be employed in private companies and vice versa. But at present, if an academic researcher leaves to work in a company and later returns, he or she cannot be promoted because they will not have published any papers during their absence from the university. A similar pattern emerges when managers take an academic post. However, there is some flexibility in this area. Some companies are sending managers to academia as part of their career development. This requires that the courses be adapted for the transition, and that industry has a model of career development that deliberately advances the capabilities of managers.

There are other barriers to mobility of knowledge workers. Pensions, social security, healthcare, and other aspects of compensation are typically tied to employment. Making these benefits portable would enable workers to seek out the best opportunities to use their skills. Moreover, social legislation in Europe is largely determined by national authorities, which implies that labor movement between member states involves plenty of complicated paperwork. Furthermore, there is an urgent need to develop a European economic immigration policy that lowers immigration barriers for a highly qualified labor force. This has proven to be a useful strategy for the US, where a continuous inflow of highly qualified labor has supported American scientific, technological, and economic strength for decades. The EU could also learn from mobility policy in China, which has adopted a number of initiatives to encourage Chinese citizens who were working abroad to return to China later in their careers. These so-called "sea turtles" bring a wealth of international business and scientific expertise with them and help rejuvenate the culture of the organizations in China that they join upon their return. However, this policy can only work when the research conditions in Europe are similar (or better) than those abroad. Top researchers will not return to their home country when the conditions for research are worse than those abroad. Finally, another area for EU reform is policy toward retirees. Yet with the continued progress in healthcare, longer life expectancies, and an aging population in most EU countries, there is too much valuable knowledge residing in the minds of retirees to be neglected. The time has come to tap into this source of "seasoned" knowledge — whether it is through coaching, mentoring, teaching, project work or other less-than-full-time employment.

In sum, labor mobility eases the tacit knowledge flow between organizations. Mobility also induces networking between organizations and knowledge spillovers (Cohen and Fields, 2000). Therefore, the productivity of a skilled workforce is determined by the quality of the skills as well as the mobility of the workforce. A fast flow of ideas generates more value than ideas that are locked into the boundaries of a single company.

15.3. Financing Open Innovation: The Funding Chain

Thus, direct incentives for R&D are ill-advised; they require public managers to make judgments about the prospects and merits of innovation at private companies. These judgments are inherently subjective and are best left to private equity suppliers, who compete to supply capital to promising opportunities. Competition enables a diversity of innovation approaches to be funded and elicits greater investment in governance by the suppliers of this capital. These owners will also be able to adapt much more readily to new information than public servants.

If highly innovative companies drive economic growth, then the EU focus should be on the economic world and the funding chain.

The funding chain conceptualizes the need to have appropriate types of financing for all stages — from research to the establishment and growth of a new venture. In each stage, the type of funding has to change, and different funding partners will be involved. Compared with the traditional innovation policy guidelines in Europe, more attention should be paid to the appropriate funding of the commercialization of new ideas into real business opportunities. A smoothly working VC market is a crucial element in the funding chain.

The size of the VC market in Europe is about one quarter that in the US. The role of VCs is to finance ventures for a number of years. These ventures then need to grow and become competitive. Accordingly, in areas where technology cycles are long (especially in biopharmaceuticals and aerospace), a venture cannot grow into a large company in just five years; 10 or 20 are needed. If there is no strong stock market, as at present, then VCs often have to sell the company prematurely to established companies. Acquisition by large companies is fine if economic reasons (such as complementary assets and global reach) drive it. But acquisitions that occur because VCs have run out of money lead to suboptimal solutions from a welfare point of view. Moreover, when the main acquirers are American companies in biotech, for instance, the result limits economic growth in Europe. It is thus a matter of encouraging more investments into these start-up firms.

Unfortunately, new regulations for banks and insurance companies are reducing their investments in the stock market, and this damages new ventures. Europe needs proactive reform. Five to seven percent of savings could, for instance, be channeled into rapidly growing and innovative companies. Europe has among the highest saving rates in the world, but these funds are invested in low-risk and underproductive areas. There is plenty for corporate and government bonds but very little for growing companies. While fiscal policy is not directly in its legal authority to control, the European Commission could use its coordinating and exhortatory powers to have member states provide new incentives for investment in R&D-based ventures. To do so, it could clearly define the target companies. They should be independent not subsidiaries of larger companies.

They should be spending 15% to 20% of their overall budgets on R&D. They should not be more than 10 years old.

With the right investments, European high-tech ventures could create more economic growth in Europe.

15.4. Adopt a Balanced Approach to Intellectual Property

However, this clarification of IP ownership should be limited in scope. In open innovation, firms invest in R&D to extend their current business models and occasionally to search for new models. These firms cannot and do not make every conceivable use of their ideas within their own walls. Innovation policies for the protection of ideas must accept the limits of what any single firm can do with its ideas and technologies and promote the recombination and reuse of the available knowledge in other companies. Direct expropriation of such ideas without compensation would be a terrible policy. But granting wide-ranging ownership rights to ideas that are not strictly controlled in their novelty, usefulness, and nontrivial nature is equally problematic. The first realization of an idea is often incomplete. Granting broad ownership rights could strangle the follow-on innovative work that enhances the value of that idea. For similar reasons, granting ownership rights to ideas for very long periods of time can be problematic. A balance must be struck between invention and diffusion. And that balance is disturbed by several factors in Europe, from the cost of patent application to the local nature of the IP market.

15.4.1. *Open innovation fostered by high-quality patents*

The European Patent Office (EPO) has the reputation of high quality, according to our interviewees. When the EPO grants a patent, it signals some embedded value when the inventor wants to license the technology, or when the startup receiving the patent seeks external financing. The EPO approach also prevents companies from becoming easily blocked (in developing or producing new products) by poor-quality patent families owned by other companies or non-practicing entities (e.g., patent trolls) as was the case in the US until recently (the strategy of the US Patent and Trademark Office has changed in the last few years in this regard).

Clear legal protection of high-quality patents is not in contradiction with an open innovation policy that strives to provide adequate incentives to undertake research and diffuse these discoveries widely. In fact, open innovation would literally be impossible without IP protection, as firms would resist sharing their ideas for fear competitors would steal them. Indeed, it can be argued that open innovation increases the need for robust IP protection. In developing a new medicine, for instance, the separate tasks of research, development, trials, and marketing may be conducted by different companies or groups — yet the overall financial return still needs to cover the costs of each step plus produce profit margins for each participant. So, there is a need to generate the same or greater returns to sustain all the parts of the R&D ecosystem — and this depends in part on robust IP. Within an open innovation framework, IP is not a fence preventing others from making use of a protected technology but rather a bridge to collaboration with other firms and organizations. Indeed, leading scholars say that a solid patent system provides opportunities for firms to overcome Arrow's (1962) "disclosure problem." However, there are still significant transaction costs in transferring technologies. Selling technologies in the marketplace is not fully leveraged and according to Gambardella *et al.* (2007), the market for technology could be 70% larger if transaction costs could be further reduced. The high percentage of unused but patented inventions could provide a ready supply of technology to the market if these costs could be addressed.

15.4.2. *Open innovation hampered by the high costs of the European IP system*

Europe has been working for almost half a century on its IP system (Van Pottelsberghe de la Potterie, 2010). However, the current system remains overly complex, opaque, and unpredictable, and it constitutes a heavy financial burden for SMEs or startup companies (Veugelers, 2009).

The European IP system is the most expensive and complex in the world due to its high level of fragmentation and translation requirements. Moreover, once a patent is granted by the EPO, it must be enforced (i.e., translated, validated, and renewed on a yearly basis) by the national jurisdictions of the countries in which the patent is applied. The London Agreement, which intends to reduce the translation requirements for patents when they are validated at national patent offices in 15 out of 34 states, has led to a reduction in the cost of patenting by 20–30% (Van Pottelsberghe and Mejer, 2010). Despite these savings, the relative cost of a European patent validated in six countries is still five times higher than in the US. These costs have a major impact on the number of potential patents that are not submitted (or withdrawn). The difference in price between the US and Europe partly explains why the USPTO attracts four times as many patent filings as the EPO (Van Pottelsberghe and Francois, 2009).

IP is increasingly embodied in business strategies, and an efficient IP system is crucial in the development of more R&D collaboration and technology transfer. A bold shift to a single European patent would drastically reduce the costs and complexity of the current system. This needs to be matched to a centralized litigation process via a single court. It is fundamental that this Pan-European Patent Court (known as the European and EU Patent Court or EEUPC) has clear rules of procedure and is run by a highly qualified group of IP judges. Otherwise, the perspective of a single patent being invalidated in any one of 27 member states after a trial of variable quality would be a significant step backward.

There is room for improvement in other areas. The EPO is currently working to reduce the time to grant a patent (currently 49 months), which compares unfavorably to the JPO (31 months) and the USPTO (27 months). In addition, Van Pottelsberghe (2010) suggests a "50% reduction in entry fees for a well-defined group of young innovative companies up to the sixth year (the average duration of the examination period). A pay-back process (of the 50% reduction) could be scheduled for companies that keep their patents enforced for more than six years." Generally, open innovation should encourage European policymakers to invigorate the European patent system. Therefore, it is interesting to notice that the EU in the last 12 months has made progress on a unified patent system.

15.4.3. *Aligning incentives of researchers and industry*

Researchers at universities and other public labs carrying out research for companies always face tension between their desire to publish early and the requirements of the contracting companies to keep inventions secret until a patent is filed. Currently, a patent application will be rejected in Europe if the invention has become publicly available before the application was filed. This includes selling the invention, giving a lecture about it, showing it to an investor without a nondisclosure agreement (NDA), or publishing it in a scientific journal. The US, by contrast, has a one-year grace period. This means that the inventor there can freely publish without losing patent rights. The European patent system would benefit from the introduction of a similar grace period.

In general, IP discussions between research institutes (or universities) and companies can be troublesome if

- Academic centers overvalue their IP and overestimate the odds of making a profit, leading to elevated expectations of royalty payments that make projects untenable, or
- Academic centers attempt to patent their work but do so badly, leading either to a lack of protection in key global markets or — worse still — creating "prior art" that invalidates patents on more useful developments of the same technology.
- These collaboration problems in research institutes or universities require professional IP management.

15.4.4. *Activating unused IP in large companies*

Multinationals have vast portfolios of patents. To protect their inventions, a company such as Philips files, via its IP and Standards organization (IP&S), an average of 1600 patent applications annually. It owned 55,000 patents in 2009 and employed 500 IP professionals and support staff worldwide. However, about 85% of all patents of large companies are never used in new products or are used to deter potential competitors. From a public policy point of view, unused patents represent a large untapped source of knowledge that could create new companies and economic growth if there were an efficient way to "activate" these unused patents in other companies.

To be sure, major companies with large patent portfolios can monetize unused technologies. Patents are frequently used as tickets in cross-licensing negotiations (mostly) with other large companies. However, licensing technologies from large companies to small firms, or creating new ventures based on the IP of large companies, is not common practice everywhere. Licensing out technology or spinning off ventures requires time and energy. And the return is likely to be small, as SMEs and startups generate insufficient revenues to seriously interest a large company that wants to monetize its unused IP. There are exceptions, however. Microsoft, for instance, has established a unit called IP Ventures, which partners with startups, venture capitalists, and government agencies to take inventions created by Microsoft Research and put them in the hands of entrepreneurs and SMEs. Microsoft is working closely with government economic development agencies such as Enterprise Ireland and the Finnish National Fund for R&D (Sitra) to transfer technology and spur the growth of small businesses (Gutierrez, 2008). Licensing out IP is also an increasing trend in pharmaceutical and chemical companies.

15.4.5. *Large-scale technology collaboration and IP agreements*

IP transfers can take more complex forms than bilateral agreements between two organizations. The growing complexity of technologies is forcing companies to team up with various types of partners in broad consortia. Examples include the IIAP programs of IMEC, CTMM, and IMI. In IMEC's Industrial Affiliation Programs, IMEC invites partners to collaborate on precompetitive research on nano-electronics and uses the so-called fingerprint IP model to deal with background IP in collaborative research and IP ownership and the use of jointly developed technologies (Leten *et al.*, 2013; Chesbrough *et al.*, 2009a, 2009b; Vanhelleputte and Reid, 2004). The Centre for Translational Molecular Medicine (CTMM) develops medical technologies that enable the design of new and "personalized" treatments for the main causes of mortality and diminished quality of life (cancer and cardiovascular diseases and, to a lesser extent, neurodegenerative and infectious/autoimmune diseases). It is a public–private consortium that comprises universities, medical centers, medical technology firms, and chemical and pharmaceutical companies. CTMM is using a similar IP model as IMEC to distribute the benefits of the joint research among the participants (including those that cannot generate patents, such as hospitals).

The Innovative Medicines Initiative (IMI) is a partnership between the European Union and the European Federation of Pharmaceutical Industries and Associations (EFPIA). The aim of IMI is to support the faster discovery and development of better medicines for patients and to enhance Europe's competitiveness by ensuring that its biopharmaceutical sector remains dynamic. Participants in the IMI (research institutes, SMEs, and large pharmaceutical companies) generate IP, which is owned by the participant(s) who generated it (or when no individual participant can be identified, the IP is jointly owned by those who have carried out the work). Participants have access to the knowledge developed in IMI before completion of the project, and they have access to IP for research purposes after the project. Beyond the research, participants may use, sublicense or commercialize the foreground they own.

These complex forms of joint research require careful thinking about ownership and the use of commonly developed IP. The pressure on universities to generate revenues from their research can exacerbate problems in some IP negotiations. In the IMI, for example, competing pharmaceutical companies agree that results of precompetitive research can be made freely available, but some university technology transfer offices want ownership over any IP generated by their work. The idea of academic centers being worried about appropriating returns, while industry at times accepts free access and runs counter to many public expectations, but it represents an important trend. These complex forms of multi-partner collaboration are shaping the future of European research; therefore, it is desirable that policymakers help in encouraging collaborative IP rules based on good practices. The current FP7 IP rules are not adapted to these complex forms of collaboration.

15.4.6. *Opening broader channels of collaboration*

Open business models have proven very effective in different parts of industry. In many cases, firms with considerable IP assets have decided to open specific parts of their IP portfolio to communities of practitioners or users. For example, IBM's IP Collaborative Innovation initiative pledged 500 patents to Open Source communities, launched an Open Innovation Network, and established an American University Summit for open collaboration. Similarly, Sony and Nokia have decided to share a portion of their patent portfolios to stimulate innovation in green technologies. Another successful collaboration is the GreenXchange, a breakthrough concept for sharing IP among companies that are working on sustainability issues in the footwear sector. And Microsoft is increasingly cooperating with major Linux software providers to enhance the interoperability of Windows and Linux through joint technology development. As customers want to use both systems to work together seamlessly and efficiently, Microsoft and Novell created an IP bridge between the worlds of Open Source and proprietary software.

15.4.7. *Promoting intermediaries to facilitate the diffusion of knowledge*

Recently, a new form of third party — innovation intermediary or "innomediary" — has emerged around the world. NineSigma,

InnoCentive, Yet2.com, and YourEncore are a few. These intermediaries facilitate collaboration across technology markets by providing innovation platforms that link companies with potential problem-solvers and facilitate the diffusion of knowledge or technologies.

There are significant transaction costs in transferring technologies. Selling technologies in the marketplace is not fully leveraged, and according to Gambardella *et al.* (2007), the market for technology could be 70% larger if transaction costs were reduced. These new intermediaries are shaping the market for technologies, and they help make the market for knowledge and IP more transparent; EU policymakers should take note. The intermediaries have been mainly focused on major companies as clients, but there is enormous potential for using their expertise to solve problems for universities, research labs, and SMEs. These cannot currently afford these innovation intermediaries, and so policymakers could analyze how costs could be lowered to an acceptable level for these groups.

15.4.8. *Extending the IP scope beyond patents*

Patents are only one form of IP protection and are very good for protecting IP that is related to a broad range of technologies. For instance, in the pharmaceutical industry, patents are used for protecting the molecular structures of medicines. But the industry has always sold more than that; value is also determined by knowledge about how these medicines can and should be used. The knowledge is generated in clinical trials, which now account for around 60% of the R&D costs (up from 50% a decade or so ago). Moreover, drug manufacturers are being asked for ever-greater amounts of data by regulators and reimbursement agencies, and these data are costly to produce. Thus, Data Exclusivity (DE) is another important form of IP protection for pharmaceutical companies; it is generating incentives for companies to collect data (particularly clinical data) on a medicine to investigate its value in treating new indications. Hence, it is important in the context of open innovation that policymakers pay attention to the increasing heterogeneity of data and information.

Similarly, trademarks, copyrights, trade secrets, and industrial design rights are important in the discussion of an open innovation policy. The emergence of the Internet is changing and will continue to change the business models that are used in many service industries (Chesbrough, 2011). Policy measures can have a considerable impact on the speed and direction of these changes — as we have seen in the music industry — but the European Commission could play a major role in proactively ensuring that IP regulation supports the conditions for business model changes in several services industries that rely on these types of IP protection.

15.5. Promoting Cooperation and Competition

15.5.1. *The locus of innovation is in the network*

Nowadays, knowledge is abundant, and the technology landscape is scattered. Therefore, policymakers have to shift their support from single firms to the innovation ecosystem that is creating and commercializing technologies. They have to look at the different nodes in the "food chain," from science to commercially viable product introductions. Innovation policy can play a crucial role in stimulating innovation systems in which universities, labs, start-ups, and SMEs and large companies jointly create new market opportunities. The locus of innovation is no longer in the firm but in the network (Powell *et al.*, 1997). An analogous shift in policymaking should redirect the policy focus from single large companies toward networks or ecosystems in which innovation partners jointly create new business opportunities.

Pharmaceutical companies, for instance, experience quick changes in their innovation process. Industry officials say their R&D productivity has declined in recent years. Attrition rates in development have remained high. At the same time, spending has increased to cover the rising demands for clinical data from regulators and payers. As a response to declining research productivity, these companies have adapted their R&D organizations. More and more stages of the R&D process are undertaken through collaboration or outsourcing. At the research level, companies deploy many different models for creating effective collaborations: contractual research agreements for specific research tasks; bilateral agreements with individual universities and research groups; collaborations with other companies on areas of precompetitive research; and bilateral agreements with other companies to progress specific research areas or specific high-cost development projects. Some companies have a venture fund and external research experts dedicated to finding partners and generating new deals and collaborations.

15.5.2. *SME formation and growth*

This shift to the network also implies that innovation public policy should seek to cultivate and strengthen SMEs. Their vitality will infuse a greater dynamism into the economy, as those companies that survive will embody new combinations of knowledge, and new business models to commercialize that knowledge. These companies will also spur greater innovation from larger companies. They provide large companies with demonstrations of the commercial viability of new approaches to commercializing ideas, and their success confronts incumbent firms with hard facts that they ignore at their peril.

Incumbents will respond to the demonstrated success of new firms with new combinations of knowledge far more rapidly than they will respond to any direct government program targeted to support them. Startups often have new technologies or are highly creative in developing new business models to commercialize knowledge; therefore, they are also great sources for large companies to in-source new technologies and business models for commercializing technologies.

To spur open innovation, policymakers should facilitate the creation of startups and encourage entrepreneurship in the European economy. They must also spur cooperation between SMEs and large companies to discover knowledge about the functioning of technologies and enact new technological ecosystems as system integrators. Finally, a new breed of managers is needed in large companies with the skills to set up new ventures such as spin-offs based on unused but patented technologies.

European VC-backed ventures should be able to grow into fully developed businesses that can compete on an international or global scale. There should be different financing schemes all the way from seed to late stage; otherwise, too many European high-tech ventures will be acquired by large American and Chinese companies. If there is sufficient money available in the VCF market, then startups can develop new manufacturing and distribution assets. The composition of the boards also plays a role in stimulating high-tech startups. These companies need directors who know the industry very well. In Europe, executives from large companies do not usually want to "waste their time" being board members in SMEs. However, large companies that do encourage their directors to sit on small boards (such as Microsoft, Novartis, GE, BP, Pfizer, and DSM) generate two effects. First, board membership gives early access to new technologies with considerable business opportunities. Second, the directors bring their experience to the startup company.

Let us take, for instance, the Novartis venture fund. When Novartis invests in startups, it shares its views on the industry with the startup and brings a great deal of expertise from the pharmaceutical industry. This is of enormous value for the startup because, while a small company may have vision and new technologies, it will probably also lack many managerial skills necessary to avoid obvious mistakes. A good board significantly increases the economic viability of startups. Governments should incentivize large companies to encourage their directors to become board members in startups.

The way in which VCs are managed is also very important. In America, VCFs are mostly managed by former entrepreneurs and former executives of large technology companies who have become investors. This approach is the right way to do it. Growing new ventures is not about how to analyze profit and loss accounts — investors have to know the field and the technology and understand the value proposition that will create competitive advantage for the venture. Too often in Europe, VC firms are headed by people with a financial background and no experience in industry or academia. Consequently, there is a high risk of making mistakes or making overly conservative decisions — creating followers instead of leading ventures. Therefore, it would be good in Europe to stimulate the formation of independent VCs that are led by people with a strong research, clinical or industrial background. The EC could, for example, launch a program through the European Investment Fund to stimulate the creation of new funds — provided that there is a new team with a broad, international background.

A final note: more than funding is required if SMEs are going to be able routinely to launch major medicines again. Regulatory and market reforms are also essential (these would benefit big and small companies). Growing needs for deep scientific knowledge, increasing sensitivities to risk (and liability), ever-greater demands for data from regulators and payers, and the need to globalize revenues to generate ROI have made launching medicines a difficult game for all, large or small. The Commission's attention to these issues — for instance, in its proposed European Innovation Partnership on healthy ageing — is needed.

15.6. Expanding Open Government

15.6.1. *Open government and open data*

Recently, there have been several "open data" initiatives around the world promoting interactive sharing of information between the government and the public. Open data refer to a practice of making data freely available online in a standard and reusable format for everyone to use (Fung and Weil, 2010). City halls collect extensive data about residents and the city. "Data" in this case refer to everything from electoral statistics to the location of schools or parking lots.

As governments realize the benefits, open data have emerged as an essential movement across the world. Many local and national governments have created their own "data portals" to list data (such as "data.gov.uk" in the United Kingdom). These open data portals allow citizens to access all public information obtained during public affairs management in standard and reusable formats. Thus, open data are the key foundation of an open government initiative. The social benefits of open government vary from citizen engagement to increased transparency and accountability, or enhanced communication channels. For instance, citizens gain greater insights into how their taxes are spent. Real-time availability of information also increases the potential to create extra services (Fung and Weil, 2010).

Open government also supports public sector innovation through diminishing bureaucracy and friction in data exchange and demolishing competitive advantages gained by proprietary access to data. Innovation is most likely to occur when data are available online in open, structured, computer-friendly formats for anyone to download (Robinson, *et al.*, 2009). Excellent examples include the USPTO and EPO databases about patents that are applied for and issued in the US and Europe, respectively. These databases have been used by thousands of researchers and have advanced our understanding of the role of innovation in creating competitive advantage at the firm level and wealth creation at the macro-economic level.

To foster innovation, government entities often use "contests" to encourage citizens to collaborate. "Apps" contests are common (such as "Apps for Democracy") to build web applications and services with open data. US government agencies have also launched challenges such as Challenge.gov or NASA Centennial Challenges Program for citizens to provide and share their solutions and innovations with the government. Other platforms for communication include "Blue Button," an online health portal where people can download their health information securely and privately, or "Federal Register 2.0," an attempt to organize articles into news sections for readers to browse by topic and by government agency, and which enables citizens to submit comments on regulatory actions.

Since government data are important for both government and citizens, a clear policy on how governments should open and distribute their data is required. Open data projects use the following principles: data should be complete, original, available online (such as in HTTP format) or in structured formats such as XML, uniquely addressable, machine readable, license-free without limitation for anyone or anything, and offered in a timely manner (Robinson *et al.*, 2009). Furthermore, governments should develop a central online portal so that data can be browsed and downloaded by citizens. There should also be a commitment by the government to regularly update data.

But there remain a number of areas where details must be worked out. Much government data are dispersed and some is still not fully disclosed. Deciding which data should be published is an important decision. Today many politicians strongly believe in the public's right to access all information — even information that is directly related to national security and privacy issues. To accomplish this, there are certain guidelines for how to ensure disclosure while protecting national security and individual privacy (Swartz, 2010). Thus, governments should strike a balance between the requirements of openness and considerations calling for nondisclosure.

15.6.2. *Extending the idea of open government*

The idea of open government can be extended to areas where the government is a monopolist. Public procurement drives demand for innovative goods and services — as analyzed previously in the Aho report (2006). Examples where public purchases play a crucial role in driving top technology are defense, aerospace, road, and railway infrastructure, and specific ICT applications. These purchases of innovative products encourage suppliers to generate top technologies that also represent interesting but untapped sources of innovations in commercial applications. There are numerous examples of how military technologies can successfully lead to commercial applications. The same holds for aerospace technology, which even leads to new products in low-tech industries — see, for example, Quilts of Denmark's functional quilts, based partly on NASA technology (Vanhaverbeke and Bakici, 2010).

However, the commercialization of technologies developed in these industries does not come automatically. On the contrary, companies that develop high-tech products for governments usually have priorities and capabilities other than those required to develop commercial products. Usually, other types of organizations handle commercialization. A few examples include MILCOM Technologies (now part of Arsenal Venture Partners) and (the early) Qinetic. Both organizations search for interesting technologies that have been developed originally for military purposes and turn them into commercial applications through licensing deals or new ventures.

Starting with the 1958 National Aeronautics and Space Act, some US federal agencies such as NASA have been required to facilitate the transfer of technology to other sectors. NASA has established 1700 spin-offs and has organized itself to actively pursue market opportunities. The transfer, application, and commercialization of NASA-funded technology occurs in many ways — knowledge sharing, technical assistance, IP licensing, cooperative research, and technology projects, and other forms of partnership (such as the NASA Open Government Plan). Similarly, the Space Foundation is a national nonprofit organization in the US that is certifying products that originate from space-related technology or use space-derived resources for consumer benefit.

Governments can further stimulate the commercialization of these technologies through funding. In the US, the Small Business Innovative Research (SBIR) program distributes $2.5 billion per year in R&D grants across 11 federal agencies, including $1.2 billion distributed by the Department of Defense. Companies whose products have high transition potential are eligible for "commercialization" funding.

In conclusion, to encourage collaboration and innovation, the old top-down model of government data management must be changed into a networked model. The scope of open data should also be expanded. Publishing data in bulk must be a government's first priority as an information provider. By publishing data in a form that is free, open, and reusable, governments will empower many innovative ideas. However, the provision of data alone will not lead to the goals of open government. Governments need to design effective legislation and policies to support this collaborative approach with citizens. Data must be processed and an open government ecosystem should be created. Open government, if implemented effectively, can improve the accountability of government, as well as boosting innovation in and beyond the public sector.

Public policymakers can also play a role in encouraging the commercialization of technologies that have been developed in industries where the government is the sole customer. Examples include the defense industry, aerospace, road, and railway infrastructure, and national security. Many of these technologies have the potential to be commercialized, but this does not happen automatically. The development of commercial applications for these technologies requires the help of specific organizations that are specialized in detecting and developing commercial applications. Governments should look at good practices and accelerate the search for commercial applications for these captive technologies.

15.7. Open Innovation Initiative for SMEs in Low- and Medium-Tech Industries

Open innovation can play a pivotal role in the innovative and commercial performance of SMEs. In SMEs, open innovation practices are part of the firm's strategy and managed essentially through an entrepreneurial process (Vanhaverbeke, 2017). As open innovation in SMEs has its own particularities, there is a need to develop a specific framework for open innovation in SMEs. Managing and organizing open innovation in SMEs are quite specific, and the lessons learned from open innovation in large firms are not readily transferable to the context of SMEs.

Most of the European GDP is generated by low-tech and medium-tech SMEs. The potential benefits of the use of open innovation are not recognized yet by most of these SMEs. There is a huge potential to spur the growth and innovativeness of the European economy if we could introduce open innovation into these companies. However, we should not be naïve: our exposure to these SMEs in the last eight years shows that they have a specific way of learning.

The challenge nowadays is to find a solution that helps SMEs to be aware of the benefits of open innovation. Furthermore, those SMEs that *are* aware of the basics of open innovation are mostly small firms that do not readily feel confident to try open innovation practices. Our experience working now for more than 10 years in open innovation with SMEs shows that they learn in a very specific way. Explanations provided by academia and consultants are usually not very fruitful. Similarly, government-led initiatives to stimulate entrepreneurship often do not provide an answer to SMEs' needs.

In contrast, SMEs learning is usually triggered by stories of entrepreneurs who have been successful in applying open innovation. Most SME managers take the attitude "if they can do it, I can do it too." Likewise, talking to other entrepreneurs with similar ambitions has been demonstrated to be very impactful. They share experiences and learn from them. Finally, local innovation (learning) networks led by a coach meet their needs and allow them to learn about OI in a shorter period of time.

To understand how open innovation can be effectively introduced into low-tech and medium-tech SMEs, we have to first to understand how SME managers get into touch with innovation and new business practices. Most of them learn new ideas in local/regional (innovation) network initiatives. The reason is that they have only a few hours during the week to do something extra beyond their operational work in the company. The action radius for such local network is 60–100 km. Some of those local networks are great initiatives where SME–managers can meet their peers, but there are also structural problems to leverage these networks:

- The coaches are only semi-professional coaches because the networks are local and small;
- There are only few good local cases, and as a result, few managers can be introduced as speakers. It is a time-consuming job for

a manager who has implemented open innovation successfully to come and tell his story 50 times a year;
- From a European perspective, we see that each region tends to reinvent the wheel, while there is few cross-regional learning (cross-regional learning is not really developed nowadays).

Setting up networks locally from scratch implies high setup costs, which can easily be avoided if the courses/guidelines/examples/cases were readily available online. So how could European innovation policy improve the situation? First, there are many excellent cases in Europe of SMEs that have been developing new businesses and gained a sustainable competitive advantage through the introduction of open innovation innovations.

As managers learn from their peers, a cross-border project should develop an online guideline with tools and inspirational videos of companies that have been successful in introducing open innovation. Therefore, such a European-wide project should start developing YouTube like videos (10–15 min) about these companies with subtitles in English. These videos could provide great practical insights on how open innovation generate benefits for SMEs and how it can be organized and managed in SMEs. A cross-border project should develop 30–40 of these short videos — each with its own focus on one or two specific topics in open innovation management. These movies should be made available on the web (YouTube) together with a syllabus/guidelines and some (open) innovation management tools. The syllabus should be set up in a modular way so that local coaches can develop their own course/sessions according to their needs or the level of open innovation maturity of the SME managers.

The guidelines and videos should empower local coaches to use great examples of best open innovation practices in European SMEs. That will significantly reduce the setup costs for their local networking actions; the quality of the coaching and discussions will significantly improve; and local managers that have been introducing open innovation successfully in their organization will no longer be overloaded by these networks. By offering convincing cases and a well-structured guideline, SME managers will be led through a top-level learning process.

The system should work as a Web 2.0 initiative where the local networks give feedback in order to evaluate and debug the system, improve the guideline, and suggest new best practices and cases, and it should allow them to relate across Europe. In this way, the initiative could be growing and improving over time. We can even think of a certification for coaches after a few years. In short, this initiative should have a strong impact on the number of companies that learn open innovation practices and durable improvement in the quality of learning.

In the long term, the course can be continuously improved through feedback from companies and coaches. In return, this will allow professionals, practitioners, and academics to better understand how to teach open innovation management to SME managers. It will allow academics to learn more about best practices in open innovation and attain new insights that in the future will help improve the methodology used. Public authorities can learn how to reach out to SMEs in an effective way, possibly at a lower cost, when providing much needed innovation training.

This should be a truly European public policy project and has to be carried out transnationally. Local coaches are now working with local examples. Introducing multimedia cases from companies across Europe will broaden the scope of the learning networks by showing management practices from different regions and countries. It will not only lead to an increase in quality of the cases but will also be an interesting way to work cross-culturally showing management practices across Europe to small firm managers, who in many cases have a rather local view on the economy and technology. Indirectly, it is also an opportunity for SME associations to link across different countries and for European public authorities to experience how this type of ICT-enabled learning offers a unique opportunity to get connected to the most difficult group of companies when it comes to innovation management.

15.8. Summary of Policy Recommendations

Many past and present innovation policies stem from a logic that is reminiscent of a closed innovation mindset. These may have been appropriate a generation ago but are no longer appropriate to the innovation needs of the EU in the 21st century. Instead, an open innovation mindset is required.

Closed innovation mindset policies	Open innovation mindset policies
Focus on developing a large domestic/European market.	Pursue global market opportunities.
Protect European companies from foreign innovators.	Invite external innovators in to spur greater competition and innovation.
Limit the number of foreign students and workers in Europe.	Encourage circulation of ideas by inviting foreigners to study and work in Europe, while also sending Europeans overseas.
Give subsidies and credits to the largest European firms to keep them innovating in their home country.	Provide the proper institutional structures for innovation and focus on SMEs.
Ensure that government funds go to domestic/European firms and avoid assistance to foreign companies.	Use government funds to stimulate greater SME formation and expansion, encourage innovation investments (whether by foreign or domestic companies) within the EU, and support export industry activities.
Focus on knowledge creation — mainly in large firms or high-tech start-ups.	Focus also on dissemination of existing knowledge mainly to low- and medium-tech SMEs. Find novel ICT-enabled ways to reach more SMEs and apply existing technologies into existing product areas.

We have summarized our recommendations as follows:

1. **Education and human capital development.** The EU is fortunate to have tremendous human capital resources at its disposal. Nonetheless there are some important changes to be made that would strengthen the excellence of research that emanates from this pool of human capital.
2. Increase meritocracy in research funding. Too many research programs within the EU sprinkle money across all the member states, with insufficient competition for these resources. The result is politically popular, but economically, the funded programs lack the excellence and scale to produce world class research and technology. Research funding competitions should move to the EU level wherever possible, to reward excellence and promote the promising ideas of new scholars. The European Research Council is a good step forward — and should be enlarged.
3. Support enhanced mobility during graduate training. EU graduate training is world class in some fields in some countries but not in others. While this condition will not change quickly, individual researchers can be given world class training if they are supported in conducting part of their training outside the EU and at the world's leading centers. In turn, EU graduate schools can broaden training by inviting the most promising scholars from outside the EU. A better ranking system for European universities would help inject much-needed transparency into the system, allowing students to make informed choices as they move. Likewise, more flexible immigration policies would also increase Europe's available brain power.
4. **Financing open innovation: the funding chain.** Funding open innovation requires a broader set of funding tools, reflecting the different financial needs at each stage of the process in which new ideas move from R&D into full commercial exploitation.
5. **Introduce the funding chain concept. Growing ideas into profitable businesses requires appropriate types of funding at each stage of the development and commercialization phase.**

A narrow focus on public subsidies for R&D inputs by firms is not in accordance with open innovation. The EU could start by encouraging member states to grant tax incentives for small, R&D-based companies.

6. **Increase the pool of funds available for VC investment. The availability of VC funding is crucial to oil the innovation engine based on the establishment and growth of new ventures. The American market dwarfs Europe's VC market, and this fact is slowing the growth and dynamism of the European economy.**
7. **Support the formation of spin-offs to commercialize research discoveries: Great technical ideas do not get commercialized because they are early stage and too risky to be privately funded. Reflection is needed on how policy can help providing funding to early-stage ventures.**
8. **A balanced approach to intellectual property.**
   One of the most powerful levers that government has to stimulate innovation is to design IP policies that reward innovative initiatives while also stimulating the diffusion of innovations throughout society. Ironically, in an open innovation world, strong IP protection is vital, to permit firms to share knowledge, but at the same time a balance must be struck to ensure rapid flow of ideas.
9. **Reduce transaction costs for IP. Current IP policy is anchored in each member country of the EU, fostering multiple filings, multiple language translations, and creating much high costs for EU patents. We need to move to a single EU patent, backed by a unified judicial process, to lower the costs of patent protection to those of rival regions. Current costs are particularly onerous for SMEs.**
10. **Foster the growth of IP intermediaries.**
    There is a growing market for IP, and the EU should encourage the expansion of this market. In addition, it should fund research into the functioning of IP markets so that future policy can be based on new and better evidence.
11. Rebalance EU policy toward universities with publicly funded research. Too many universities are focused on maximizing the

royalty income they receive from publicly funded research. The focus on royalty income, encouraged by governments trying to capture as much value as possible from their funding, may limit the flow of knowledge to industry which, in turn, hampers the technological progress and competitiveness of the industry. A more balanced approach would be to give greater weight to the overall social and economic impact of publicly funded research, with particular emphasis on broadly diffusing the research output within society.

12. **Promoting cooperation, competition, and rivalry.** Competition is vitally important to innovation. It enhances the willingness of firms to take the risks that advance new thinking, new processes and new markets in an innovative society.
13. Shift support from national champions toward SMEs and startup companies. SMEs are powerful agents of innovation diffusion within a society. Even when large firms remain at the top, the presence of striving SME firms in their industries forces large firms to innovate more rapidly to keep ahead. Policies should support SME formation, expansion, and exports outside the EU.
14. Promote spin-offs from large companies and universities. Many innovative ideas start small, too small to be of interest to large companies. Many other ideas start inside a university lab but require risk capital and entrepreneurial management to move into the market. Government can help facilitate these spin-offs by encouraging the transfer of IP to these spin-offs (perhaps providing tax incentives for large companies) and supporting the invested risk capital.
15. Focus on innovation networks. The locus of innovation is no longer in single large companies but in innovation networks involving a mix of partners: universities, labs, startup companies, multinationals, and governments. The relationship between these players largely determines the overall performance of an innovation system. The success of large firms hinges increasingly on their ecosystem.

16. **Expanding open government.**
    Government is not a bystander in the innovation system. It possesses a wealth of information distributed through a myriad number of databases that are often difficult to access. Government also buys innovation from many suppliers in society, and its opportunities to foster innovation through its procurement activities also deserve more attention.
17. **Accelerate the publication of government data wherever possible.** Citizens and companies can often spot wonderful innovation opportunities if given the necessary information. This has already been demonstrated through mashing data from different sources and developing applications to analyze and interpret public data.
18. **Utilize open innovation in government procurement.** When buying new technologies, create and employ open innovation intermediaries to seek out solutions from anywhere in the world, vs. the usual suppliers to the government. The U.S. Department of Homeland Security, for example, has created a government organization, SECURE, to procure defense and security-related technologies using open innovation.
19. **Foster commercial application of technologies developed for the government.** Public policymakers should encourage the commercialization of technologies that have been developed for military, aerospace, road, and railway infrastructure, and national security. Many of these technologies can be turned into interesting commercial applications, but this process will not happen automatically without government incentives.
20. **Improving open innovation in low- and medium-tech SMEs**
    Vanhaverbeke (2017) shows that open innovation has a huge potential in low- and medium-tech SMEs. Yet, most small firms in these industries are not innovating at all, while there is an urgent need to do so, because of the growing globalization and intensified competition.

    Most policymakers assume that an explanation of the need and urgency to innovate will incentivize SME managers to start innovation in their organization. This assumption is erroneous. Observing SMEs for eight years, we experience that entrepreneurs and SME managers only start to innovate when other SME managers show that it can be done. SME managers learn from the experience of other managers. As they have limited time to learn this can only be done in local networks organized by local organizations supporting SMEs.

These local networks should be/remain at the core of the European innovation policy for SMEs as these networks understand the needs of and are trusted by the SMEs in a particular region. The problem is that good local open innovation examples are rare and managers who have successfully implemented open innovation get too many requests to speak. Moreover, many coaches do not have access to professional guidelines to instruct open innovation to SME managers.

We claim that a set of short videos/multimedia cases about best practices of open innovation in European SMEs (to be distributed freely on YouTube) and a concise guideline would strongly empower local coaches in their attempt to initiate SME managers into open innovation.

The production of the videos and the guideline should not be expensive, and given its potential impact on changing the mindset of SME managers, it is potentially one of the most powerful initiatives policymakers can take at the EU level to stimulate (open) innovation in SMEs.

References

Aho, E. (2006). *Creating an Innovative Europe: Report of the Independent Expert Group on R&D and Innovation*, Office for Official Publications of the European Communities, Belgium, ISBN 92-79-00964-8.

Arora, A. and Gambardella, A. (2010). Ideas for rent: an overview of markets for technology. *Industrial and Corporate Change*, 19, pp. 775–803.

Arrow, K. (1962). Economic welfare and the allocation of resources for invention. In: National Bureau of Economic Research (ed.), *The Rate and Direction of Inventive Activity*, Princeton University Press, Princeton, pp. 609–625.

Bird, B. J. (2002). Learning entrepreneurship competencies: the self-directed learning approach. *International Journal of Entrepreneurship Education*, 1, pp. 203–227.

Chesbrough, H. (2003). *Open Innovation; The New Imperative for Creating and Profiting from Technology*, HBSP, Harvard: MA, USA HBSP: Harvard Business School Press.

Chesbrough, H. (2006). *Open Innovation Business Models; How to Thrive in the New Innovation Landscape*, HBSP, Harvard, Boston, MA. USA.

Chesbrough, H. (2011). *Open Services Innovation: Business Rethinking Your Business to Grow and Compete in a New Era*, Jossey, Bass, San Francisco, CA.

Chesbrough, H., Vanhaverbeke, W. and West, J. (eds.) (2006). *Open Innovation: Researching a New Paradigm*, Oxford University Press.

Chesbrough, H.W., Vanhaverbeke, W. and Odusanya, L. (2009a). Ineriniversity Microelectronics Center (IMEC) (A). Berkeley-Haas Case Series, B5712, p. 17.

Chesbrough, H.W., Vanhaverbeke, W. and Odusanya, L. (2009b). Ineriniversity Microelectronics Center (IMEC) (A). Creating value through a new approach to IPR Management, Berkeley-Haas Case Series, B5713, p. 3.

Cohen, S. S. and G. Fields (2000). Social capital and capital gains: an examination of social capital in Silicon Valley. In: M. Kenney (ed.), *Understanding Silicon Valley: Anatomy of an Entrepreneurial Region*, Stanford University Press, Stanford, CA.

Cooke, P. (2005). Regionally asymmetric knowledge capabilities and open innovation: Exploring "Globalisation 2" — A new model of industry organisation. *Research Policy*, 34(8), pp. 1128–1149.

De Jong, J. P. J., Vanhaverbeke, W., Kalvet, T. and Chesbrough, H. (2008). Policies for Open Innovation: Theory, Framework and Cases. Research Project Funded by VISION Era-Net, July 2006.

Doz, Y., Santos, J. and Williamson, P. (2001). *From Global to Metanational: How Companies Win in the Knowledge Economy*, Harvard Business Press, Boston, MA.

Ebersberger, B., Herstad, S. J., Iversen, E., Som, O. and Kirner, F. (2011). Open Innovation in Europe, Preliminary Draft 5/1/2011.

Fung, A. and Weil, D. (2010). Open government and open society. In: Lathrop, D. and Ruma, L. (eds.), *Open Government: Collaboration, Transparency, and Participation in Practice*, O'Reilly Media, Inc, Sebastopol, CA, USA, pp. 105–112.

Gambardella, A., Giuri, P. and Luzzi, A. (2007). The market for patents in Europe. *Research Policy*, 26, pp. 1163–1183.

Gutierrez, H. (2008). Microsoft's Collaboration Imperative, *IAM Magazine*, April/May.

Leten, B., Vanhaverbeke, W. and Roijakkers, N. (2013). IP models to orchestrate innovation ecosystems: IMEC, a public research institute in nano-electronic. *California Management Review*, 55(4), pp. 51–64.

McEvoy, N. (January 16, 2011). Level five open government innovation. *Cloud Computing Journal*, Available from: http://cloudcomputing.sys-con.com/node/1677330 [Accessed 10th March, 2011].

Milcom Technologies. Available from: http://www.milcomtech.com/ [Accessed 11th March, 2011].

O'Doherty, D. and Arnold, K. (2003). Understanding innovation: the need for a systemic approach. The IPTS Report 71. Seville: IPTS.

Robinson, D. G., Yu, H. and Felten, E. W. (2009). Government data and the invisible hand. *Yale Journal of Law & Technology*, 11.

Swartz, A. (2010). When is transparency useful? In: Lathrop, D. and Ruma, L. (eds.), *Open Government: Collaboration, Transparency, and Participation in Practice*, O'Reilly Media, Inc., pp. 267–272.

Vanhaverbeke, W. (2017). *Managing Open Innovation in SMEs*, Cambridge University Press, Cambridge, UK.

Vanhaverbeke, W. (2016). *Quilts of Denmark: Managing Open Innovation*, Ivey Cases, 9B16M213, 13 pages.

Vanhaverbeke, W. (2016). *Patient Room of the Future: User-Oriented Innovation*, Ivey Cases 9B16M195, 12 pages.

Van Helleputte, J. and Reid, A. (2004). Tackling the paradox: can attaining global research excellence be compatible with local technology development? *R&D Management*, 34(1), pp. 33–44.

Van Looy, B., Ranga, M., Callaert, J., Debackere, K. and Zimmermann, E. (2004). Combining entrepreneurial and scientific performance in academia: towards a compounded and reciprocal Matthew-effect? *Research Policy*, 33(3), pp. 425–441.

Van Pottelsberghe de la Potterie, B. (2010). Europe should stop taxing innovation, *Breugel Policy Brief*, 2010/02.

Van Pottelsberghe de la Potterie, B. and Franois, D. (2009). The cost factor in patent systems. *Journal of Industry, Competition and Trade*, 9, pp. 329–355.

Van Pottelsberghe de la Potterie, B. and Mejer, M. (2010). The London Agreement and the cost of patenting in Europe. *European Journal of Law and Economics*, 29, pp. 211–237.

Veugelers, R. (2009). A lifeline for Europe's young radical innovator's. *Breugel Policy Brief*, 2009/01.

Online resources:

NASA Open Government Plan (April 7, 2010), Available from: http://www.nasa.gov/pdf/440945main_NASA_OpenGov_Plan_v17[508].pdf [Accessed March 12, 2011].

Obama, B. Transparency and Open Government, Available from: http://www.whitehouse.gov/the_press_office/TransparencyandOpenGovernment/ [Accessed March 10, 2011].

QinetiQ, Advanced Sensors Innovation Project, Available from: http://www.qinetiq.com/home/capabilities/sensors.html [Accessed March 10, 2011].

Index

L

M

N

O

P